6th International Symposium on the Science and Technology of Light Sources

Budapest, Hungary
30 August – 3 September 1992

ISBN: 978-1-5108-4110-9

Printed from e-media with permission by:

Curran Associates, Inc.
57 Morehouse Lane
Red Hook, NY 12571

Some format issues inherent in the e-media version may also appear in this print version.

Printed by Curran Associates, Inc. (2017)

For permission requests, please contact Foundation for the Advancement of the Science & Technology of Light Sources (FAST-LS) at the address below.

Foundation for the Advancement of the Science & Technology of Light Sources
FAST-LS
Belmayne House
99 Clarkehouse Road
Sheffield, United Kingdom
S10 2LN

www.fast-ls.org

Additional copies of this publication are available from:

Curran Associates, Inc.
57 Morehouse Lane
Red Hook, NY 12571 USA
Phone: 845-758-0400
Fax: 845-758-2633
Email: curran@proceedings.com
Web: www.proceedings.com

THE 6TH INTERNATIONAL SYMPOSIUM ON THE SCIENCE & TECHNOLOGY OF

LIGHT SOURCES

BUDAPEST – HUNGARY

30 AUGUST – 3 SEPTEMBER 1992

Invited Papers
and
Abstracts

LS 6

TECHNICAL UNIVERSITY OF BUDAPEST

6th International Symposium on the Science & Technology of

Light Sources

(LS:6)

Proceedings of the
Sixth International Symposium on the
Science & Technology of Light Sources (LS:6)
Budapest - Hungary
30 August - 3 September 1992

Invited Papers
and
Abstracts of Landmark
and Poster Papers

Edited by:
L. Bartha
Academy of Sciences, Hungary
and
F. J. Kedves
TUNGSRAM Co. Ltd.. Hungary

PREFACE

This volume contains both the full texts of the invited papers, and the abstracts of the accepted Contributed papers, that are to be presented at the Sixth International Symposium on the Science and Technology of Light Sources (LS:6) being held in Budapest, Hungary, from the 30th of August to the 3rd of September, 1992. The Budapest Symposium has shown a decrease in terms of the number of delegates (about 150) and the number and range of the contributed papers (84) than the earlier Symposia. The previous Symposia were held in Loughborough (1975; 90 delegates), Enschede (1979; 126), Toulouse (1983; 150), Karlsruhe (1986; 198) and, most recently in York (1989; over 200).

The seven main sessions contained in the scientific programme of LS:6 comprise a long poster session bracketed by one, or more, invited and/or "Landmark" papers presented for the full Symposium. The Landmark papers included in the programme were selected from the Contributed papers by the Organising Committee for presentation to the full Symposium because they met one, or more, of several criteria, for example they were regarded as being either novel, timely, or of general importance in their particular session. In addition special consideration was given to the papers of younger scientists.

Each paper in the volume has been assigned a number according to its position in the running order as planned in the programme and a further letter designation according to which of the three possible categories it belongs, i.e. I (Invited), L (Landmark), P (Poster).

László Bartha

Ferenc J. Kedves

July 1992

Editors

RADOMIR LAGUSHENKO
(1926-1991)

Dr. Radomir Lagushenko died on October 20, 1991 due to a massive heart attack. He worked at GTE Electrical Products Group Research and Development Laboratories at Danvers, Massachusetts, U.S.A., since 1979 as a theoretical physicist. He spent his career with GTE conducting research in such fields as novel fluorescent and compact fluorescent light sources, gaseous electronics, plasma physics, radiation transfer, modeling of low pressure discharges, electron energy distributions, etc.

Dr. Lagushenko, who has published more than 80 articles in technical journals internationally and who regularly was invited as a guest speaker to a variety of conferences, earned master's and doctorate degrees in physics from Leningrad University in his native Russia. Prior to joining GTE, he was professor of Theoretical Physics at Leningrad State University.

Radomir was a pleasure to talk to. In spite of his profound knowledge, he was totally unassuming, humble, and patient. He would not undertake an extensive calculation unless it was really necessary. His insight into the nature of electrical discharges allowed him to often give a quick, "back of the envelope" estimate of the desired parameters to help out with experimental work. I do not know of any person who came into contact with Radomir and was not impressed with his knowledge and pleasant personality.

Prior to his death, Radomir was in the process of assembling a major study of his treatment of electrical discharges over the years. Furthermore, the following week, he was scheduled to give a treatment of the negative glow in the 44th Gaseous Electronic Conference. He will be sorely missed by many of his colleagues and friends.

Dr. Lagushenko is survived by his wife, Olga, and son, Tim, of Brookline, Massachusetts.

Jakob Maya
GTE Electrical Products

Maurice Cayless 1926-1989

Maurice Cayless died shortly after LS5 in York. He, together with Prof. Raffle, started this series of conferences. I had hoped that he would be able to come to the York banquet as a guest; alas illness prevented his attendance and caused his death shortly afterwards.

Many people attending LS6 will have met him, and many more will be aware of his pioneering work. He is best known for his theoretical work: it is less well known that he was a first rate experimental scientist.

In a major contribution to low pressure discharges, which formed the basis of much subsequent work, Cayless was the first to use digital computers to solve the diffusion and energy balance equations. He was amongst the first in civilian science to appreciate the power of computers for tackling problems of this complexity. He also understood the enormous insight that is gained if this type of theoretical problem can be tackled with minimal approximations. This process, now known as computer modelling, is now commonplace. The computer he used had a tiny 48 words of fast store: to overcome its limitations, the problems had to be programmed in machine code using numerical techniques that he, himself had devised. The predictions agreed so well with experiment that new lamp types could be designed by computer.

His interest turned now to high pressure (HP) lamps. Drawing analogies with mercury lamps, his theory predicted that as the pressure was raised, the light should become whiter. Lacking other materials, his collaborator Clarke was able to make a metal lamp with a sapphire window using automobile spark plugs as electrodes (there is a picture of it in the book by de Groot and van Vliet). This 'lamp' ran for only a short time, long enough to see that the light was white, but not for the predicted reason. Cayless' understanding of radiation transport and processes was so refined that he was immediately able to formulate the theory that is used to this day.

In 1968 Maurice joined Thorn Lighting, to set up a new Lamp Research department, and where he did much to encourage and develop a new generation of lamp scientists.

David Wharmby

LS:6 Organizing Committee

Prof. Dr Bartha, L.(Chairman)	Academy of Sciences, Hungary
Prof. Dr Kedves, F. J.(Secretary)	TUNGSRAM Co. Ltd., Hungary
Dr Berta, I. (Treasurer)	Technical University, Hungary
Dr Devonshire, R.	University of Sheffield, UK
Dr Wharmby, D. O.	Thorn Lighting Ltd., UK
Dr Hoerster, H.	Philips Research, Germany
Dr Jack, A. G.	Philips Lighting B.V., Holland
Prof. Popp, H. P.	University of Karlsruhe, Germany
Dr Maya, J.	GTE Lighting Products Group, USA
Dr Pitchford, L.	University of Paul Sabatier, France

LS:6 Local Organizing Committee

Prof. Dr Bartha, L. (Chairman)	Academy of Sciences, Hungary
Prof. Dr Kedves, F. J. (Secretary)	TUNGSRAM Co.Ltd., Hungary
Dr Berta, I. (Treasurer)	Technical University, Hungary
Dr Gaál, I.	Academy of Sciences, Hungary
Ms Jetzin, M.	Technical University (Conf. Off.)
Ms Somlyodi, Zs.	Technical University (Conf. Off.)
Ms Stoll, Á.	Technical University (Conf. Off.)

LS:6 Corresponding Members

Dr Bergman, R. S.	GE Lighting Business Group, USA
Dr Berman, S.	Lawrence Berkeley Laboratory, USA
Prof. Blanc, D.	University of Paul Sabatier, France
Prof. Cai, Z.	Fudan University, China
Dr Kamiya, S.	Matsushita El. Corp., Japan
Prof. Karabourniotis	University of Crete, Greece
Dr Kossakowski, J.	CIE, Poland
Dr Waymouth, J. F.	Consultant, USA

Financial Guarantors

The LS:6 Organizing Committee gratefully acknowledge the important role of the following companies who have acted as financial guarantors:

GE Lighting

GTE Products Corporation

Philips Lighting

Hungarian Academy of Sciences

Technical University of Budapest

Aug 30., Sunday

18.30 - 19.00	OPENING		*TUB K building*
19.00 - 20.00	**M. Vukcevich:** The 3rd Century of Incandescence Lamp *informal talk*		*Main Hall*
20.00 - 22.00	WELCOME PARTY		"

Aug. 31., Monday

8.30 – 10.10	I:1, I:2	**Phosphors**	*Main Hall*
10.10 – 11.50	P:3-P:10	"	"
11.50 – 12.40	L:11, L:12	"	"
12.40 – 14.00	LUNCH		*(University cafeteria)*
14.00 – 15.45	I:13, L:14, L:15	**LP Discharges**	*Main Hall*
15.45 – 17.30	P:16-P:29	"	"
17.30 – 18.15	I:30	"	"
18.30 – 20.30	SIGHTSEEING TOUR		

Sep. 1., Tuesday

8.30 – 9.15	I:31	**HT Chemistry**	*Main Hall*
9.15 – 11.00	P:32-P:39	**+**	"
11.00 – 12.15	L:40-L:42	**Materials**	"
12.15 – 14.00	LUNCH		*(University cafeteria)*
14.00 – 14.45	I:43	**Incandescent lamps**	*Main Hall*
14.45 – 16.45	P:44-P:52	"	"
16.45 – 17.35	L:53-L:54	"	"
20.00	GET TOGETHER PARTY AT HOTEL GELLÉRT *Given by TUNGSRAM*		

Sep. 2., Wednesday

8.30 – 9.15	I:55	**Modelling/Diagnostics**	*Main Hall*
9.15 – 11.10	P:56-P:64	"	"
11.10 – 12.00	L:65-L:66	"	"
12.00 – 14.00	LUNCH		*(University cafeteria)*
14.00 – 14.45	I:67	**HID**	*Main Hall*
14.45 – 17.10	P:68-P:73	"	"
17.10 – 18.00	L:74-L:75	"	"
20.00	BANQUET		

Sep. 3., Thursday

8.30 – 9.15	I:76	**RF Discharge/Novel**	*Main Hall*
9.15 – 10.05	L:77-L:78	"	"
10.05 – 11.30	P:79-P:83	"	"
11.30 – 12.15	I:84	"	"
12.15 – 12.30	CLOSING		
12.30 – 14.30	LUNCH		*(University cafeteria)*
15.00	VISITS		

Sep. 4., Friday

POST CONFERENCE TOURS

Session 1

Chairman: R. Devonshire

Phosphors

Session 2

Chairman: J. Maya

LP Discharges

Session 3

Chairman: R. Bergman

HT Chemistry + Materials

Session 4

Chairman: S. Kamiya

Incandescent lamps

Session 5

Chairman: D. O. Wharmby

Modelling/Diagnostics

Session 6

Chairman: A. G. Jack

IIID

Session 7

Chairman: J. F. Waymouth

RF Discharge/Novel

Scientific Programme

Texts of Invited, Landmark and Poster Papers

Session 1

Chairman: R. Devonshire

Phosphors

1:I **J. Schanda**
Energy efficient light sources and applications

2:I **S. Kamiya**
Recent progress in lamp phosphors

3:P **V. Bobay, J. Osán, L. Balázs**
Thermoluminescent study of Seca blue phosphor samples

4:P **X. M. Chen**
Manufacturing of the photoluminescent glass doped with $B_2O_3-La_2O_3$ and its application in fluorescent lamp

5:P **C. Caretti, P. Manini, A. Renzo, E. Rabusin**
Sodum radiation-induced hydrogen emission from a H_2-doped barium film

6:P **L. L. Hope**
Ray-tracing estimation of Kubelka-Munk parameters and application to phosphor layers and plaques

7:P **K. Iwama, T. Higashi, T. Matsuda, S. Kamiya**
The luminescence properties of Ce^{3+} - and Mn^{2+} - activated cerium magnesium aluminate phosphors

8:P **L. Pogány, L. Balázs**
Characterization of ALON layer on the inner bulb surface

9:P **E. Tombácz, J. Szigeti, L. Balázs, G. Sajó**
Interpretation of pH-dependent colloid stability of aqueous phosphor suspensions and their stabilization by polyelectrolyte adsorption

T. M. Zhou, Y. J. He, H. Shao, L. M. Wang
Studies on luminous features of phosphor coating of compact fluorescent lamps

H. Ito, Y. Yuge, A. Taya, M. Tamatani, K. Terasima
Charging tendency of phosphors and fluorescent lamp performance

H. C. G. Verhaar, B. J. Roelevink
Mercury-analysis of fluorescent lamps

ENERGY EFFICIENT LIGHT SOURCES AND APPLICATIONS

J. Schanda
CIE Central Bureau, Vienna and Res. Inst. Techn. Phys. HUS, Budapest

1 Introduction

The development of electrical lighting reaches nowadays its fourth phase. The early days - roughly the first 50 years - were dominated by incandescent light sources that developed from the carbonated and osmium filament lamp till the double spiral and krypton filled tungsten incandescent lamp. During this *tungsten lighting* phase, the efficacy increased continuously, lamps based on the same principle were built in a wide wattage range.

In the application area, a tremendous step was undertaken when instead of kerosine and some gas-lights electrical lighting could be used. As the supply of electricity became cheaper, higher and higher task illuminances were produced, till finally the heat dissipation in the illuminated areas prevented to produce indoor illuminations that would be comparable with the open air day-time illumination levels. Glare problems and daylight similar colour were not so much of a concern.

The second phase of lighting started with the introduction of low and high pressure gas discharge lamps and thus will be referred to as *discharge lamps lighting* phase. Their development dominated the next 25 years of light source evolution.

At this stage, illuminating engineers thought that with the three to four times more efficient light sources they can provide ideal lighting conditions. Concern about light colour and colour rendering became obvious, and light source manufacturers started to diversify their production.

During the same period, classical incandescent lamps reached a phase of maturation and technologists had time to check out details of their profession and started to understand the physics of their lamps. Progress, however, was much slower.

Then the energy crisis of the seventies brought about a new situation. The first reaction was to switch the light off. Lamp and luminaire manufacturers responded quickly, and introduced *energy saving lighting*. Many lighting codes have set the prescribed values back to lower values. Nevertheless with modern light sources and lighting gear now one can produce lighting that is equally effective like that designed 20 years ago, but with half the energy input.

With this we have reached the end of phase three, and now step into phase four: *demand lighting*. During phase two and three manufacturers showed their technology know-how by building more and more complicated lamps. When one manufacturer brought onto the market a 50 W high pressure sodium lamp, the other had to show that he could produce one with 30 W rating, and so on, irrespective whether the lamp was really efficient or whether it filled an application area gap or not. By now we know the advantages and limitations of the different lamp constructions and can start a diversification in accordance with the existing demand.

2. Demand lighting

Firstly we have to specify what should be understood under demand lighting. As lighting practice developed, the use of different lamp types for different applications was started. When new lamp types were developed, they were tried in a number of applications and only practice showed in which application they could survive. During these years also lighting practice has learned a lot, and now we can - more or less - prescribe the characteristics of a light source we would like to use in an application. Thus, the lighting engineer can put his demand to the lamp developer describing the requested characteristics of the desired lamp, and then the lamp constructor can tailor the lamp to suit these demands. This we would like to term as demand lighting.

Demand lighting should ease the work of both the lamp technologist and the applied lighting engineer. In a short survey paper we are unable to deal with all the different applications of light sources to give an overview of the existing demands, especially as - to a large extent - this new approach is still in its infancy and only the years to come will make a detailed analysis possible.

Before some of the applications are highlighted, we will have to deal with those basic concepts of lamp characteristics that will have to be changed to reach clear categorisation of the demands.

3. Lamp characteristics

3.1 Efficiency versus efficacy

The energy crisis produced a cut-back on oil production and export in the mid 70s. By the second half of the 80s, mankind had to realize that neither fossil nor atomic energy can be used without environmental impair and considerable risk. There seems to be no way out of using energy more consciously, and this applies for lighting to use efficient lamps and lighting systems.

Up to now lamp and luminaire technologists have worked in parallel, trying to optimize their own product individually. With modern lamps, luminaires, and lighting gear it becomes obvious that the main goal is to optimize system performance. It is useless to develop a high efficacy lamp if one looses a lot of energy in the ballast or in the production of the luminair or that of the ballast. A lower lamp efficacy can be counter balanced by a more efficient luminaire.

It will probably be a long way to go to change from the basic lumen/watt catalogue data to performance related characterization, especially as it needs the close cooperation of lamp and luminaire manufacturers and will need, in the future, the involvement of the design engineer as well.

To show this on an example: For an office desktop lighting system, at present halogen incandescent lamps and compact fluorescent lamps are competing with traditional incandescent lamp luminaires.

The user should be interested in quality lighting, i.e. a good contrast rendering factor [1,2], even illumination of the working area, economic optimum for the total life of the system and also aesthetics of the office furniture. And then we have not mentioned items as UV safety, best colour characteristics, etc. As we cannot anticipate that the end user is an expert in lighting, it is obvious that he needs guidance. Luminaire manufacturers often concentrate on one family of sources, i.e. they use in their constructions either incandescent lamps or compact fluorescent ones. That means that guidance has to come from the lamp manufacturer, or better-say, the lamp manufacturer has to know all the implications of his technological solution and has to optimize the construction to meet all the requirements.

On the market there are computer programs for lighting calculations which perform optimization calculations of some sort, producing overall efficiency and cost/benefit relation figures, but these programs are often based on assumptions not fulfilled in practice, the outcome of the calculations has to be made better understood and analyzed. In the US, the Environmental Protection Agency started a "Green Lights Program" [3] that should help producers to apply and users to understand these concepts. Similar moves - according to the knowledge of your reporter - are still missing in Europe and other continents. One of the tasks of lamp manufacturers should be to state clearly the preferred application for a given class of lamps.

3.2 Colour and colour rendering

The method of calculating colour rendering indices was, to my understanding, highly misused by lamp manufacturers. The concept [4] was worked out at times when only standard and deLuxe fluorescent tubes - and a very limited range of high pressure discharge lamps - were on the market. It was a genuine idea to determine the spectral power distribution to optimize efficacy and colour rendering simultaneously. But finally it was a misuse of the concept and new starts are needed to get a better understanding of what colour rendering really means, and how it should be applied [5,6].

I would like to question already the very first step of calculating the colour rendering index, i.e. the concept that a Planckian radiator of low colour temperature should have a colour rendering index of 100.

The human visual system, developed during the millenia, and the chromatic adaptation system adjusted themselves to compensate for the different phases of daylight. It is a real marvel of our visual system that it can adapt so well to incandescent lamp colour temperature too.

But colour rendering is more than colour adaptation. During the periods prior to electric lighting, we could hardly speak of colour perception in case of artificially light scenes. The state of adaptation was practically always mesopic. Thus, it is just too understandable that people who saw colour under incandescent lamp light for the first time, were content with its quality. At the invention of fluorescent lighting, it was still the first generation of scientists who grew up with the benefits of electrical lighting, who had to compare the colour rendering of these new sources with those of the already established incandescent lamps. But an established light source does not need to mean "optimum" performance as regards colour rendering.

With our present-day knowledge about the poor colour rendering of blues and greens under Standard Illuminant A, I would much more contest for a colour rendering system that is task dependent.

Results of many psychological investigations that analyzed the mood and behaviour of people under different lighting situations are available. Thus, we know that we need a different light colour for our sitting room than for our office.

One and the same lamp, if used in a restaurant or in an open space office, will produce different visual impression of the colour quality of the source. This should be reflected by a new colour rendering system.

Lamp manufacturers will have to be open to the request of the applied lighting engineer to describe the characteristics of their products in a meaningful form and will have to keep in mind the end user's requirements when designing a lamp for a given application.

3.3 Integration of luminair parts

It started probably in the US with the introduction of sealed beam automotive lamps that lamp manufacturers incorporated parts of the luminair business into their task. Nowadays almost all lamps with a confined luminous body are also available with integrated reflectors. As lamp manufacturers know little about the wish of the end user, all lamps are produced with two to three light distribution characteristics from narrow beam to broad beam or even batwing light distribution. A better cooperation between lamp and luminaire manufacturers could help to limit the choice of lamps to the amount really needed.

In selecting optimum solutions the overall energy efficiency has to be considered. Thus it is not worth integrating a heavy glass reflector (PAR 38 type) to filament constructions designed for a short life, as total energy consumption (energy used by the lamp plus energy used to produce the lamp and reflector) will be too high. For such lamps, retrofits in multiple use reflectors are more economical.

The small tungsten halogen lamps with their very confined luminous body and the small, well defined reflector, have - on the other hand - filled a market niche in a short time and made efficient decorative lighting possible.

3.4 Wattage range of lamps

It is most important to find the proper application areas for the different light source types and then to develop the wattage ranges according to the application.

Thus, e.g, there are applications where full light output is needed within a very short time after switch on. For such an application it is not worthwhile to develop metal halide constructions, as they will not be accepted by the user.

At present we have a wide variety of lamps based on different principles and built in different sizes, from low wattages to very high ones. Within the limited space and time it is not possible to deal with this subject in detail, we have to confine ourselves to some examples:

3.4.1 Lowest wattage range, lumen output of a few hundred to thousand lumen

These lamps are used either for emergency lighting, where a rapid start and good efficacy is of paramount importance, or are used in applications where the source is relatively near to the illuminated object (desktop or reading lights, etc.). In the latter case, a warm white or neutral light colour, good colour rendering, instant full light output and flicker free operation are important characteristics. The easy directing of the light onto the task is usually not of paramount importance. Thus, sources with a relatively large luminous surface can also be used.

In practice this means tungsten halogen lamps with or without reflector and compact fluorescent lamps of low wattage.

3.4.2 Medium wattage range

For the general lighting in a living-room or an office, we need sources with somewhat higher ratings. Depending on the atmosphere we would like to create, sources with warm or neutral, perhaps cool-white colour temperature can be used. In this area the available sources are more than abundant.

It will probably take some time before the general purpose incandescent lamps will disappear from this application, but your reporter does not believe in the substitution of these lamps with tungsten halogen ones. The lumen saving of such a change or the total energy efficiency saving does not support such a move.

At this place we do not want to go into detail with shop window and other small area decorative lighting, where the past years have shown a boom of halogen incandescent lamps. It should only be mentioned that both in these applications and others where the lamp-target (human or material) distance is low, the UV emission of the lamp has to be controlled (7). The higher efficacy of the compact fluorescent lamp will certainly gain in importance in this area, especially if the electronic ballasts used become even smaller, lighter, and, last but not least, cheaper.

On the other hand, in such applications, dimming is often required. Lamp construction has to consider this requirement.

For decades now we used fluorescent tubes in office lighting, and incandescent lamps in home lighting. Compact fluorescent lamps challenge both applications. While for home lighting these lamps can bring a considerable saving in energy, it is more questionable how well this will be accepted in office lighting.

The trend of producing more and more compact lamps by folding the discharge tube many times is questionable. If the light loss due to self-absorption by the parallel standing tube parts cannot be counterbalanced by a luminair construction of higher efficiency and inherent advantages of the smaller size (cheaper luminaire with better light distribution), then this solution cannot be favoured.

In office lighting, one has to take into consideration also the general use of modern office equipment. This will mean in the future that lighting has to be optimized not only for cathode ray tube video display units (VDUs) - and written, printed paper - but also for liquid crystal displays, where the surface curvature and the orientation of the surface compared to the normal working plane is different. This means flexible, easily and individually directed light with intensity regulation.

A growing concern of ophthalmologists is flicker and beat frequencies between flicker of the light source and the VDU (8). The obvious answer of the lamp manufacturer and illuminating engineer will be electronic high frequency operation.

3.4.3. High wattage range

In industrial high bay applications, street lighting, flood lighting, etc., high efficacy lamps with small compact sources are requested.

An interesting debate has erupted recently on efficacy versus lamp spectrum: A group in the US questioned the use of the photopic luminosity curve in calculating visibility, and introduced the concept of "pupil lumens" (9). According to this concept, light colour can be optimized for seeing by providing more blue-green energy exciting the rods of the visual system, as this has as a consequence a smaller pupil diameter that should lead to smaller aberrations in the eye. This concept has still to be checked in practical and laboratory studies. If accepted, it could lead to a shift of using metal halide lamps instead of the high pressure sodium lamps. But it is most unlikely that this idea will meet general acceptance (10).

There is, however, also another aspect that has already been touched and cannot be neglected in these applications, namely colour rendering. Preliminary calculations based on different colour models and visual investigations show that the colour rendering indices of high pressure sodium lamps are underrated by the present method of calculating the general colour rendering index (11). Proper evaluation of the colour characteristics of these lamps might lead to new developments and, consequently, to new applications.

4. Application of energy efficient light sources

In the application area, we will deal only with lighting applications that consume large amounts of energy, and will not consider areas as sports lighting, flood lighting, etc., although they are of high importance for the lamp manufacturers.

4.1. General indoor lighting

The concept of modern building design is system optimization. Thus, the architectural construction will already take optimum daylighting - with minimum heat load in hot climates and optimum utilization of natural radiation heating in cold climates - into consideration. Electrical lighting has to be used as supplementary lighting and for providing good visibility in the hours of darkness.

Older concepts called for an even illumination both time wise and space wise. Now we already know that some change in light level is tolerated, even favoured by the observer as it provides an arousal. It is a challenge for the illuminating engineer

- to provide supplementary lighting during the day that compensates for major changes in day-lighting, without abrupt steps in illuminance and
- to design a night time lighting that is both efficient and pleasant.

For the lamp- and lighting gear manufacturer this means dimmable sources.

For the lighting engineer, the system concept has to mean also total energy optimization: energy used to build the lamp, luminaire, and control gear and energy use during the life of the equipment to produce the best performance lighting in case of work place illumination. We will have to go even one step further and a diversification of the concepts will be necessary, depending on the geographic location of the building:

In cold climates, the heat dissipated during the use of the light might not be of such primary importance as in hot climates, where with every extra input Watt for lighting several Watts have to be consumed to cool the room. Thus, in a cold climate the energy to produce the lighting system has to get a higher weight in the total energy balance than in hot climates.

4.2 Domestic and flexible task lighting

It is expected that home lighting in the near future will bring a shift in favour of compact fluorescent lamps. At present, this is probably more a challenge to the luminaire manufacturer than to the lamp designer.

In task lighting, whether reading lamps, desktop lamps in the office or the home, or work bench lamps, the incandescent tungsten halogen lamps will probably keep a large percentage of the applications, sharing the market with the compact fluorescent lamps of smaller ratings.

4.3. Outdoor and street lighting

We expect a further advance of high pressure sodium lamps in this field. The system concept will call also here for dimmable solutions, although two-three level switching is acceptable (normal lighting, reduced lighting for low traffic periods and increased lighting for special circumstances, as e.g. accidents).

The urban area lighting will be more and more diversified as well (12):

Residential areas with mainly pedestrian traffic and a high demand for "security lighting" will need aesthetic lighting with good efficacy and colour properties, probably using metal halide lamps.

In shopping districts, the street lighting has to blend with the shop window lights and street lights have to provide already an adaptation that makes the goods in the shop window appealing: If the eye has been adapted to the yellowish colour of high pressure sodium light and the observer looks at the shop window illuminated with sources of only 3200K colour temperature, his first impression will be as seeing a scene illuminated with cool-white lights. Not all goods will look appealing under this situation. If the casual observer does not take the necessary time to reaccomodate to the shop window light, he will, perhaps, neglect the displayed goods. Is system approach necessary for the lighting designer of the shop-window and for the street lighting engineer? And what happens if the possible customer enters the shop? Lighting levels, colours and colour rendering have to be harmonized outside the shop and inside. For shopping areas this would mean street lighting lamps with colour properties adjusted to those used in the shops (i.e. metal halide lamps with lamp chromaticity - not colour rendering- in the range of compact fluorescent lamp chromaticities).

5. Demand of developing countries

In the demand areas and applications we have concentrated very much on the demands of industrialized nations who certainly have to be concerned about their energy consumption and have to do all they can to lower their energy demand. Something they would like to do without lowering their standard of living. In large parts of the world, the major questions are not the provision of an optimum visual environment with reasonable energy input, but that of acceptable visibility with affordable energy use. This is also a demand lighting. And here it has to be even more important to provide this light with minimum energy used.

Optimum light distributions might be different in case of lowest level acceptable illumination and optimum visual comfort conditions, as e.g. glare to illumination level is not a linear function of illumination.

Little has been done in this area, and as it is certainly not the most lucrative occupation for an R&D department of a lamp or luminaire manufacturer to deal with these questions, it has also been largely neglected. Projects sponsored by environmental and other agencies concerned with third world development programmes will have to start to shed light into areas and make their explorations attractive to lighting research groups.

6. Conclusions

Light source development has to go in the future towards higher efficacy viewed from a system-viewpoint. This means for the lamp manufacturer

- compact sources to be able to provide high efficiency luminaires with better overall light distribution design;
- improved colour characteristics based on new colour rendering concepts and better blending of sources using different light generation concepts, and
- sources with easy and wide range control of luminous output, to be able to make full advantage of a system concept.

Besides these demands, the request of users, where not optimum viewing comfort but acceptable viewing with optimum energy balance is in the foreground, has to be considered and inexpensive lamps have to be provided. Inexpensive in a sense that the overall cost of lamp, luminaire installation and use is minimized, calculating with real energy costs (available plus new electric plants).

Finally, there is a further aspect the lamp manufacturer has to keep in mind when designing a new light source: his lamp has a finite life and after use as much of the lamp as possible should be recycable, and the non-recyclable parts have to be disposed safely. The total (energy and labour) costs of this disposal have to be also a part of the deliberations when a new light source is constructed and have to be considered when the total energy balance of the lighting is drawn.

References

1 CIE Publ. 29/2, 1986: Guide on interior lighting, 2nd ed., CIE Central Bureau, Vienna.

2 LiTG-Fachausschuß "Innenbeleuchtung", Arbeitsgruppe "Blendung", 1989: Der Kontrastwiedergabefaktor CRF - ein Gütemerkmal der Innenraumbeleuchtung.

3 U.S. Environmental Protection Agency, Global Change Division: Green Lights Program.

4 CIE Publ. 13/2, 1988: Method of measuring and specifying colour rendering of light sources, 2nd ed., corrected reprint. CIE Central Bureau, Vienna.

5 Schanda, J: A combined colour preference - colour rendering index. Lighting Res. & Technol., 17/1, 31-34, 1985.

6 CIE TC 1-33: Colour Rendering

7 CIE TC 6-18: Evaluation of potential optical hazards associated with "desk top" quartz halogen lamps.

8 Willkins, A: Intermittent illumination from visual display units and fluorescent lighting affects movements of the eye across text. Human Factors, 28/1, 75-81, 1986.

9 Berman, S: Energy efficiency consequences of scotopic sensitivity. IESNA Conf., 1991.

10 Rhea, M: Are we seeing eye to eye on pupil lumens? El. Design, March 1992.

11 Schanda, J: Chromatic adaptation and colour rendering. CIE-Journal, 1/2, 30-37, 1982.

12 CIE Publ. 92, 1982: Guide to the lighting of urban areas. CIE Central Bureau, Vienna.

RECENT PROGRESS IN LAMP PHOSPHORS

S. Kamiya
Matsushita Electronics Corp.
R & D Center Light
Takatsukishi, Osaka, Japan

1. Introduction

It was well known that rare earth ions have peculiar emission spectra since rather old. Practical application for fluorescent lamps, however, started in 1966. Eu-activated Sr-orthophosphate was applied for photocopying fluorescent lamps as the first practical use of rare earth phosphor. Since after, research and development of these phosphors were accelerated and the decade of 1970s was golden ages of rare earth phosphors development. Numerous charateristic phosphors were succesively developed and put into practical use. However, after Ce, Tb and Mn co-activated Gd-Mg borate phosphor reported in third LS in 1983, worthnoticing new phosphor did not appear. On the other hand, remarkable progress was achieved during this period in fluorescent lamp technologies both quality improvements and new products development. Out of those innovations, development of new compact fluorescent lamps was essencial. These lamps are forced to higher wall loading than that of conventional design lamps because of their compactness. High wall loading results the temperature rise on tube wall as well as increasing of 185 nm radiation which has influence for deterioration of phosphors.

Indeed most of rare earth phosphors have superior characteristics than conventionalphosphors even under such strict conditions. Nevertheless, further improvements were requested to meet with new lamp type requirements. Therefore the efforts of phosphor engineers were concentrated to solve these problems by means of optimization of chemical composition, finding of effective additives and improving the manufacturing processes in last few years. This efforts is still continued.
In this paper, the results of recent activities are reviewed and author's personal view is added.

2. Emission spectra of rare earth ions

Peculiar characteristics of rare earth ions are caused by electrons in imperfect 4f shell. Emission of rare earth ions are originated different two energy transfer mechanisms, namely transition within 4f shell and transition from 5d to 4f shell. Emission spectra generated by transition within 4f shell are little influenced by the crystal structures where rare earth ion exist. 4f shell is not outermost shell of the ion and surrounded by 8 electrons outside. These electrons works as the shield to protect the influence of crystal field. Therefore, emission spectrum almost depends on the kind of rare earth ion itself. Typical example is Tb^{3+} emission. Spectral distribution curves of Tb^{3+}in Y_2SiO_5,$CeMgAl_{11}O_{19}$,$La(PO)_4$, and $GdMgB_5O_{10}$ are quite similar. Only slight difference in fine structure of spectra and intensity ratio of sub bands is observed . 8 rare earth ions have emission in visible wavelength region. Out of these, Eu^{3+} and Tb^{3+} are being widely used for practical phosphors. Dy^{3+} had been used for light white colour lamps. Spectral power distribution curves of these emissions are shown in Fig.1 compared to those of Eu^{2+} emission. Emission spectrum of Sm^{3+} seems attractive but efficient phosphor is yet found. Contrary to the emissions generated by transition within 4f shell, emissions by the transition from 5d to 4f are influenced directly by crystal field because 5d has no outside shield as different from 4f shield. This means emission spectrum depends on strongly the surrounding of rare earth ion existed. Eu^{2+} emission is typical example. Peak wavelength of Eu^{2+}emissions are distributed from 360 nm till 500 nm depending on the matrix crystals as shown in Fig.1 and Fig.2.

$Ce^{3+}$is also belonging in this category and emission spectra also varies from near ultra violet till near infrared depending on the matrix structure. On practical lamp phosphor, however, Ce^{3+}'s function as sensitizer for Tb^{3+} and Mn^{2+} is more important than Ce^{3+} emission itself.

3. Phosphors for 3 bands lamps

First commercialized 3 bands fluorecsent lamps were realized by utilizing the combination of $BaMg_2Al_{16}O_{27}$:Eu^{2+}, $CeMgAl_{11}O_{19}$:

(Ce^{3+}):Tb^{3+} and Y_2O_3:Eu^{3+} phosphors.[1] Emission peaks of these phosphors locate around at 450 nm, 540 nm and 611 nm respectively. In spite of various efforts, practical alternative of Y_2O_3:Eu^{3+} phosphor is not found yet. As the alternatives of $CeMgAl_{11}O_{19}$:(Ce^{3+}):Tb^{3+}, $LaPO_4$:Ce^{3+}:Tb^{3+} [2] and $GdMgB_5O_{16}$:Ce^{3+}:Tb^{3+} phosphors [3] are being used. These three phosphors have quite similar emission spectrum as mentioned section 2. Only small difference in fine structures of spectrum and intensity ratio of sub-bands are observed. This small difference gives some variation of Ra value within around plus or minus one in 3 bands lamps. Highest Ra is obtained by GdMg-borate and lowest by La-phosphate whereas lamp efficacy is the opposite. There is no significant difference in light output maintenance between these two phosphors. Therefore, as far as applied for standard fluorescent lamps, there is little difference to choose between the two. As for blue component phosphors, Eu^{2+} activated BaMg-aluminate and alkaline earth metals chloro-apatite are being used. Extensive studies have been made to clarify the influence of chemical composition especially Sr,Ba and Ca ratio for emission spectrum.[4,5,6] Increasing of Ca and Ba ratio gives wider spectra trailed to longer wavelength side. Precise analysis of the spectra is still expected to understand the emission mechanism. For further improvement of colour rendering properties, addition of deep red emissive phosphor such as $3.5MgO\ 0.5MgF_2$:Mn^{4+} is effective. This measure, however, accompanies decrease in lamp efficacy. On the other hands, it has been proved that providing the emission in the region of 490 nm is also effective to improve Ra value with the small sacrifice of lamp efficacy. Fig.3 shows spectral power distribution curves of blue green emitting phosphors [7,8,9] utilizing for this purpose. The 3 bands fluorescent lamps of Ra 87 - 88 commercialized in Japan are this application.

4. Phosphors for compact fluorescent lamps

General requirements of emission characteristics for compact fluorescent lamps are similar to those of normal 3 bands lamps. Special attentions ,however, should be paid additionally for following two points. First point [illegible]igh emission efficiency as well as good physical and chemical stability under the increased radiation of 185 nm which is generated distinctly in compact fluorescent lamps. Second point is less emission quenching at high temperature ambient. As far as at wall temperature zone of compact fluorescent lamps, temperature quenching of blue and red emissive phosphors is not so significant. Tb^{3+} emission of Ce^{3+}-Tb^{3+} co-activated phosphors, however, is influenced by ambient temperature and is observed temperature quenching phenomenon at above 100°C. In case of $La(PO)_4$, degree of temperature quenching depends on Ce content. With increasing in Ce content, quenching starts at lower temperature. Therefore, Ce content in phosphors should be decided considering maximum wall temperature of the lamps to be applied.[10] There is another report on improvement of temperature quenching of $La(PO)_4$.[11] According to this report, temperature quenching causes by existence of Tb^{4+}. Transfered energy to Tb^{4+} disappears through non-radiative process at higher temperature. Generally higher the Ce content, there is more chance to formate Tb^{4+}. To prevent or suppress the formation of Tb^{4+}, addition of BO^{3-} ion or Th metal is reported as effective measure. This consideration seems quite interesting. Further support researches are expected.

5. Expectation for new phosphors and subjects to be studied

From the results of research and development up till now and the recent research trend, author's private view on expectation for new phosphors and subjects to be studied are summarized.

5. 1 Rare earth - Mn co-activated phosphors

As mentioned section 2, spectrum of Eu^{2+} emission is sensitively influenced by structure of matrix crystal. It is well known that spectrum of Mn^{2+} emission also widely changes depending on matrix crystal. In Fig.4, some examples of the spectral power distribution curves of Mn^{2+} emission on practically used phosphors (present and past) are illustrated. Peak wavelength locates at the range from 515 nm to 630 nm. Co-activation of rare earth ions and Mn^{2+} seems interesting study subject.

As far as Eu^{2+}- Mn^{2+} co-activation, only BaMg-aluminate phosphors are being practically used of which emission peaks locate at around 450 nm and 515 nm. However, from the research made in before 1970,[12,13] several phosphors can be found of which peaks of emission spectra locate longer wavelength than 600 nm. If new matrix crystal which gives efficient Mn emission of longer wavelength region is found, it might be interesting and useful.

Whereas various Ce^{3+}-Tb^{3+}co-activated phosphors were developed, Tb^{3+}emission shows almost same spectral distribution as the reason mentioned section 2 . On the other hand, Ce^{3+}- Mn^{2+} co-activation has a possibility of variety of spectral distributions. For instance some of aluminate emit green light of which peak wavelength is around 516nm [14], and peak of yellow emission generated by Sr-fluorophosphate is reported as 577nm.[15] GdMg-borate emits red light of which peak locates 630 nm.

Matrix crystals which can be co-activated both by Ce^{3+}- Tb^{3+} and by Ce^{3+}- Mn^{2+} generate both $Tb^{3+}$and Mn^{2+} emissions when Ce, Tb and Mn are doped. Emission intensity ratio of Tb^{3+}/Mn^{2+} depends on the mole ratio of additives. Spectral power distributions of such phosphors are shown in Fig.5 and Fig.6 as example. Combinations of GdMg-borate co-activated by Ce^{3+},Tb^{3+} and $Mn^{2+}$and $Sr_4Al_{14}O_{25}$: $Eu^{2+}$are able to offer Ra 95 fluorescent lamps of colour temperature throughout 3000 - 5000 K . Because Mn^{2+}emission in CeMg-aluminate is too close to that of Tb^{3+}, this phosphor is not so attractive for general lighting purpose.

Nevertheless, it is useful to improve the colour purity of green component of colour disply pixel based on fluorescent lamp principle.[16] Further research for energy transfer mechanism as well as discovery of new phosphors of Ce^{3+}- Tb^{3+}- Mn^{2+} are expected.

5. 2 Phosphors for rare gas discharge fluorescent lamps

Drawbacks of fluorescent lamps are slow rising of light output after switching on and temperature dependence. These drawbacks restrict to use the fluorescent lamps in some cases, for instance, light sources for office automation equipments such as facsimiles and LCD displays. For such applications rare gas discharge fluorescent lamps are being popularized. Excitating source for phosphors is mainly Xenon's resonance line at 147nm of wavelength which is far shorter than mercury vapour lines of 254 and 185 nm. Under such excitation, conventional phosphors which have high efficacy in normal fluorescent lamps are not always efficient. According to latest report[17], for green colour emission, $(Y,Sc)_2SiO_5:Tb^{3+}$ phosphor gives far better result than Ce^{3+}- Tb^{3+} co-activated phosphors which are most efficient for normal fluorescent lamps. For red colour emission, $Gd_2O_3:Eu^{3+}$ is preferable than $Y_2O_3:Eu^{3+}$ from the view to temperature dependence. It is also reported that $BaMgAl_{10}O_{17}:Eu^{2+}$ has highest efficacy among the Eu^{2+} activated phosphors but light output maintenance is worse than other two colour emissive phosphors. On the other hand, $Y(VP)O_4:Dy^{3+}$ was proposed as the white colour emitting single component phosphor. Spectrum of this phosphor consists of blue green VO^{-3}band and blue green and yellow Dy^{3+} bands as shown in Fig.7.[18]

Since the intensity of resonance lines of rare gas are far weaker than those of Hg, application of rare gas discharge type fluorescent lamps will be limited. Nevertheless, research on the behavior of the luminescence under such wavelength excitation as neither the region of electron beam nor UV generated by Hg discharge are very interesting. Development of these kinds of phosphors is more closely related to actuarization of colour plasma display unit.

5.3 Low Hg consumption phosphors

Hg is absolutely necessary to operate fluorescent lamps. However, since dumping of used fluorescent lamps will gives influence to environment, mercury in fluorescent lamps is being the object of public attention. From environment of view, Hg content in fluorescent lamps must be minimized. In order to realize this minimization, accurate Hg consumption mechanism in fluorescent lamps during burning have to be clarified. Due to the complicated complex factors, details of mercury consumption mechanism is not known yet and only few reports can be found. Presentation by P.W.C.Verhees [19] in York symposium, was the first one essentially dealt with this subject. However,it is well known fact either by experience or experiment that consumption of Hg in fluorescent lamps depends quite deal on the sort of phosphor. According to above mentioned report, the consumption of Hg in standard 40 watt type lamp of 1.2 m length with 26 mm dia., during 5000 hrs burning is 3 - 8 mg in case of conventional Ca-halophosphate phosphor whereas 0.5 - 4 mg in case of rare earth phosphors blend. The courses of this large difference are estimated as Na diffusing speed from glass tube and formation probability of Na-Hg amalgam and mercury diffusing speed into glass by exchanging Na ion and Hg ion are depend on the properties of different phosphors. To prevent or suppress these phenomena, application of protective layer on the inner surface of glass tube has been proposed by which diffused Na ions can be prevented to direct contact with phosphor. Transparent metal oxides layer such as Y_2O_3,[19] CeO_2,[20] Al_2O_3 and etc. have been reported as effective materials.

Meanwhile very interesting reserch works on relationship surface potential and adsorption quantity of mercury has been presented[21,22]. According to the report, Hg and HgO adsorption quantity is depending on electric charging tendency as shown in Fig.8. Adsorption became large when surface charge of phosphor turned to negative. Although each phosphor has normally peculiar surface potential, modification can be possible by special surface treatment.

Surface charge of phosphors might have influence for coagulation of particles in blended phosphor system.

From the industrial view point, development of low mercury consuming phosphors might be one of the most important project for the time being ,whichever by new material or surface treatment.

References

1)J.M.P.J.Verstegen : J.Electrochem.Soc.,121-12, 1623, (1974)
2)S.Nakajima et al : Proc. Japanese Phosphor Research Soc. 186th meeting (1981)
3)J.Th.W.de Hair et al : Proc.LS 3 (1983)
4)M.Nakamoto et al : Proc. IES meeting Atranta(1982)
5)H.Sumoto et al : Proc. Annual Meeting Illum.Engng.Inst. Japan (1990)
6)T.Hisamune et al : Proc.Annual Meeting Illum.Engng.Inst. Japan (1991)

7)J.T.C.Van Kemenade et al : Spring Meeting of Electrochem.Soc
Abstruct No.607 (1983)

8)A.Taya et al : Proc. Japanese Phosphor Research Soc.
192th Meeting (1982)

9)K.Murakami et al : MITSUBISHI DENKI GIHO 52, 701 (1978)

10)K.Ichinomiya et al : Proc. Japanese Phosphor Research Soc.
233th Meeting (1990)

11)N.Hashimoto et al : MITSUBISHI DENKI GIHO 61,609 (1987)

12)T.L.Barry : J.Electrochem.Soc.,117 381 (1970)

13)C.C.Lagos : J.Electrochem.Soc. 115 1271 (1968)

14)W.Liu et al : J.Luminescence 40&41 877 (1988)

15)R.G.Pappalardo : J.Electrochem.Soc. 130 2087 (1983)

16)T.Higashi et al : Proc. Annual Meeting of Illum. Engng.
Inst.Japan (1990)

17)N.Hashimoto et al : Proc.Annual Meeting of Illum.Engng.
Inst.Japan (1991)

18)K.Nakagawa et al : Proc.Annual Meeting of Illum.Engng.
Inst. Japan (1988)

19)P.W.C.Verhees : Proc. LS 5 (1989) Abs. No 105L

20)M.Myojyo : Proc.Annual Meeting of Illum. Engng.
Inst. Japan (1988)

21)H.Ito et al : Proc. Annual Meeting of Illum.Engng.
Ins. Japan (1990)

22)H.Ito et al : Proc. Annual Meeting of Illum.Engng.
Ins. Japan (1991)

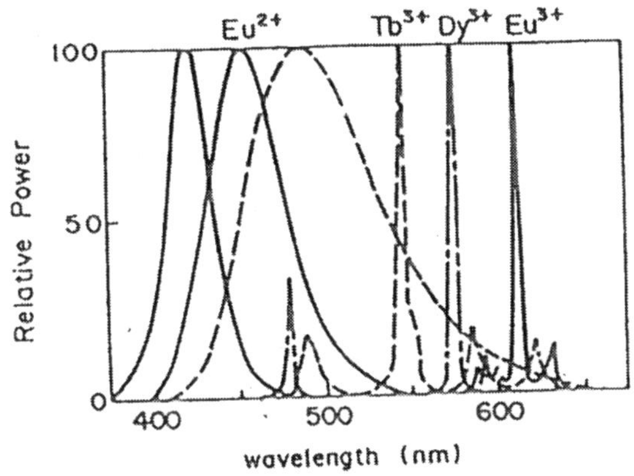

Fig. 1 Spectral power distributions of various rare earth metal ions

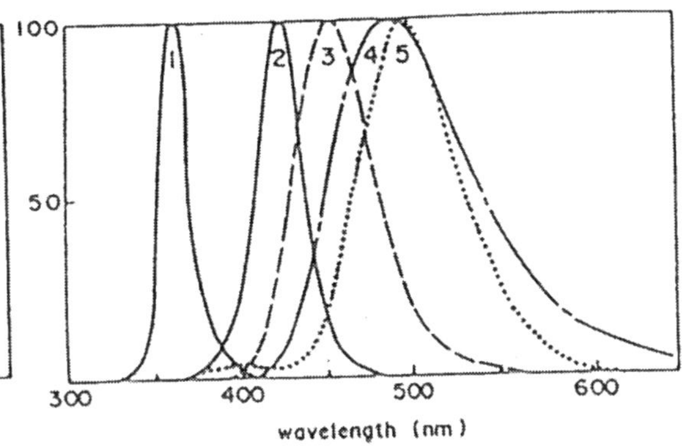

Fig. 2 Spectral power distributions of Eu^{2+} emission in different matrix crystals
1. $SrB_4O_7F:Eu^{2+}$ 2. $Sr_2P_2O_7:Eu^{2+}$ 3. $BaMg_2Al_{16}O_{27}:Eu$
4. $(Ba,Ca,Mg)_5(PO_4)_3C$: Eu^{2+} 5. $Sr_4Al_{14}O_{25}:Eu^{2+}$

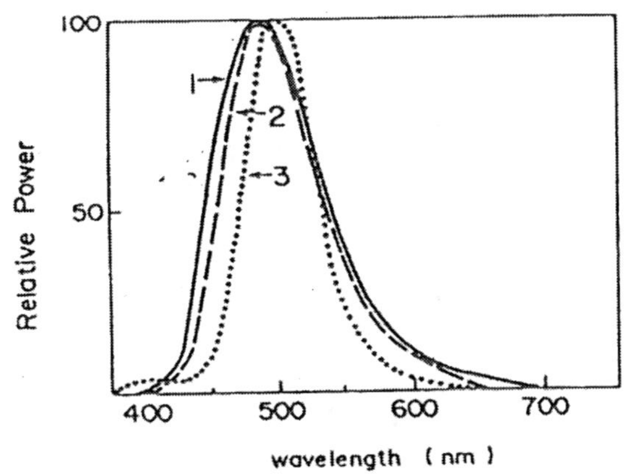

Fig. 3 Spectral power distributions of Eu^{2+} emission in Blue-Green region
1. $(Ba,Ca,Mg)_{10}(PO_4)_6C\ell_2:Eu^{2+}$ 2. $2SrO \cdot 0.84P_2O_5 \cdot 0.16B_2O_3:Eu^{2+}$
3. $Sr_4Al_{14}O_{25}:Eu^{2+}$

Emission Colour	Chemical Composition
Blue	$(Sr,Ca,Ba)_5(PO_4)_3Cl:Eu^{2+}$ $BaMg_2Al_{16}O_{27}:Eu^{2+}$
Green	$CeMgAl_{11}O_{19}:(Ce^{3+}):Tb^{3+}$ $LaPO_4:Ce^{3+}:Tb^{3+}$ $GdMgB_5O_{10}:Ce^{3+},Tb^{3+}$
Red	$Y_2O_3:Eu^{3+}$

Table 1. Phosphors for 3 bands fluorescent lamps

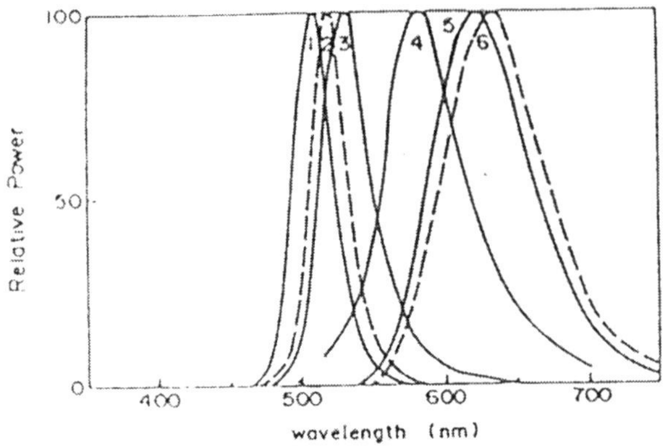

Fig. 4 Spectral power distributions of Mn^{2+} emission in various matrix crystals
1. $MgGa_2O_4$ 2. $BaMg_2Al_{16}O_{27}$ 3. Zn_2SiO_4
4. $3Ca_3(PO_4)_2 \cdot Ca(F,Cl)_2$ 5. $Cd_2B_2O_5$ 6. $GdMgB_5O_{10}$

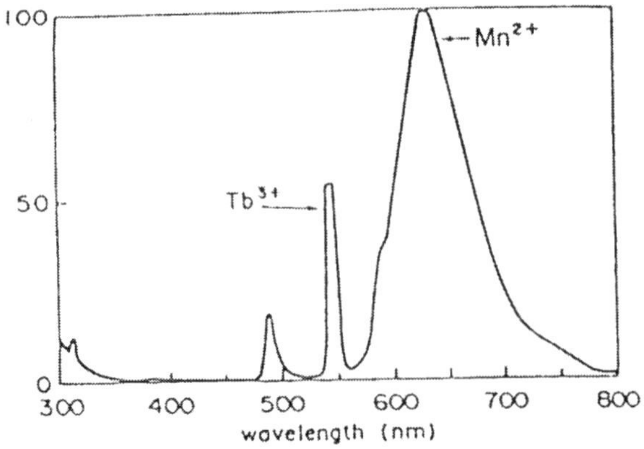

Fig. 5 Spectral power distribution of $GdMgB_5O_{10}$: Ce : Tb : Mn

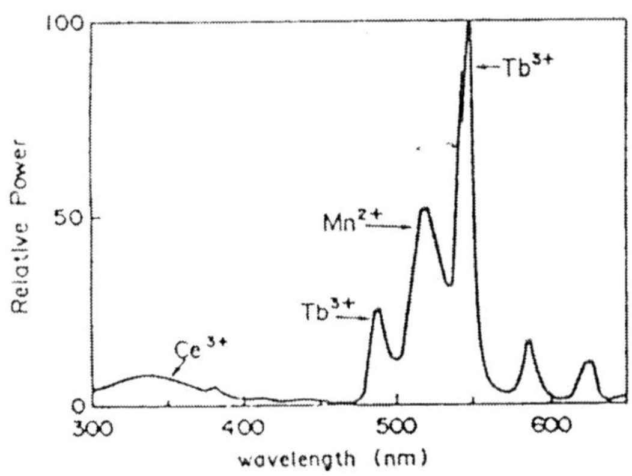

Fig. 6 Spectral power distribution of $CeMgAl_{11}O_{19}$: (Ce) : Tb : Mn

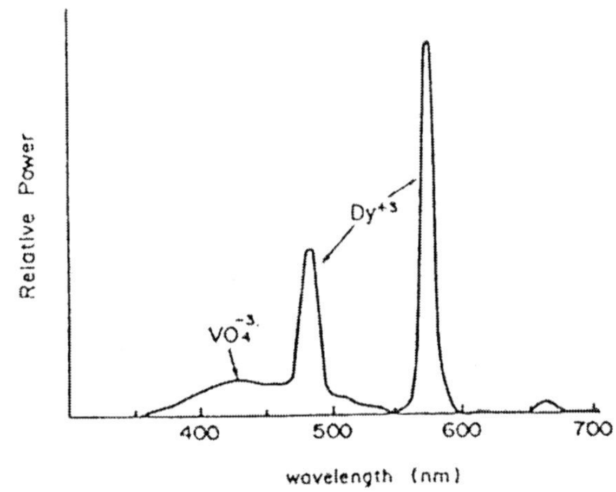

Fig. 7 Spectral power distribution of $Y(PV)O_4$:Dy

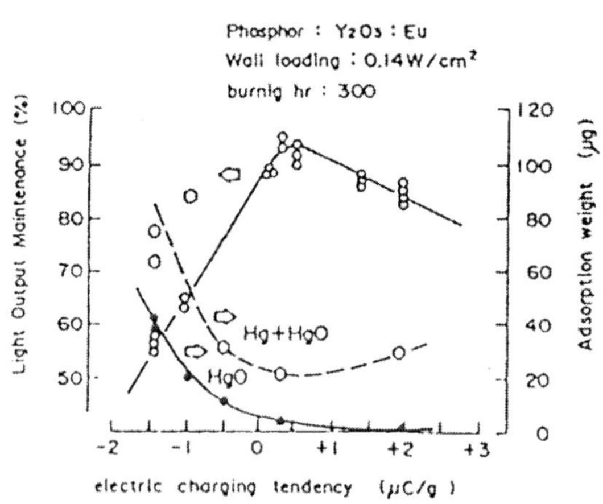

Fig. 8 Light output maintenance, adsorption of Hg and HgO as a function of electrical charging tendency of phosphor

THERMOLUMINESCENT STUDY OF SECA BLUE PHOSPHOR SAMPLES

Veronika Bohay, János Osán, László Balázs

TUNGSRAM BIKK, Phosphor and Coating ATC, H-1340,
Budapest, Vaci ut 77, Hungary.

The $(Sr,Ca,Ba)_{10}(PO_4)_6Cl_2:Eu^{2+}$ chemical composition blue phosphor component (SECA) were studied. Two different composition samples were measured:

Sample 1 - near stochiometric
Sample 2 - cation excess to anion

The TL-curve shows that the intensity of the 417 K glow peak of the sample 2 lower than that of sample 1.(Fig. 1)
It must be noted, that the 350 and 417 K TL-peaks are due to Eu^{2+} emission. The peak above 480 K can be explained by O^{2-}-vacancies [1].
The thermoluminescent properties of sample 1 were studied in detail. The excitation of samples was done with 185 + 254 nm emission of low pressure mercury lamp in cleaned N_2 atmosphere.
Sample 1 and its dried suspension were heat treated for 10 min at the temperature of 573, 773, 873, 973 K. (Figs. 2, 3)
One of the important effects is that the intensity of the 350 and 417 K peak decreases with the increasing of "lehr" temperature. This tendency is in correlation with 450 nm emission intensity.
In the case of dried suspension series less decrease was found, but the 973 K heat treatment caused very high drop in glow intensity. It is presumably due to europium valence change, which is shown by ESR-results.Miscellaneous phase was not detected by X-ray diffraction method after the heat treatment.
The trapped electrons can be released not only by heating, but by visible irradiation too [2]. At the bleaching the visible reflexion increases, the intensity of 350 and 417 K TL-peaks decreases (Fig. 4) and we could detect blue light at 450 nm (Fig. 5).
For understanding the stability of three band fluorescent lamp studying only the three phosphors is not sufficient, also their interaction must be examined.

References

[1] A. Meijerink & G. Blasse, J Phys Condens Matter, 2, 3169 (1990)
[2] Mian Zeng Su et al., J Less-Common Metals, 148, 241 (1989)

BLUE PHOSPHOR SAMPLES

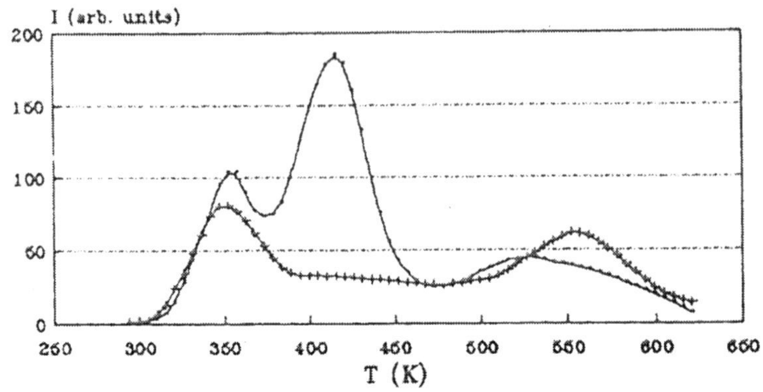

Effect of composition

— Sample 1 —+— Sample 2

Figure 1

SAMPLE 1 BLUE PHOSPHOR

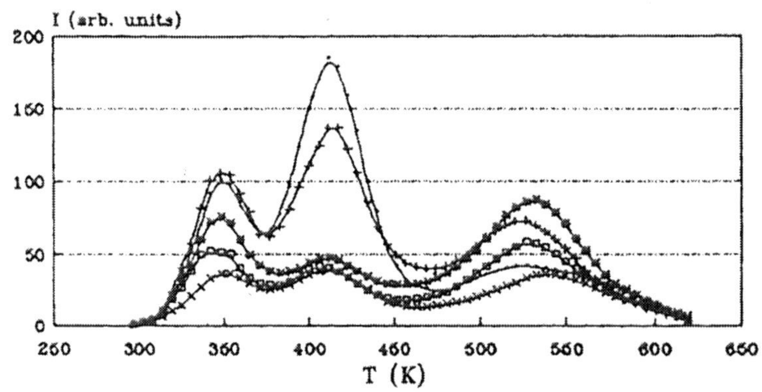

Heat treatment

— Base sample —+— 573 K —*— 773 K —□— 873 K —×— 973 K

Figure 2

DRIED SUSPENSION OF SAMPLE 1

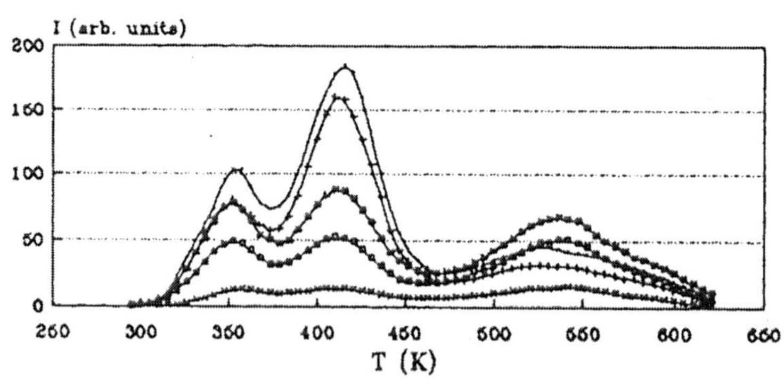

Heat treatment

— Base sample —+— D. susp. 573 K —*— D. susp. 773 K

—■— D. susp. 873 K —×— D. susp. 973 K

Figure 3

SAMPLE 1 BLUE PHOSPHOR

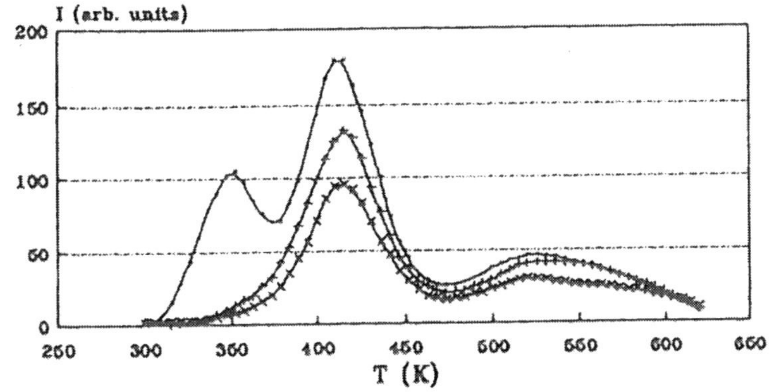

"Bleaching" effect

— Base sample —+— 632,8 nm 30 min —*— 543 nm 30 min

Figure 4

BLEACHING WITH He-Ne LASER

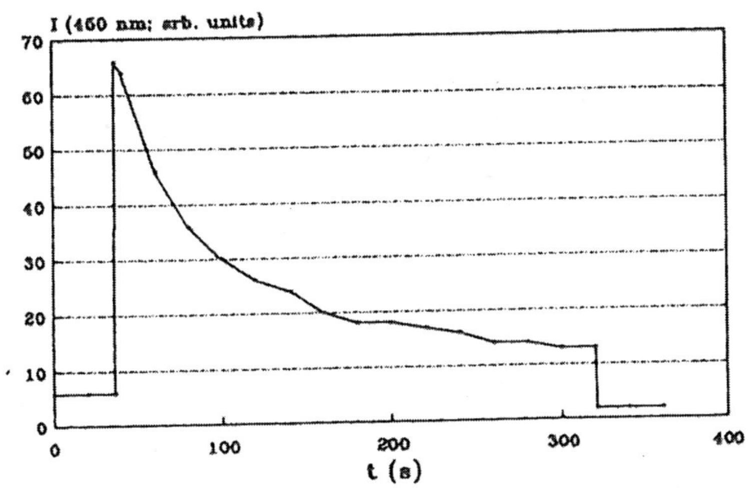

Figure 5

Manufacturing of the photoluminescent glass doped with $B_2O_3-La_2O_3$ and its application in fluorescent lamp

CHEN XUE MING
(Chang Sha Lamp Plant, Hunan China)

Summary:

A kind of photoluminescence glass doped with $B_2O_3-La_2O_3$ has been explored in this paper. Its optimum composition and skills manufacturing have also been explored. With its stability and radiation under the bombardment of UV (2537 Á), it can replace partial phosphor and increase light-efficiency of the fluorescent lamp 20% higher and postpone its life-time more than one thousand hours.

Key words: photoluminescence glass manufacture silicate doped with $B_2O_3-La_2O_3$.

SODIUM RADIATION-INDUCED HYDROGEN EMISSION FROM A H_2-DOPED BARIUM FILM

Corrado Carretti, Paolo Manini, Antonio Renzo, Elio Rabusin
SAES Getters S.p.A., Via Gallarate 215, 20151 Milan - ITALY

1. Introduction

Barium is widely used as a getter in vacuum devices such as television picture tubes, electronic tubes, lamps.
Many experiments have therefore been carried out in past years to determine sorption characteristics for specific gases and to elucidate the effect of deposition parameters on the structure and properties of the barium film.
As far as HPS lamp are concerned, a number of studies have been addressed at investigating the behaviour of barium under various operating conditions (1)(2)(3).
However, the effect of the arc tube sodium radiation on the gettering properties of the barium film seems to have received up to now little attention.
This can be of particular interest in the case of hydrogen which is sorbed by barium in large amounts during the life of the lamp (3).
In the present paper a few preliminary results aimed at clarifying this point are presented. We have found that hydrogen-doped barium film can release some hydrogen when irradiated with the sodium light.

2. Experimental

A properly designed pyrex bulb, containing a sodium arc tube of 400W nominal power, was used as a test vessel (fig. 1).
The bulb was connected to a bench equipped with a quadrupole mass spectrometer for partial pressure measurements, as per fig. 2.
For each test a couple of ring type barium getters were mounted each yielding about 8 mg of Ba. After mounting and before starting with the test, the bench and the bulb were baked under vacuum at 573 K for one night to reduce the background pressure in the 10^{-6} Pa range. The arc-tube was also conditioned by switching it on for 4 hours prior to each test. The barium film temperature was measured by means of a K-type thermocouple directly positioned on it. Hydrogen of 99.9% nominal purity was used, for sorption/release studies of the gas under irradiation.

3. Results

In a first set of tests the barium film was made to sorb hydrogen (0.133-0.665 Pa m^3/g) and then exposed to the arc tube radiation.
The sudden hydrogen pressure increase which takes place at switching on the arc tube is clearly visible in fig. 3. The rise in temperature of the barium film because of the heat radiated by the arc-tube is relatively slow (only 50 K during the first two minutes). Hydrogen pressure increase can therefore be essentially ascribed to the arc-tube irradiation, rather than to the known equilibrium pressure of H_2 in the Ba-H_2 system. After the pressure has reached a peak value it slowly decreases with time. This is mainly due to the pumping action of the filament of the mass spectrometer, as studied in separate experiments.

In a second set of tests a barium film was progressively dosed with up to 12 Pa m^3/g while the arc tube was switched on (film temperature was 510 K). The hydrogen pressure over the film was measured 20 minutes after the sorption of each dose, this time being considered sufficient to reach a steady state condition. The experiment was repeated on another barium film kept at the same temperature but in absence of radiation (arc-tube switched off). Heating was accomplished by means of an external heating tape. Results are shown in fig. 4.
They confirm that sodium radiation promotes hydrogen outgassing from the barium film, leading to H_2 pressure in the bulb higher than those expected on a pure thermal basis.
This result can be of technological interest in HPS lamp manufacturing, where efficient hydrogen gettering must be assured for a good thermal insulation. Further work is in progress aimed at a better understanding of this phenomenology.

References

(1) P.C. Ward, J. Illum. Eng. Soc., 9 (1980) p. 194
(2) V.D. Meyer, E.F. Wyner and P.L. Fortucci, J. Illum. Eng. Soc., 9 (1980). p. 175
(3) G.H. McGrath and M. Connolly, Illum. Eng. Soc. Conference (1980)

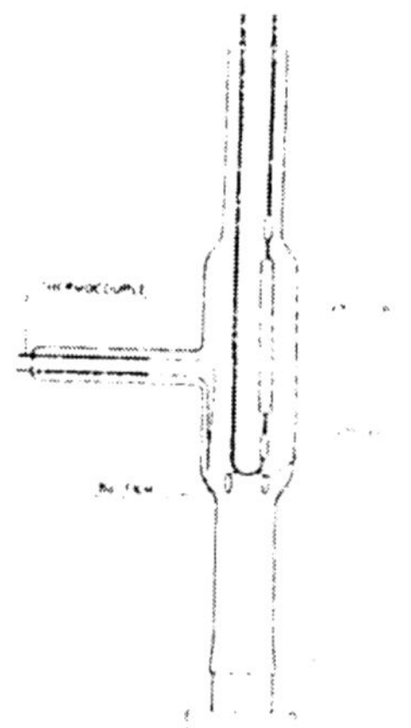

Fig. 1 - Lamp bulb

Fig. 2 - Experimental set-up

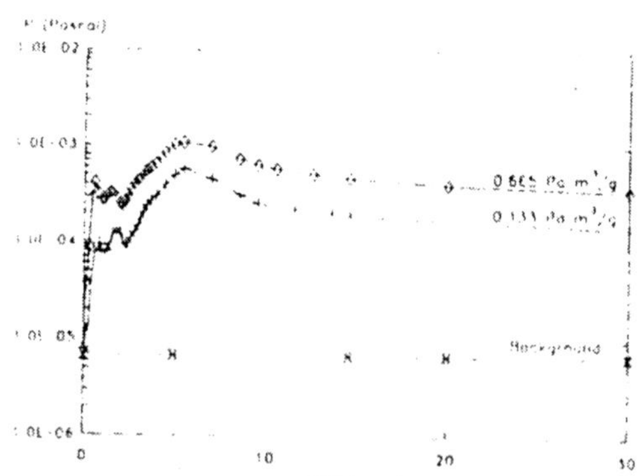

Fig. 3 Hydrogen pressure evolution after the lamp switch on

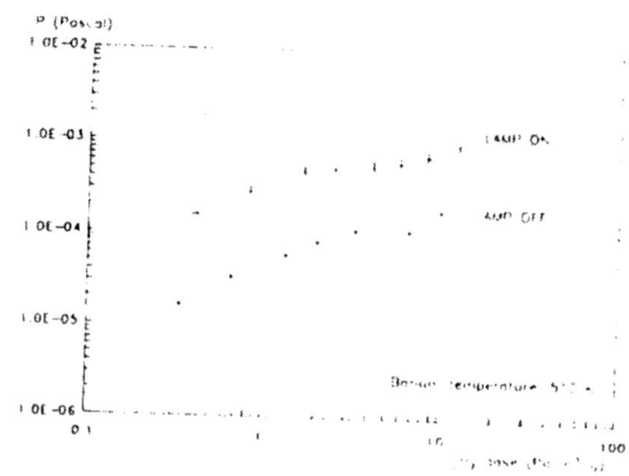

Fig. 4 Hydrogen release as a consequence of sodium arc-tube irradiation

RAY-TRACING ESTIMATION OF KUBELKA-MUNK PARAMETERS AND APPLICATION TO PHOSPHOR LAYERS AND PLAQUES

Lawrence L. Hope
GTE Electrical Products
60 Boston Street
Salem, MA 01970

The Kubelka-Munk theory is widely used in the lighting industry to predict optical behavior in particulate phosphors, e.g. those in fluorescent lamps. Customary methods for estimating the Kubelka-Munk scattering parameter β and absorption parameter μ cannot deal well with two issues of interest: applicability to thin layers and effects of surface roughness. Thin layers are the rule in fluorescent lamps with a typical thickness of three particles; it has never been clear that a single value for β and μ can describe the entire range of layer thicknesses from very thin to plaques. Effects of surface roughness on layer reflectance are also important determinants of lamp performance.

Reflectance and transmittance values for particulate phosphors are presented here that were computed by ray tracing of vector waves representing ultraviolet flux in explicitly defined close-packed arrays of particles. The particles are characterized by radius R, internal Beer's law absorption coefficient μ_s, and index of refraction. Parameter configurations considered include weak, intermediate, and strong absorption. For each configuration, it is found that a single pair of Kubelka-Munk parameters can reproduce the ray tracing results for the full range of layer thicknesses, from one particle to plaques.

A typical comparison of ray tracing data (points marked *) and Kubelka-Munk estimates (curves) for transmittance and reflectance is shown in Figure 1. The value of the dimensionless parameter $\mu_s R$ is 1.3 and the index of refraction is 1.5. Layer count varies from one to six.

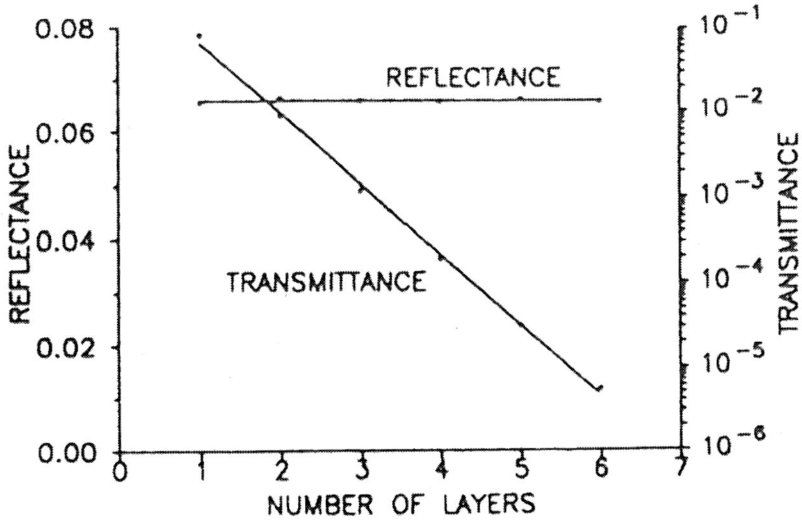

Figure 1. Ray Tracing Data for Transmittance and Reflectance with Estimates from Kubelka-Munk Theory

Figure 2 shows the dimensionless ratios β/μ_s and μ/μ_s as functions of $\mu_s R$. Best fit points are marked *. As expected from scattering theory, in the limit of large $\mu_s R$, β/μ_s varies inversely with $\mu_s R$. It is found that μ/μ_s also varies inversely with $\mu_s R$ in this limit. For this and other examples run, μ exceeds μ_s for small $\mu_s R$. The transition from values below μ_s to those above μ_s is smooth as scattering and light trapping are increased by reduced particle size.

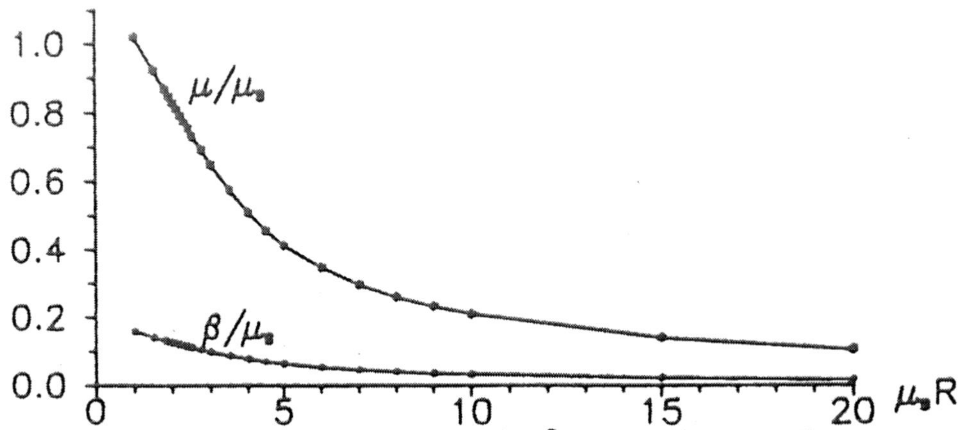

Figure 2. Dimensionless Ratios β/μ_s and μ/μ_s vs $\mu_s R$

Figure 3 shows the effect of surface roughness on plaque reflectance. Total thickness for this case was sufficient to absorb all incident radiation. Particles were removed at random from the top layer on the incidence side. For non-random particle removal, the computed reflectance data varies with the pattern of particle population on the top layer. Moderate roughness is found to reduce reflectance of ultraviolet to the discharge.

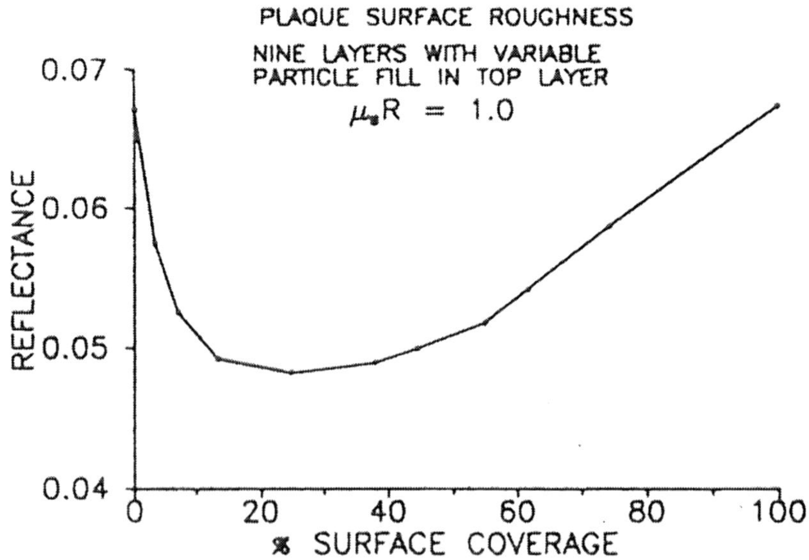

Figure 3. Plaque Reflectance vs per cent Surface Coverage

THE LUMINESCENCE PROPERTIES OF Ce^{3+}- AND Mn^{2+}-ACTIVATED CERIUM MAGNESIUM ALUMINATE PHOSPHORS.

K. Iwama, T. Higashi, T. Matsuda and S. Kamiya
Light R&D Center, Matsushita Electronics Corp.
Takatsuki, Osaka, Japan.

Although $La(PO_4):Ce^{3+}:Tb^{3+}$ or other Ce^{3+} - Tb^{3+} coactivated phosphors have efficacy green emission, colour purity is not satisfactory in such applications of large colour display system utilizing fluorescent lamp principle. Higher the colour purity brings better colour reproduciability in wide range.

Out of Ce^{3+}-Tb^{3+} coactivated phosphors, $CeMgAl_{11}O_{19}:(Ce^{3+}):Tb^{3+}$, generates additional emission peaked at 515 nm by dosing Mn^{2+} ion. Combination of Tb^{3+} and Mn^{2+} gives improved colour purity whereas the light out put efficacy goes down. Investigation has been made to improve the efficiency of Mn^{2+} emission by means partial substitution of Mg by different alkaline earth metals such as Ba, Sr, Ca and Zn. These substitutions were effective to increase in Mn^{2+} emission as shown in Fig. 1. Out of these 4 metals, improvement by Zn substitution was dominant. Fig. 2 shows the excitation spectra of Mn^{2+} emission as a function of Zn substituting ratio y for $Ce(Mg_{0.82-y}Zn_y)Al_{11}O_{19}:Mn_{0.18}$. Excitation spectraof Mn^{2+} emission extend to longer wavelength side and more widely overlap the u.v. luminescence of Ce^{3+} ions with increasing in Zn content. Increase in intensity of Mn^{2+} emission by Zn substituion can be explained as the results of efficient energy transfer from Ce^{3+} to Mn^{2+}. Measurements of emission spectra suports this assumption. Fig. 3 shows the relative intensities of Ce^{3+} and Mn^{2+} emission in $Ce(Mg_{0.7-x}Zn_{0.3})Al_{11}O_{19}:Mn_x^{2+}$ as a function of Mn content x. When Mn content x incresed above 0.5, additional band was observed in 600-700 nm region. Chromaticity coordinates (x,y) of $CeMg_{0.52}Zn_{0.3}Al_{11}O_{19}:Mn_{0.18}$ under 254nm excitation is shown in Fig. 4, compared to other phosphors. Comparison of light out put efficacy and colour coordinates is summarized in Table 1.

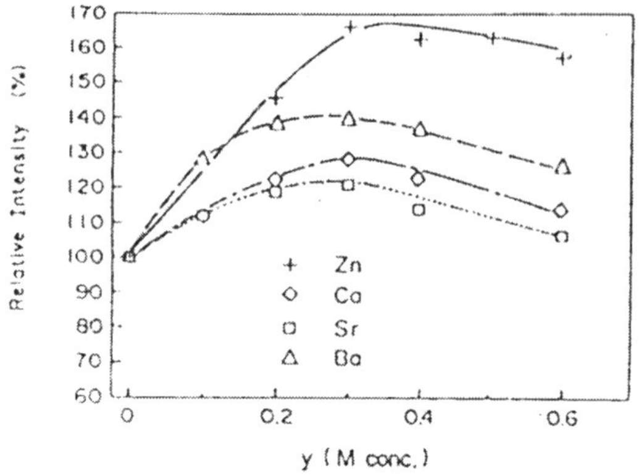

Fig. 1. Mn^{2+} emission intensity as a function of M substituting ratio y in $Ce(Mg_{0.82-y}M_y)Al_{11}O_{19}:Mn_{0.18}$, (M=Zn, Ba, Sr, Ca)

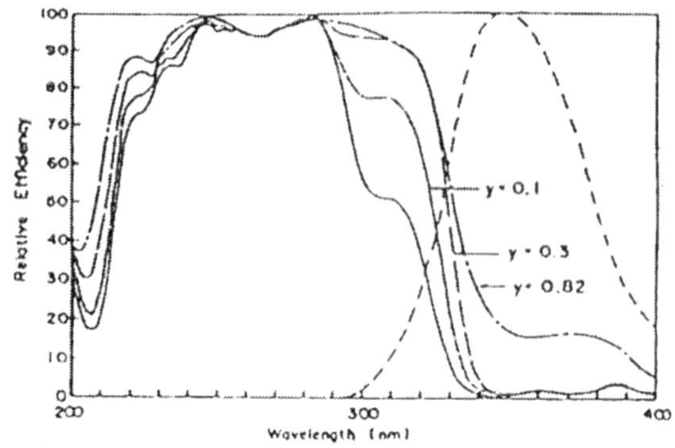

Fig. 2. Excitation spectra of the 518nm in $Ce(Mg_{0.82-y}Zn_y)Al_{11}O_{19}:Mn$. The dashed line represents the u.v. emission band in $CeMgAl_{11}O_{19}:Mn$.

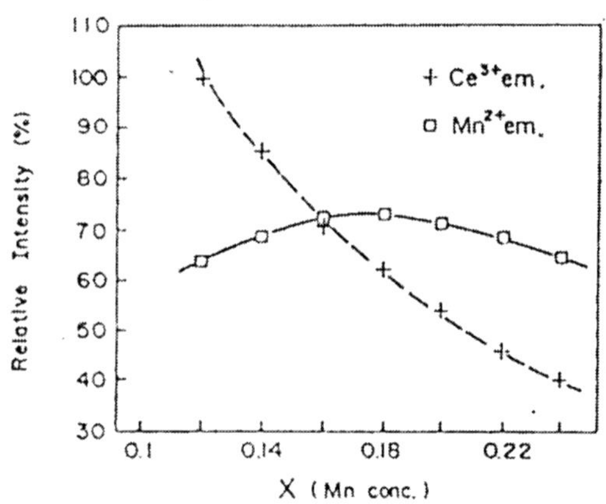

Fig. 3. Ce^{3+} and Mn^{2+} emission intensity as a function of x in $Ce(Mg_{0.7-x}Zn_{0.3})Al_{11}O_{19}:Mn_x$.

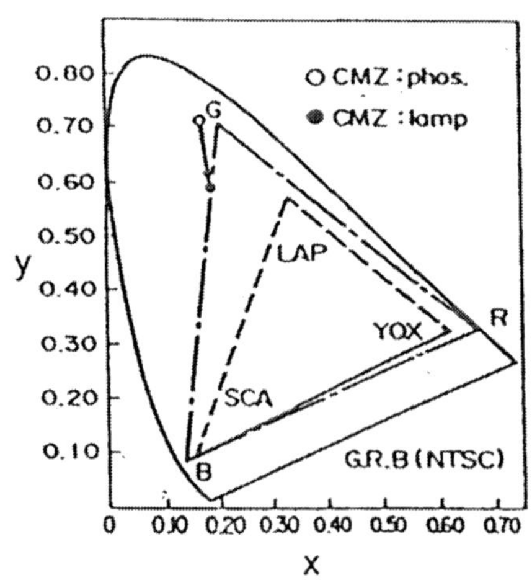

Fig. 4. Chromaticity coordinates (x, y)

Table 1. Characteristics of single-phosphor lamps. (40W Fluorescent lamps)

phosphor	efficacy (lm/W) 100h	efficacy (lm/W) 1000h	chromaticity coordinates x/y
$CeMg_{0.82}Al_{11}O_{19}:Mn_{0.18}$	47,7	45,1	0,170/0,715
$Ce(Mg_{0.52}Zn_{0.3})Al_{11}O_{19}:Mn_{0.18}$	90,3	87,2	0,170/0,705
$LaPO_4$:Ce,Tb (conventional)	134,0	129,4	0,334/0,572

Characterization of ALON layer on the inner bulb surface

L. Pogány, L. Balázs
Central Research Inst. for Physics H-1525 Budapest POB 49
*TUNGSRAM Co. BIKK H-1340 Budapest váci út 77

The luminescent lamps loss their Mercury content during the operating time. Part of the mercury builds amalgam with the Na diffusing out from the glass. To prevent this process a thin alumina layer - called ALON - has been applied on the inner surface of the glass bulb. The aim of this investigation is to characterize this ALON layer, to evaluate its ability for preventing the diffusion of the Na. It is important to know how much percent of the glass surface will be covered by the alumina, and how reliable is the technology for depositing the Alumina layer to the glass surface. A semiquantitative method using Energy dispersive X-ray measurements will be shown to evaluate the coverage and the thickness of the ALON. Morphological pictures shows that the layer consists of evenly distributed Alumina grains.

Fig 1 One layer, tilt 30°
Magnification = X2000

The layer thickness was measured on the broken glass edge by Scanning Electron Microscope. Pictures were taken at surface tilting angle of 85°-87° to the electron beam. The morphological investigation don't shows layer "peel off" phenomenon on the glass surface. For the thickness measurement samples were made with one, two and three layers of alumina. The measured layer thickness was 0.1-0.4 μm in the case of one deposition process.

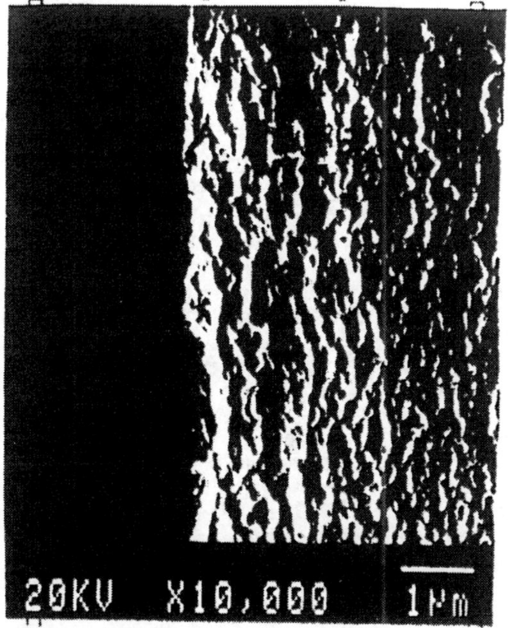

Fig 2 One layer, tilt 85°
Magnification = X10000

X-ray measurement was used for more effective characterization of the ALON layer. X-ray spectra were taken from ALON layers of different thickness. By changing the thickness, the ratio of the excitation volume changes. If the glass and the layer composition don't varies, the change comes only from the thickness fluctuation of the Alumina layer. The semiquantitative thickness evaluation uses these phenomenon. The Al2O3 content of the glass substratum were used as standard. The spectra were taken at different accelerating voltages and the net peak integrals of the Na, Mg Si, K and Ca were normalized to the Al net peak integral. The Si/Al ratio dependences from the accelerating voltage and shows different shapes for the different layer thickness.

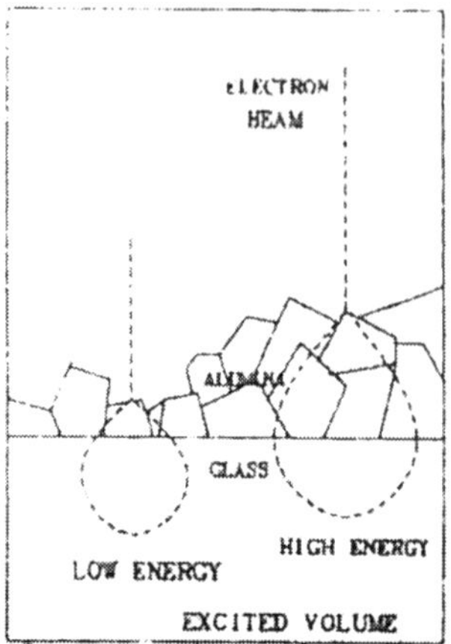

Fig 3 Change in the excitation volume by the primer electron beam energy

Comparison of the curves taken from the pure glass surface and taken from glass surfaces covered by Alumina of different thickness shows that the high energy part is characteristic to the layer thickness.

By this curves a comparison can be made between the different Alumina layers. If the electron beam scans surface area - for example 0.1 mm x 0.1 mm - during the spectrum measurement, the resulting Si/Al ratio voltage dependence will contain informations about the surface coverage of the Alumina layer. If the electron beam finds uncovered glass surface, the Na/Si curve will approximate to the pure glass curve. The inflexions point of the curve is characteristic to the surface covering ratio.The electron accelerating voltage were 4, 5, 6, 7, 8, 9, 10, 15, 20, 25 keV. The net peak integrals were computed and all peaks were normalized to the Al peak in the same spectrum. The Na diffusion can be investigated by investigating the Na/Si curves of the samples made by different annealing time. Samples were made 1. by the usual annealing process (1 minute at 400 C°), 2. two times annealed 3. tree times annealed. The Na/Si curves have very similar shape to the curves of Fig 3. It means that the diffusion of the Na through the ALON layer does not exist.

Conclusions

The qualitative comparison of the pure glass surfaces to the ALON covered glass surfaces shows that the Alumina covers the glass surface almost continuously.

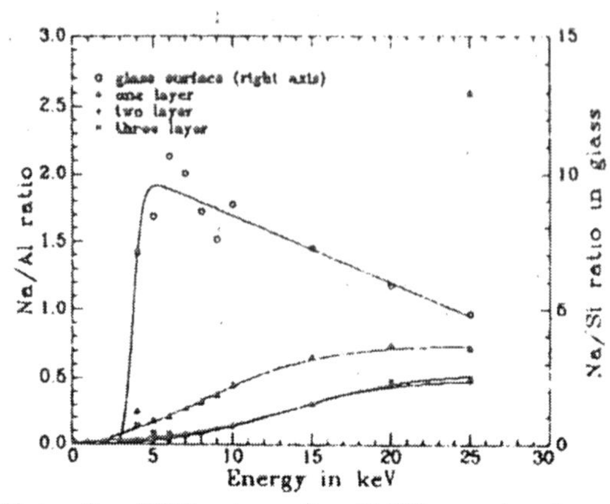

Fig 4 Effect of different layer thickness

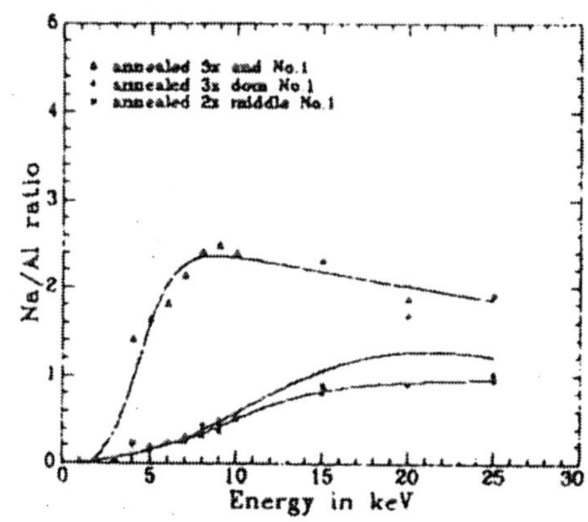

Fig 5 Effect of the annealing time on Na diffusion through ALON covered surface.

Comparison of curves of Fig 3 show without any computation that ALON layer covers the glass surface almost perfectly. The free glass surface without Alumina layer must be smaller then the 1-5 percent of the whole surface. The thinnest layer is about 5 times smaller then the thickest. The absolute layer thickness can be measured with the X-ray method semiquantitatively. The quantitative measurement needs a known standard Alumina layer. Curves of Fig 4 show that the ALON layer effectively prevents the Na diffusion from the glass to the inner layer surface.

INTERPRETATION OF pH-DEPENDENT COLLOID STABILITY OF AQUEOUS PHOSPHOR SUSPENSIONS AND THEIR STABILIZATION BY POLYELECTROLYTE ADSORPTION

Etelka Tombácz*, Judit Szigeti**, László Balázs**, Gábor Sajó**

*Attila József University, Department of Colloid Chemistry H-6720 Szeged Aradi Vértanúk tere 1. HUNGARY

**TUNGSRAM, BIKK, Phosphor and Coating ATC, H-1340, Budapest Váci ut 77, HUNGARY.

The fluorescent lamp phosphor layer quality is determined by the suspension properties, specifically the phosphor and suspension surface chemistry.

Different phosphor suspensions samples ($Y_2O_3:Eu^{3+}$; $LaPO_4:Ce^{3+},Tb^{3+}$ and Sr,Ca,Ba-chlorapatite:Eu^{2+}) were investigated using the methods of microcalorimetry, potentiometric titration, solution adsorption and rheology. The smallest values of heat of immersion [Fig.1.] and approximately zero net charge [Fig.2.] forming on the proton selective surface sites were observed at neutral pH. Under such conditions the coulombic repulsive forces have no effect and the solid particles stick together and coagulated structure develops in suspensions between pH 6 and 8. Relatively high yield stress values ($\tau \sim$ 10-20 Pa) characteristic for structure formation were measured in concentrated suspensions . In both acidic and alkaline suspensions charges can form in the interfacial layer due to the specific adsorption of H^+ and OH^- ions. The charged particles repel each other and the yield stress values of suspensions decrease significantly as the pH becomes more acidic or alkaline [Fig.3.]. Yttria-containing suspensions could not be measured below pH 6 because of dissolution of yttria. Newtonian flow behavior characteristic for the well-stabilized systems was observed in the strongly alkaline suspensions of La-phosphate [Fig.3.] and Ca,Sr-phosphate. The structure formation in the critical range of pH between 6 and 8 can be destroyed by additives which can modify surface properties [Fig.4.]. The adsorption of highly charged polyelectrolyte results a high excess in the surface charge of particles. Relatively small amount (some tenth mg/g) of ammonium salt of polyacrylic acid adsorbed can cause a big change in the stability of suspensions. Using well-stabilized suspensions with Newtonian flow behavior in which the single particles of phosphors are suspended smooth fluorescent coat will produced on the glass bulb surface.

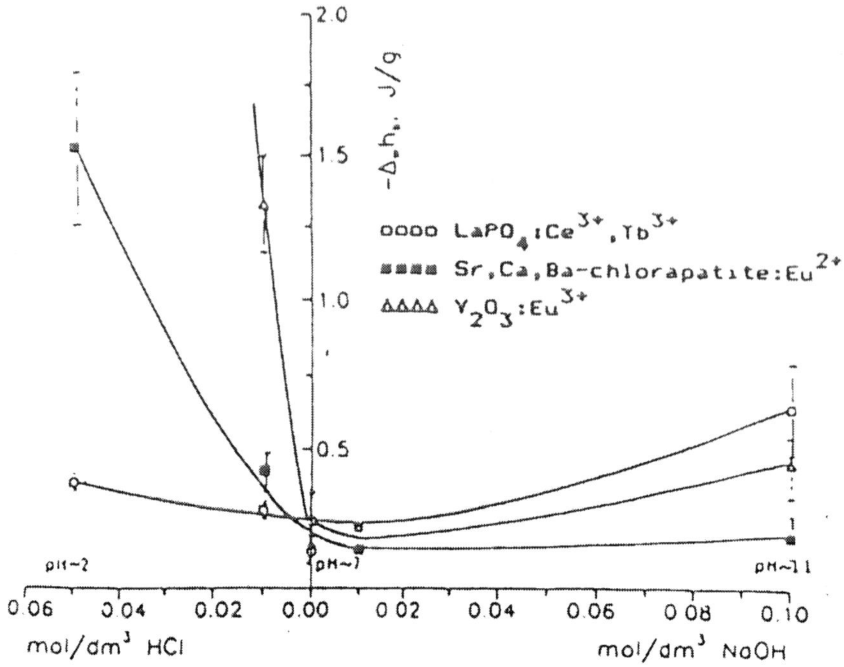

Fig.1. Dependence of specific heat of immersion on the concentration of immersion HCl and NaOH solutions

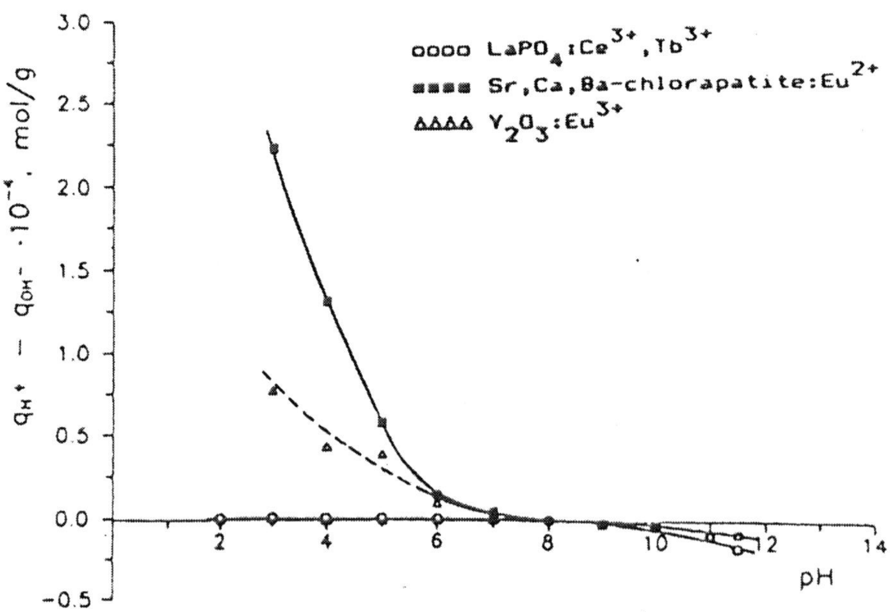

Fig.2. Dependence of net amount of surface charge on pH

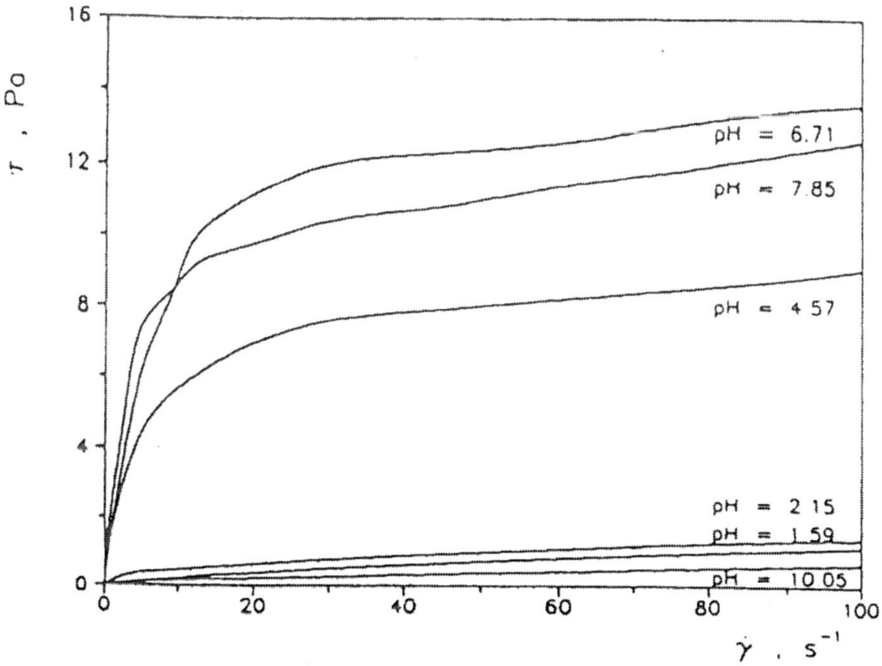

Fig.3. Characteristic flow curves for 70.4 mass % aqueous suspensions of $LaPO_4$:Ce,Tb at different pHs and at temperature 23°C

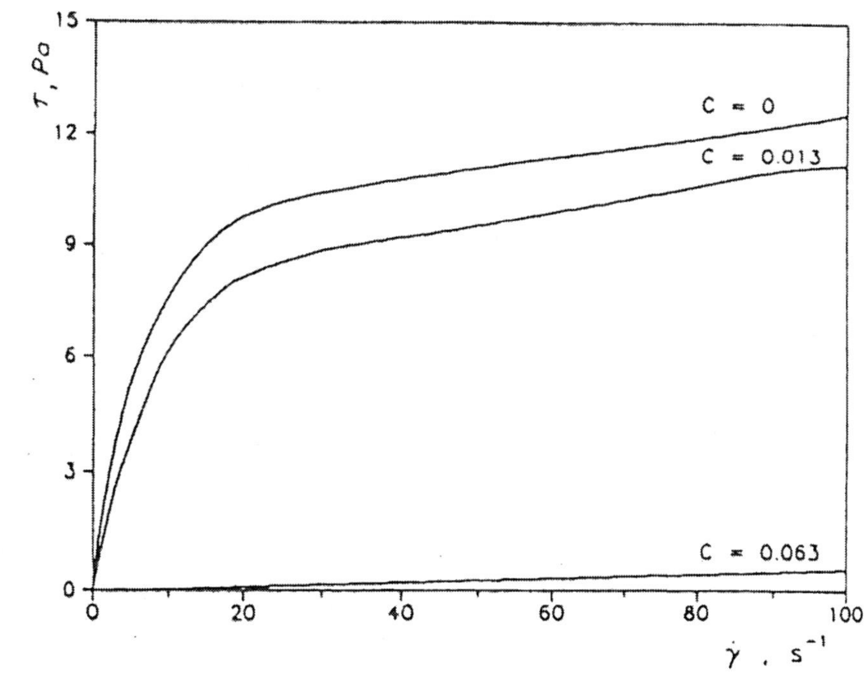

Fig.4. Effect of polyelectrolyte on the flow properties of 70.4 mass % $LaPO_4$:Ce,Tb suspensions. The up curves of some characteristic flow curves are plotted. (The specific amount of added polyelectrolyte is indicated with c in unit mg/g.)

Studies on Luminous Features of Phosphor Coating of Compact Fluorescent Lamps

T. M. Zhou, Y. J. He, H. Shao, and L. M. Wang

Institute For Electric Light Sources, Fudan University
Shanghai, P. R. China

1. Introduction:

Although the new three band phosphors opened the way for compact fluorescent lamps which have high luminous efficacy and good color rendering, the expensiveness of them somewhat limits the popularization of this kind of lamps. It becomes an important subject how to use these phosphors rationally and effectively.

In this paper, the effect of phosphor coating weight and particle size on the luminen output of lamps is illustrated both from the theory and experiments, and the technique of double coatings is presented.

2. Theoretical analysis:

The transmission T and the reflection R of diffuse radiation by a powder layer are given by the well known Schuster-Kubelka-Munk relations [1]-[4]:

$$T=(1-R_\infty^2)/(e^{pw}-R_\infty^2 e^{-pw}) \quad (1)$$

$$R=R_\infty(e^{pw}-e^{-pw})/(e^{pw}-R_\infty e^{-pw}) \quad (2)$$

where R is the diffuse reflection of a semi-infinitely thick layer, w is the layer thickness in mg/cm^2, and $p=S(1-R_\infty^2)/2R_\infty$, in which S is the scattering coefficent of the layer in cm^2/mg and can be calculated [5] through:

$$\log S=\log K-\log d_{op}-1.1513\log\sigma_g \quad (3)$$

where d_{op} is the optical diameter of phosphor particle, $\log\sigma_g$ is the logarithmic standard deviation, and K is a constant.

In a lamp with a single phosphor coating on the bulb, the luminous flux leaving the lamp can be derived, resulting equ.(4):

$$\Phi=\{\phi^{(lum)}\tau^{(lum)}+[I_f+I_b(1-\alpha^{(lum)})\tau^{(lum)}]\tau^{(uv)}/T^{(uv)}\}Km\cdot\int_{380}^{780}P'(\lambda)V(\lambda)\,d\lambda/[\int_{380}^{780}P'(\lambda)d\lambda] \quad (4)$$

where $\phi^{(lum)}$ is the visible mercury radiation flux from the discharge of the lamp, $P'(\lambda)$ the relative spectral power distribution of the lamp, $V(\lambda)$ the relative photopic luminous efficiency function, α the fraction of the reflected radiation absorbed by the discharge, τ=T/[1-(1-α)R], Km=683lm/w, and I_f, I_b is the forward and backward luminescence radiation flux, respectively, which can be obtained with the formulae (5) and (6) in the reference [6].

In a lamp with two different phosphor coatings, as shown in Fig.1, the total luminous flux can be derived [7], with the use of the results of a single layer.

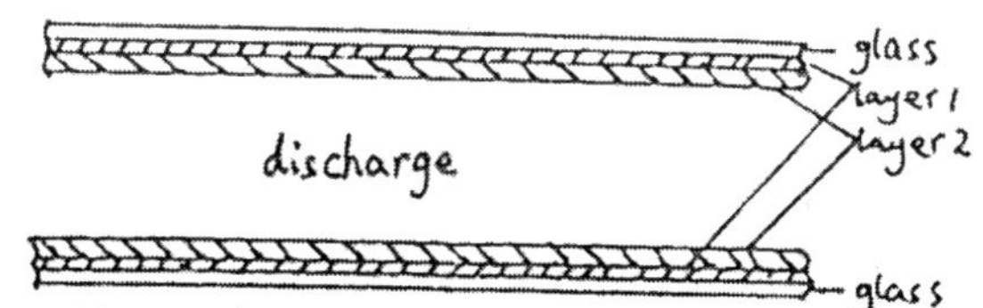

Fig.1 Lamp tube coated with two layers

This total luminous flux is most simpl expressed in the form:

$$\Phi=\nu^{(lum)}\nu^{(uv)}\tau_1^{(lum)}\{I_{f1}/(\nu^{(lum)}T_1^{(lum)}+\phi_2^{(lum)})\nu^{(uv)}+I_{b1}\delta^{(lum)}(R_2,T_2)+R_1^{(uv)}\delta^{(uv)}[\delta^{(lum)}(I_{b2},I_{f2}),\delta^{(lum)}(I_{f2},I_{b2})]\}Km\int_{380}^{780}P'(\lambda)V(\lambda)d\lambda/[\int_{380}^{780}P'(\lambda)d\lambda] \quad (5$$

where the suffices 1 and 2 refer to th values for layer 1 and 2, respectively $\delta(x,y)=x+y(1-\alpha)\tau_2$, $\nu=1-R_1\delta(R_2,T_2)$, and $\phi_2^{(lum)}$ i the visible radiation flux leaving the lamp i layer 1 is absent, equal to $\phi^{(lum)}\tau_2^{(lum)}$ $\delta^{(lum)}(I_{f2},I_{b2})\tau_2^{(uv)}/T_2^{(uv)}$.

3. Experimental results:

To check whether the theory can b applied, the luminous efficacy of PL-9 lamp coated with a single rare-earth activate phosphor was measured and calculated. Fig. shows the luminous efficacy of a set of suc lamps varying only in coating thickness.

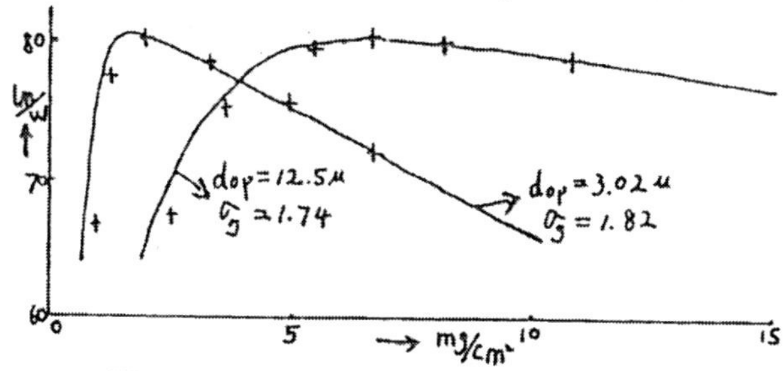

Fig.2 Luminous efficacy of lamps vs. coating weight
phosphor: $CeMgAl_{11}O_{19}:Tb^{3+}$,
$R_\infty^{(uv)}$=0.086 $R_\infty^{(lum)}$=0.960
—— calculated values + measured value

Fig.3 shows the luminous efficacy of a set of lamps varying only in phosphor particle size.

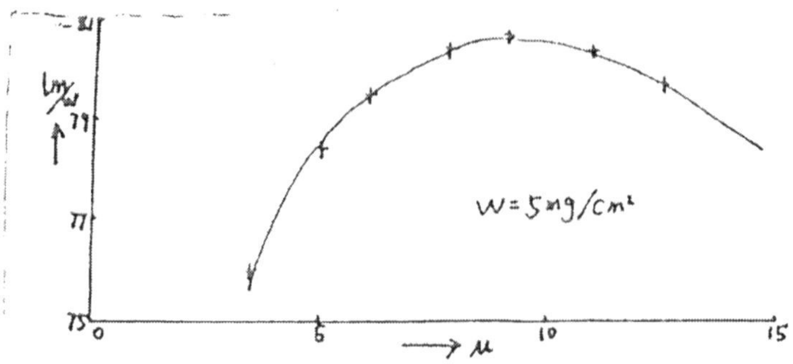

Fig.3 Luminous efficacy of lamps vs. particle size
phosphor: the same as in Fig.1
—— calcated values + measured values

Finally, the caculations and experiments of the Pl-9 lamps coated with two different phosphor layers was carried out, where layer 1 is the warm white halophosphate phosphor and layer 2 is the three-band phosphor. Table 1 gives the results and compares them with the lamps coated only with the three-band phosphor.

Table 1. Property comparisons of PL-9 between single coating and double coatings

	Single coating ($W=5mg/cm^2$)		Double coatings ($W_1=2mg/cm^2$ $W_2=2.9mg/cm^2$)	
	Calculated	Measured	Calculated	Measured
Tc	2930k	2950k	3000k	3060k
Lm/w	54.0	52.2	53.1	51.0
Ra	82.0	82.2	82.5	82.6
Flux Maintaince after 2000 hrs		77.7%		79.1%

4. Conclusions:

It has been shown that the luminous efficacy of a compact fluorescent lamp varies with coating thickness or with particle size in a manner predictable from the present theory. It can also be seen from Fig.2 that with the use of superfine phosphor, the optimum thickness of the coating is decreased without any decreases of the efficacy of the lamp. It is a way of reducing the price of a lamp.

Another way for this purpose is the use of double coating technique. The lamp efficacy only decreases by about 2%, while considerable savings of 3-band phosphors can be achieved.

Reference:

[1] A. Schuster; Astrophys. J., 21, 1 (1965)
[2] P. Kubelka and F. Munk; Z. Tech. Phys., 12, 593 (1931)
[3] P. Kubelka; J. Opt. Soc. Am., 38, 448 (1948)
[4] J. W. ter Vrugt; Philips Res., 20, 23 (1965)
[5] K. H. Butler and H. H. Homer; Illum. Eng., 55, 396 (1965)
[6] J. W. ter Vrugt et al; Lgt. Res. & Technol., 7, 23 (1975)
[7] T. M. Zhou et al; Elec. Lgt. Srcs, 5, 20 (1990)

CHARGING TENDENCY OF PHOSPHORS AND FLUORESCENT LAMP PERFORMANCE

Hidenori Ito*, Yoji Yuge*, Akira Taya*, Masaaki Tamatani**and Kenji Terasima**
* Toshiba Lighting and Technology Corporation , Yokosuka 237, Japan
** Toshiba Corporation, Kawasaki 210, Japan

1. INTRODUCTION

Decrease in lumen maintenance of phosphor screens in high-load compact fluorescent lamps is thought to be caused by degradation of phosphors themselves and mercury laydown on the phosphor surfaces . Several mechanisms have been proposed for the degradation of phosphors themselves. The adsorption mechanism of the mercury laydown, however, is not fully understood. The purpose of the present paper is to find a contribution of the mercury laydown to the lumen depreciation for rare earth phosphor screens and to verify the assumption that some kinds of phosphor surface properties govern the mercury adsorption.

2. ASSUMPTION

It is well known that electric charges are generated on a material surface by contact of the surface with a dissimilar material. The charging tendency for a phosphor powder can be measured from the polarity and amount of the electric charge generated by using a so-called blow-off apparatus when it is blended with a reference such as iron powder. We assumed that the adsorption degree of mercury or its oxide on phosphors depends on electrostatic force working at the interface between them and therefore depends on their charging tendencies.

3. EXPERIMENTAL

The charging tendency of phosphors can be controlled by a treatment which attaches colloid particles on the phosphor particle surfaces. Fig. 1 shows some examples. The amount of the charge is represented in microCoulomb per 1 g of the sample. The phosphors attached by MgO and SiO2 colloids have (+) and (-) charge against iron powder reference, respectively.

High-load fluorescent lamps (10mm of inner diameter, 0.14 watt per cm2 of inner tube area, and about 80C of tube temperature) with a cold spot as a mercury reservoir were prepared by coating these phosphors.

4. RESULTS

Lumen maintenance after 500-hour operation was evaluated. Fig. 2 a shows the results for $Y2O3$:Eu phosphors. Lumen maintenance has a maximum value at a certain charging tendency value at the (+) polarity side. The decreasing rate of lumen maintenance at the (+) side of the chage is much less as compared with that at the (-) side. Fig. 2-b shows the elative brightness to the initial value for phosphors which were recovered from 500-hour operation lamps and were heated at 500°C to remove mercury and its oxide. The brightness does not depend on the phosphor charging tendency, and returns to almost the initial value. This means that degradation of phosphor itself is not so large and that the charging tendency does not affect the phosphor degradation.

Fig. 3 shows the amount of HgO adsorbed on phosphor screens in fluorescent lamps after 500-hour operation. The amount is minimum at a charging tendency value which coincides with that for the maximum of lamp lumen maintenance. From Fig. 2 and 3 it is concluded that the lumen depreciation is governed by mercury oxide (which has brown-black color hue) adsorption on the phosphor layer.

The charging tendency of HgO is determined by making a contact electrification series comparing it with various standard metal plates, since it is not determined by the blow-off charging amount which depends on the difference in particle size from phosphors. It was found that the charging tendency of HgO is close to that of the phosphor which has the maximum value in lumen maintenance of lamps after 500-hour operation.

5. CONCLUSION

These results show (1) that HgO adsorbed on the phosphors significantly affects the decrease in lumen maintenance of fluorescent lamps, and (2) that the adsorption degree is determined by the difference in the charging tendency between HgO and phosphor surfaces.

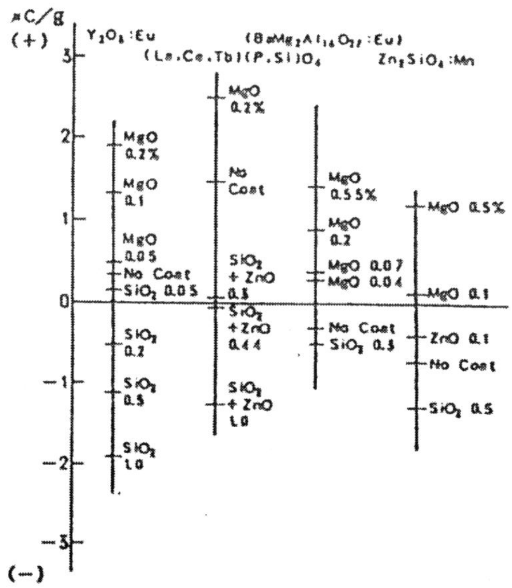

Fig. 1. Charging tendency of phosphors against a reference iron powder measured by the blow-off method.

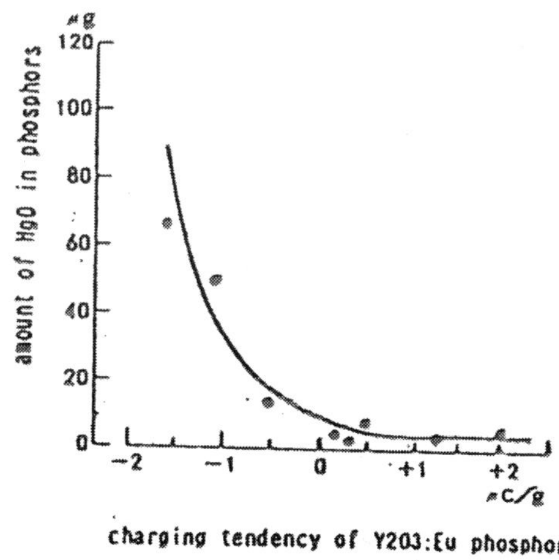

Fig. 3. Relation between the charging tendency of Y203:Eu phosphor and the amount of HgO in phosphors.

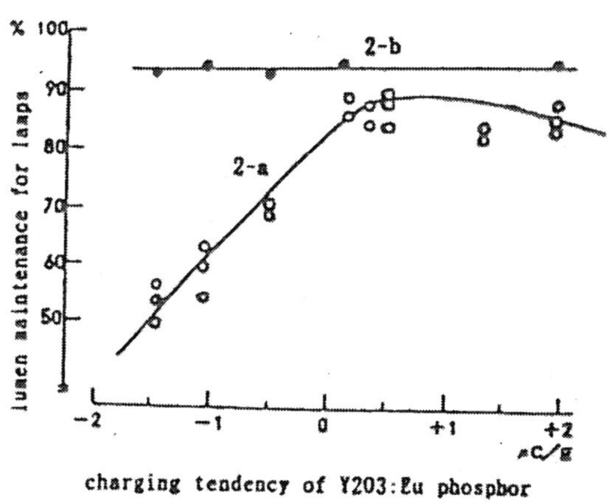

Fig. 2-a Relation between the charging tendency of Y203:Eu phosphor and the lumen maintenance of fluorescent lamp.

Fig. 2-b Relation between the charging tendency of Y203:Eu phosphor and the relative brightness of phosphor after 500-hour operation.

MERCURY-ANALYSIS OF FLUORESCENT LAMPS

H.C.G.Verhaar*, B.J.Roelevink**
* Philips Lighting B.V., Eindhoven, The Netherlands
**Philips Lighting B.V.,Roosendaal, The Netherlands

All fluorescent lamps contain more mercury than is needed for optimal luminous flux alone. The gradual disappearance of the element, when aging the lamp must be compensated with an overdose Hg in lamp production. Generally four major mercury-consumers are identified, viz. the glass-wall; electrode- and emitter-material; the phosphor-layer and gaseous impurities. The processes behind these phenomena have been the subject of many studies (see e.g.[1]).
Since not many quantitative data are known, their determination is a major issue, as a basis for further action. In order to come up with these data, the development of a set of analytical methods and procedures has been carried out.
These tools must enable the quantitative determination of both free and bound (i.e. consumed) mercury, to obtain a reliable Hg mass-balance, necessary for drawing valid conclusions on Hg-consumption processes and their impact on lamp-life.
Three interconnected methods have been developed for this purpose, viz.:

1. Wet-chemical profiling, for the destructive analysis of both bound and free mercury.
2. Tracer-analysis, for the non-destructive determination of primarily bound mercury,
3. Cataphoresis-analysis, for the non-destructive determination of free mercury,

These three tools will be discussed for 36W four feet T8 lamps. Of all methods results will be presented.

For the more regular analysis of the total amount of mercury in common fluorescent lamps a simple but adequate destructive wet-chemical procedure was developed. The information on the amounts of both bound and free mercury is obtained, by segmenting the lamps in a number of smaller parts after collecting the free mercury on a cold spot. These segments are on their turn submerged into a nitric acid solution for 3 hours at 60°C. Subsequently mercury is analysed by the cold-vapour technique (hydride-AAS),which is known for its suitability in mercury analysis. Data will be presented of Hg-consumption profiles of 36W T8 halophosphate lamps.
A method that offers attractive opportunities in non-destructive analysis of bound mercury in fluorescent lamps is based on the use of the radioactive mercury isotope ^{203}Hg.

When the ^{203}Hg isotope, which emits at 279eV with a half-life of about 46 days, is present in a lamp (with predetermined activity), it is possible to measure Hg-consumption profiles of (a part of) the lamp of interest. Since the destructive analysis yields the more straightforward data, the latter technique is particularly usefull in the more detailed research of Hg-consumption processes. We have used tracer mercury in the evaluation of the effects of variations in electrode operation on mercury consumption. Figure one gives the profile of bound mercury of a 36W four feet T8 lamp, that has been operated with an electrode heating current of 500mA. The activity was scanned through a 1cm lead diafragm.
In case one is mainly interested in the amount of free mercury, which is still present in the lamp (e.g. to ensure that specified lamp-life can be reached), use can be made of the so-called cataphoresis effect, which occurs when lamps are operated on a DC voltage. When operated on a DC voltage, the positive Hg^+ ions move towards the cathode of the lamp and will condensate as mercury on a cold spot at the cathode end. Eventually a mercury deficiency at the anode side of the lamp occurs, visible as a dark area. The time it takes before this effect is observable is a measure for the amount of free mercury.
It will be shown that the combination of these analytical methods offers a powerful tool in the quantitative research of Hg-consumption processes.

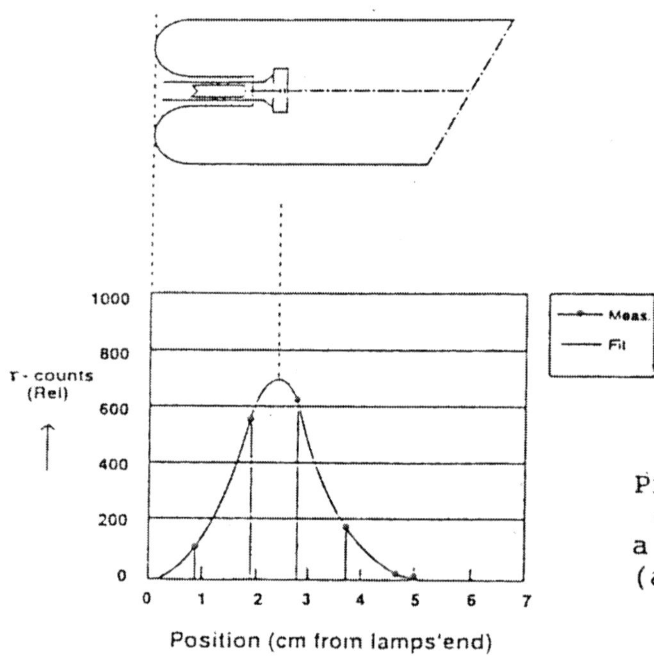

Figure 1.
Profile of bound mercury, detected with ^{203}Hg, for a heating current of 500mA (after 150 burning hours).

[1] B.J. Mulder, S. van Heusden, J. Electrochem. Soc., 130 (1983) 440

Session 2

Chairman: J. Maya

LP Discharges

13:I **M. Neiger**
Dielectric barrier discharges: an unusual new light source

14:L **Y. Watanabe**
Cathode fall characteristics at high frequency operation

15:L **R. Itatani, M. Kubo, M. Aono, M. Ryoko**
Variable luminous color discharge lamp

16:P **E. Arnold, K.-J. Dietz, G. Gnass, S. Reber, W. Schwarz**
A low pressure mercury lamp with enhanced UV–C radiation flux

17:P **D. Chen, G. Ma, L. Yang, Z. Cai**
Improvement of compact fluorecent lamps by filling different mixture gases

18:P **S. Coe, J. A. Stocks, A. J. Tambini**
Experimental and theoretical study of the negative glow and Faraday dark space regions of a fluorescent lamp

19:P **J. J. de Groot**
Miniaturisation in electronic compact fluorescent lamps

20:P **M. Duncan, R. Devonshire, J. A. Whitby**
Laser diagnostics of electrode operation in low-pressure Hg--Ar lamps

21:P **D. Y. Fang**
The axial electric field of compact fluoescent lamps

22:P **J. Ingold**
Cathode-plasma interface in the fluorescent lamp

23:P **X. Li**
A novel low wattage fluorescent lamp

24:P **F. A. S. Ligthart**
General scaling law for the luminous flux of fluorescent tubes

25:P **P. E. Moskowitz**
Fluorescent lamp electrode diagnostics by laser induced fluorescence

26:P **T. Sakurai, T. Saikatsu, O. Myodo**
VUV radiation of the rare-gas discharge fluorescent lamp at pulsed discharge

27:P **L. Wang, G. X. Jia, Lixi Yang, J. Guo**
Circuit parameters measurement of high frequency electric discharge lamps and electronic ballasts

28:P **L. Wang, Lixi Yang, G. Jia, Laishun Yang**
Frequency property of low pressure electric discharge lamps

29:P **S. Zhu, W. Zhou, S. Li**
Studies of the VUV xenon resonance radiation sources

30:I **F. Beleznay**
Semiconductor light sources

DIELECTRIC BARRIER DISCHARGES: AN UNUSUAL NEW LIGHT SOURCE

M. Neiger
LTI / University of Karlsruhe, FRG

INTRODUCTION

Dielectric barrier discharges (DBD's) are a special kind of capacitatively excited electrodeless discharges operating at the lower end of the high frequency regime (mostly 1 kHz to 1 MHz) and at relatively high pressures of 0.1 to several bar. While this type of discharge has been known for well over a century [1] and found wide spread application for Ozon production in so-called "Silent Discharge Ozonizers" at low frequencies (0.1 to 1 kHz) [2], its potential for generating uv and visible light with fairly high efficiency and reasonable radiative flux has become apparent only recently [3,4]. In addition to the known advantages of conventional capacitatively excited high frequency discharges at low operating pressures (< 100 mbar), namely virtually no chemistry constraints of the gas fill and considerably less geometric constraints due to the inherent series capacitance of the coupling scheme, DBD's produce a highly non-equilibrium cw discharge even at pressures beyond atmospheric. Such a discharge consists of a series of individual discharge pulses with nonconducting time intervals in between. The latter characteristic adds an additional degree of freedom, because electric energy input and gas kinetic processes (including production of radiating species) often are seperated in time and can therefore be optimized independently.
Most investigations of DBD's as possible light sources have concentrated on discharges with rare gas and rare gas / halcgen fills for the production of incoherent excimer radiation in the uv and vuv spectral range [5-7]. This is due mainly to the fact that excimer molecules either have no stable ground state at all or only a very weakly bound one which eliminates self- absorption problems even at very high pressures. Another important point is the fact that narrow bandwidth uv sources, including excimer lasers, with radiative efficiencies exceeding 5% are basically limited to two wavelengths, Hg 254 nm and Hg 185 nm from low pressure Hg / rare gas discharges. Contrary to this the large number of narrow band excimer (and trimer) wavelengths available from rare gas, rare gas halide and rare gas metal excimers (see Fig.1) are rather interesting for applications such as medical uv sources, uv lamps for photochemical applications and new types of fluorescent lamps. Another interesting aspect of these lamps is the absence of mercury in the lamp fill.

DIELECTRIC BARRIER DISCHARGE OPERATION

Figure 2a/b shows two different discharge arrangements made from quartz glass, which were used in our laboratory, a flat pill box shaped one (see Fig.2a) and a coaxial one (see Fig.2b). In both cases the discharge gap is separated from the metallic mesh type electrodes by dielectric quartz walls. Discharge radiation can be emitted into all directions, or by using one reflecting electrode, e.g. coaxially inward or outward only. Electrical excitation of DBD's is simple and done by applying mostly sinusoidal high voltage (1-15 kVpp) to the outside electrodes. Figure 3 shows photographs of DBD's operated at different frequencies in discharge vessels according to Fig.2a (40 mm diameter, 5 mm discharge gap) and viewed end-on facing one flat side [7]. The photographs at the left are discharges in Xenon, those at the right side Xe/O_2 (1000:1) discharges, both at a total pressure of 0.2 bar. Two basic modes of DBD discharges can be seen, a fairly homogeneous one at low frequencies and a filamentary one at higher frequencies. The bright spots of the latter mode show the size of the footprints of the discharge at the dielectric rather than the filament diameter, which is much smaller. Figure 4 shows examples of current and uv radiation (Xe_2^* 172 nm) signals versus time for a single discharge puls in Xenon [7]. At low frequencies of 5 kHz (Fig.4a) current and uv pulse are well seperated in time, indicating discharge build-up and quenching well before the production of the radiating Xe_2^* species. At higher frequencies of 22 kHz at the same pressure (Fig. 4b) plasma formation and radiation emission overlap considerably. As our measurements show, the latter behavior is accompanied by a decrease in radiative efficiency due to quenching of excimer molecules by plasma electrons. However, increasing the operating pressure with operating frequency offsets this efficiency loss, indicating a f/p - relationship.

UV EMISION AND RADIATIVE EFFICIENCY OF DBD EXCIMER LAMPS

Examples of interesting excimer wavelengths of prospective DBD-lamps are $XeCl^*$ / 308 nm (e.g. phosphor excitation, skin treatment, uv curing), $KrCl^*$ / 222 nm (e.g. phosphor excitation, photochemical water treatment) and Xe_2^* / 172 nm (e.g. phosphor excitation, uv-induced surface modification). Figure 5 is an example of the uv emission spectrum of a $KrCl^*$ DBD-lamp [6] for different fill pressures, normalized at 222 nm ($KrCl^*$ peak), where detector sensitivity was only 15% of that at 325 nm (Kr_2Cl^*-peak). Figure 6a/b shows results of measurements of uv flux (Fig.6a) and external uv efficiency (Fig.6b) of flat, 10 cm^2 area lamps. External efficiencies exceeding 10% in the 210 - 240 nm range ($KrCl^*$ excimer band) can be obtained for optimized operating conditions. From goniometric measurements we deduce internal (plasma) efficiencies larger than 15% for $KrCl^*$ [6].

The kinetics of excimer production in a lamp with Kr/Cl_2 - fill is fairly simple. The current pulse excites and ionizes Kr-atoms and creates negative Cl^- ions

(1) $Kr + e \rightarrow Kr^* + e$

(2) $Kr + e \rightarrow Kr^+ + 2e$

(3) $Cl_2 + e \rightarrow Cl + Cl^-$

During the afterglow phase and after quenching of the discharge current radiating $KrCl^*$ species are formed:

(4) $Kr^* + Cl_2 \rightarrow KrCl^* + Cl$

(5) $Kr^+ + Cl^- + Kr \rightarrow KrCl^* + Kr$

Reaction (4) dominates at lower pressures. With increasing Krypton pressure reaction (5) takes over. At Krypton pressures approaching atmospheric and beyond, Kr_2Cl^* trimer formation becomes increasingly important (see Fig.6a):

(6) $KrCl^* + 2\,Kr \rightarrow Kr_2Cl^* + Kr$

Radiative lifetimes of the order 100 ns for rare gas halide dimers require 3-body collision rates of the order 10exp7 per sec before trimer formation according to reaction (6) becomes important.

The uv emission spektrum of a DBD with a Xenon only fill is shown in Fig.7, normalized at the peak of the 172 nm Xe_2^* excimer band [7]. The spectral width is only weakly dependent on operating pressure in the range 0.1 to 1 bar. It is important to note that all other spectral features from 220nm to 800 nm are more than 3 orders of magnitude weaker than the peak of the excimer emission. The Xe_2^* uv flux and uv efficiency versus pressure are shown in Fig.8a/b for a 10 mm discharge gap and a 10 cm2 area flat cell lamp. For low lamp power (low operating frequencies / 1kHz / e.g. 10 W/m2 uv power) external efficiencies up to 15% are possible. At about 1 kW/m2 uv power external efficiencies are down to 10%. As in the case of the $KrCl^*$ excimer lamp internal plasma efficiencies seem to be at least 50 % higher.

Also as in the case of $KrCl^*$ the efficiency decrease is mainly caused by two factors: (i) increasing overlap of plasma phase and excimer build-up phase with increasing operating frequency, which causes increased electron collisional quenching of excimer radiators, and (ii) gas heating effects reducing 3-body reaction rates and thus excimer build-up. As increased operating pressure leads to shorter current pulses, i.e. plasma life time, and at the same time higher 3-body reaction rates, increasing both operating frequency and pressure significantly offsets those efficency losses. Additional lamp cooling keeps efficiencies up even for frequencies well beyond 100 kHz.

SUMMARY

A new uv light source using a dielectric barrier discharge (DBD) has been developed recently. The discharge is c.w. and operates at medium to high pressures either in a homogeneous mode (low power) or a multi-filamentary mode (higher power) as a rapid succesion of individual current pulses. Each of these pulses constitutes a highly non-equilibrium high pressure glow discharge plasma. The time averaged light output is spatially steady and stable. Due to the capacitive electrical coupling the choice of discharge geometries is very flexible, making possible e.g. large area flat panel sources with atmospheric pressure gas fill. Incoherent narrow band excimer radiation can be produced with higher than 10% external (higher than 15% internal) efficiency with specific uv fluxes of the order 0.1 W/cm2 from the lamp surface. Further studies will show whether direct production of visible light with similar efficiency is possible using different lamp fills.

ACKNOWLEDGEMENT

The author greatly appreciates support and funding by the German Federal Ministry of Research and Technology (BMFT) and by the Philips Research Laboratory, PFA Aachen / FRG.

LITERATURE

[1] W.v.Siemens: Ann. Phys. u. Chem.,Vol.102 (1857).
[2] B. Eliason, M. Hirth, U. Kogelschatz: J. Phys. D 20, p. 1421 (1987).
[3] B. Eliason, U. Kogelschatz: Proc. 40th GEC, Atlanta, p.174 (1987).
B. Eliason, U. Kogelschatz: Appl. Phys. B 46, p. 299 (1988).
[4] V. Schorpp, K. Stockwald, M. Neiger: Proc. 41st GEC, Minneapolis, p.74 (1988).
H. Müller, M. Neiger: Proc. 41st GEC, Minneapolis, p.55 (1988).
[5] B. Gellert, U. Kogelschatz: Appl. Phys. B 52, p. 14 (1991).
[6] V. Schorpp: PhD Dissertation, Univ. Karlsruhe / LTI (1991).
[7] K. Stockwald: PhD Dissertation, Univ. Karlsruhe / LTI (1991).
[8] K. Beneking (Philips PFA / Aachen): priv. comm. (1988).
[9] U. Kogelschatz: Appl. Surface Sc. 54, p. 410 (1992).

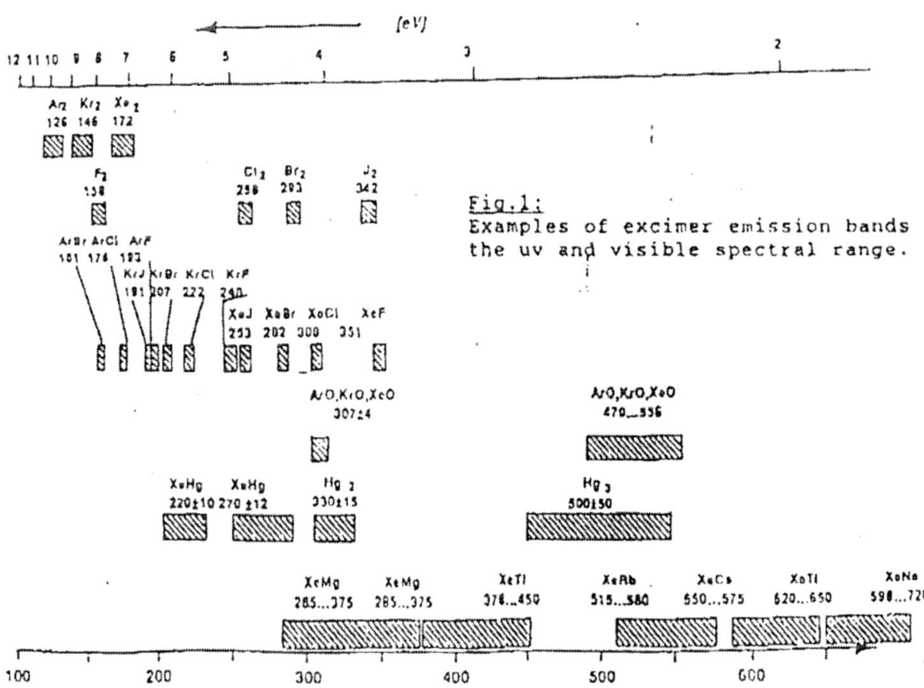

Fig.1:
Examples of excimer emission bands in the uv and visible spectral range.

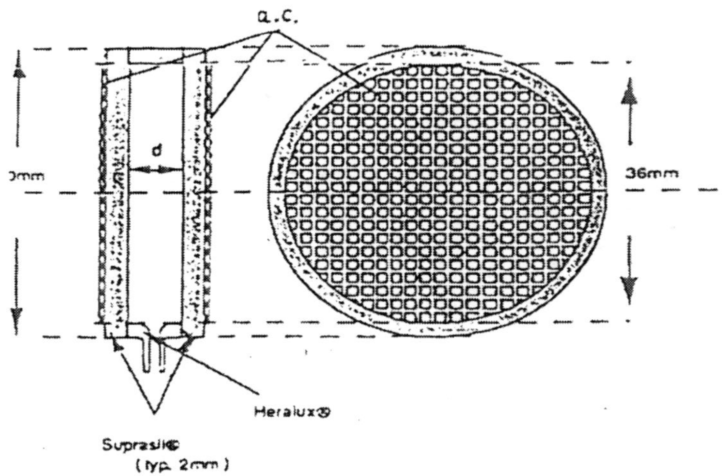

Fig. 2:
Construction of flat (top, Fig.2a) and coaxial (bottom, Fig.2b) quartz vessels.

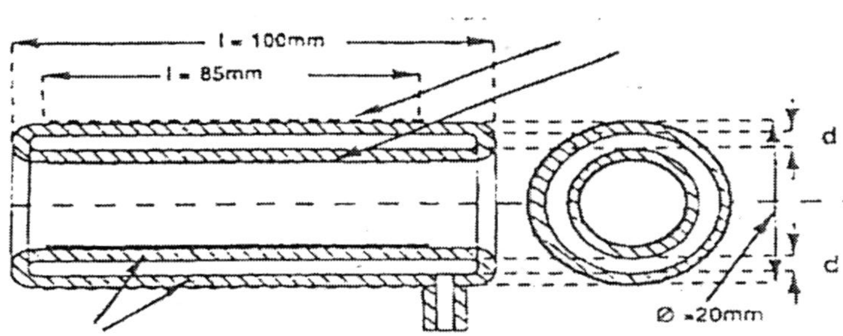

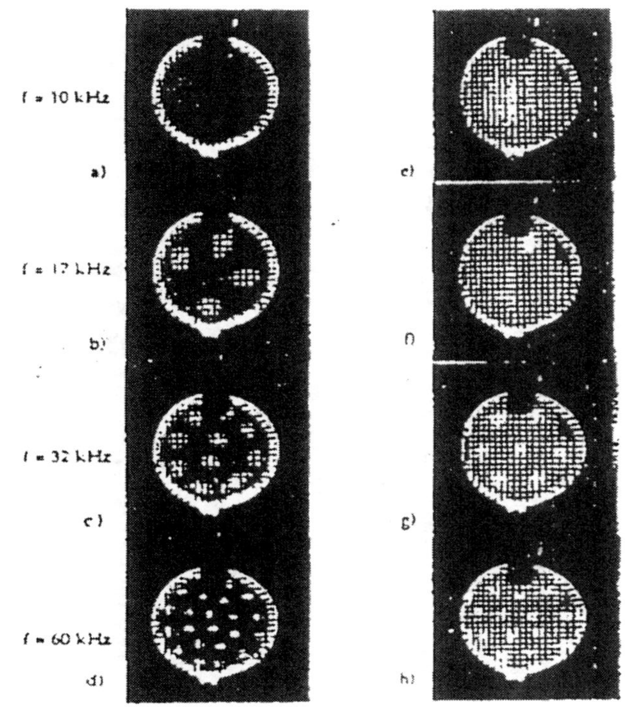

Fig. 3:
Photographic exposures of flat panel DBD's operating at different frequencies:
Left (Fig.3a): Xenon DBD (0.2 bar).
Right (Fig.3b): Xe/O2 DBD (0.2 bar).

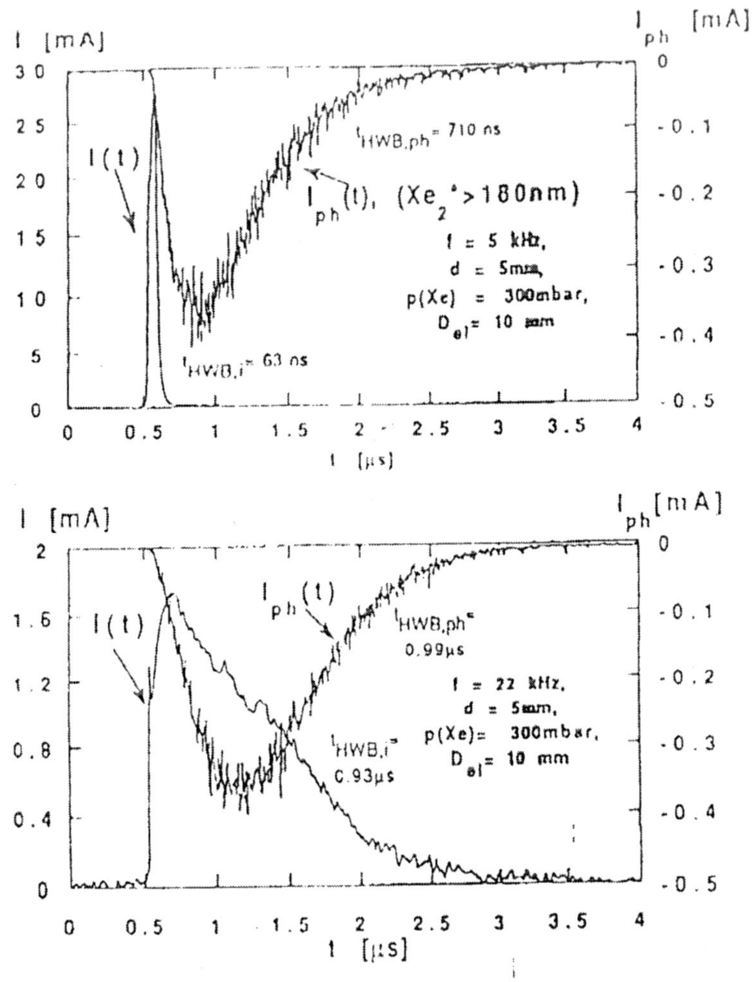

Fig.4:
Oscilloscope traces of current and uv (172 nm) output for a single discharge pulse in a Xenon DBD at 2 different operating frequencies:
Top (Fig.4a): 5 kHz.
Bottom (Fig.4b): 22 kHz.

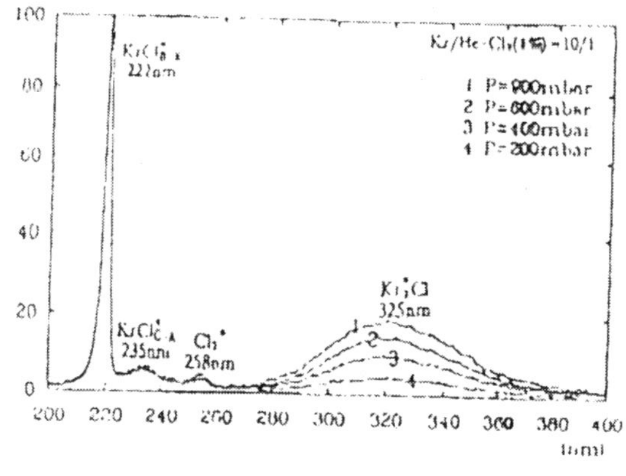

Fig.5:
UV emission from a Kr/Cl2 DBD lamp at different pressures

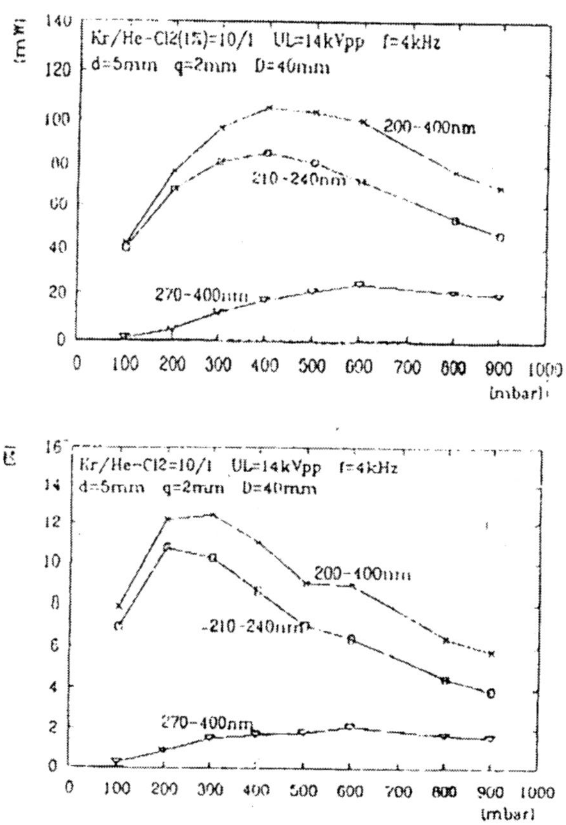

Fig.6:
UV flux (top, Fig.6a) and uv efficiencies (bottom, Fig.6b) for excimer radiation from a Kr/Cl2 DBD as a function of operating pressure.

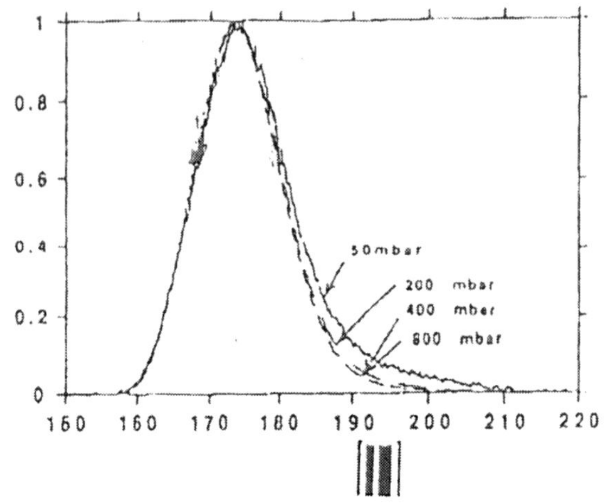

Fig.7:
UV emission from a Xenon DBD lamp.

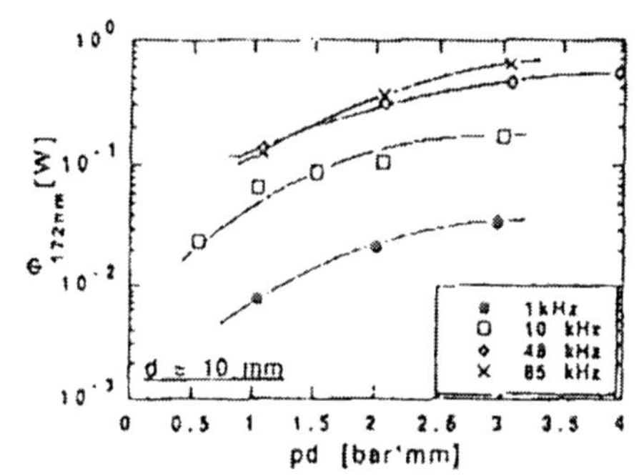

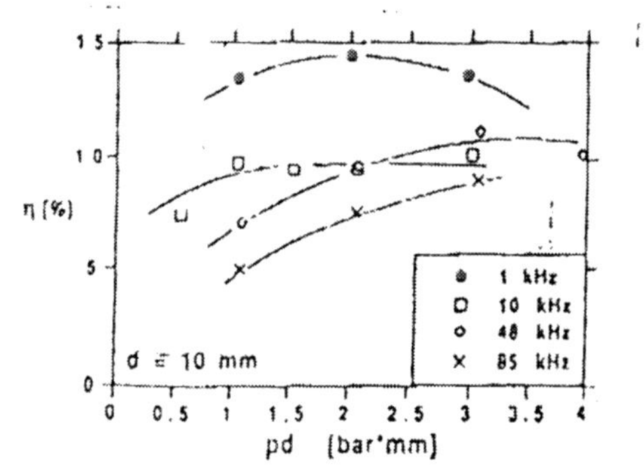

Fig.8:
UV flux (top, Fig.8a) and uv efficiencies (bottom, Fig.8b) for Xe_2^* (172 nm) excimer radiation as a function of pressure for different operating frequencies.

CATHODE FALL CHARACTERISTICS AT HIGH FREQUENCY OPERATION

Yoshio Watanabe
Kanagawa University, Yokohama, Japan

1. Introduction The main mechanism of electron emission from an electrode of a fluorescent lamp is thermal emission. The electrode heating period is divided into anode- and cathode-cycle. As an anode fall will vanish at high frequency operation, the electrode temperature at high frequency operation must be lower than that at low frequency operation[1]. Thus, cathode fall characteristics should vary according to the operation frequency.

2. Experiment Cathode fall voltage, V_K, is measured in two manners, which are hereafter called 'direct measurement (Fig.1)' and 'probe measurement (Fig.2)'. The experimental tube has a pair of electrodes used in an ordinary 20W lamp. The distance between the electrodes is 100 mm. Argon of 400 Pa and mercury are filled. The probe is located near to the measured electrode. The distance between them is about 2 mm.

[V_K at DC] The measured probe voltage, V_P, as a function of the probe current, I_P, in probe measurement is shown in Fig.3, where the discharge current, I_D, which is the anode current plus I_P, is kept constant.

[V_K at 43kHz (a sinusoidal wave)] The V_P waveforms at each I_D in direct measurement are measured. I_D comprised a non-negligible DC current component in direct measurement even if the opposite electrode was heated by a current of 0.4 A through an additional source. The V_P waveforms at each I_P in probe measurement are shown in Fig.4, where I_P is a constant direct current.

[V_K at 50Hz] The V_P waveforms at each I_D in direct measurement without an additional electrode heating source are shown in Fig.5. I_D comprised a small DC current component. This fact suggests that anode fall loss contributes

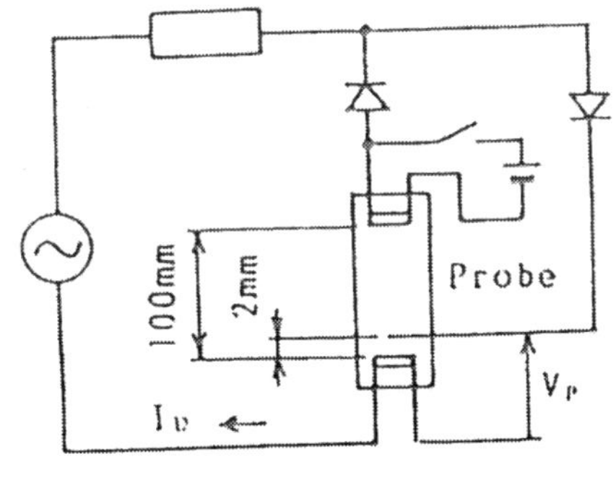

Fig.1 Direct measurement.

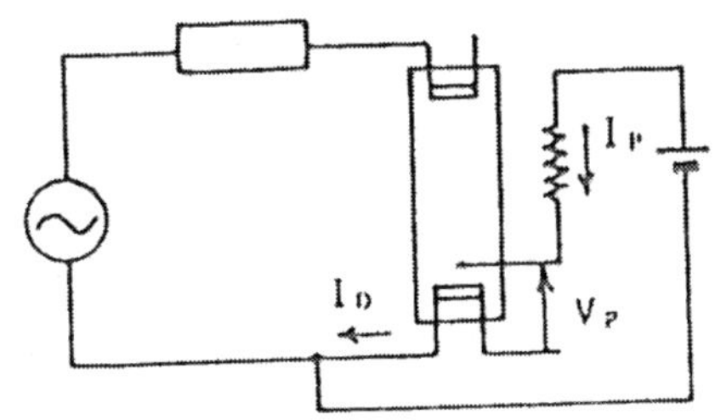

Fig.2 Probe measurement.

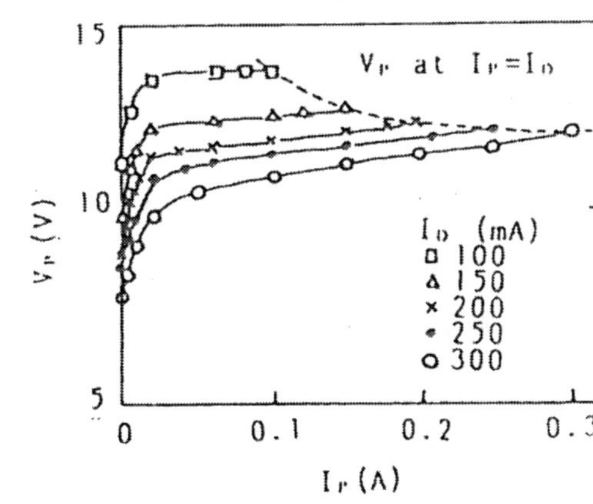

Fig.3 V_P-I_P at DC operation.

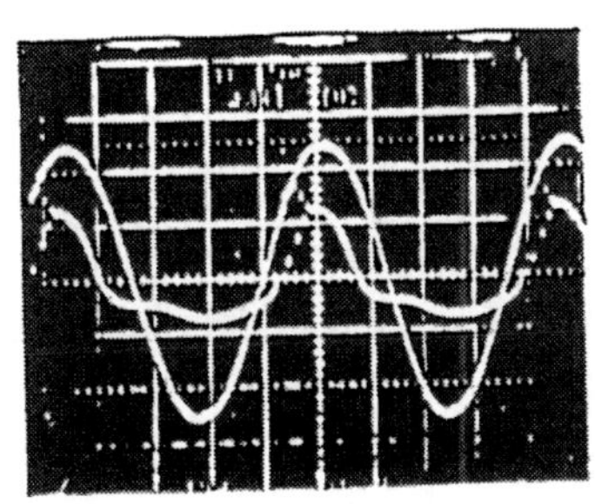

to the electrode heating as well as in 43 kHz operation, but the contribution is not so large compared with 43 kHz operation. The V_P waveforms at each I_P in probe measurement are shown in Fig.6. MHz oscillations are observed in the two portions of V_P waveform.

3. Discussion

[V_K at DC] V_P at $I_P=I_D$ are taken as a cathode fall voltage, V_K, since anode fall and positive column will not be formed because of the short distance between the cathode (electrode) and the anode (probe).

[V_K at 43kHz] V_K at 43kHz operation is obtained by subtracting V_{P0} from V_P in probe measurement, where V_{P0} is the V_P value at the phase when I_D intersects the zero line. The waveforms of $V_P - V_{P0}$ give the same waveform for all values of I_P, including $I_P=0$.

[V_K at 50Hz] To eliminate an anode fall voltage and an ion sheath potential fluctuation in front of the probe, the V_P waveform at the suitable I_P should be chosen to get V_K waveform. V_P waveform obtained at $I_P=3$ mA is taken in this paper. The maximum of V_K is determined from the maximum of the V_P in direct measurement, since V_P will approach to the V_K value after the negative glow formation.

[Comparison of V_K at each operation frequency] Using the V_K waveforms, the effective values of V_K are calculated. V_K as a function of I_D in effective value is shown in Fig.7. Fig.7 shows that V_K at 50Hz operation is higher than V_K at 43 kHz operation and nearly equal to V_K at DC operation.

4. Conclusion

It is shown that the cathode fall voltage at high frequency operation is lower than the one at 50Hz operation. One of the reason why the luminance efficacy of a fluorescent lamp increases at high frequency operation is due to the decrement of the cathode fall voltage as well as the vanishment of the anode fall.

Reference 1) R.R.Verderber, O.Morse and F.Rubinstein, Cont. Rec. IEEE IAS Annu. Meet. (Ind. Appl. Soc.) 20th (1985) pp.1689-1693

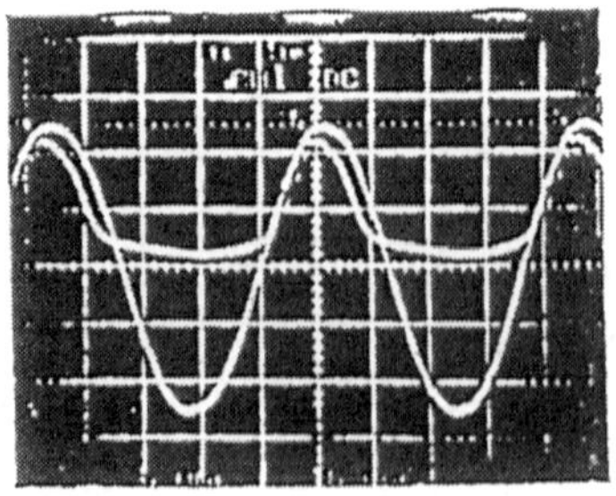

Fig.4 V_P in probemeasurement at 43kHz. I_D=0.35A.

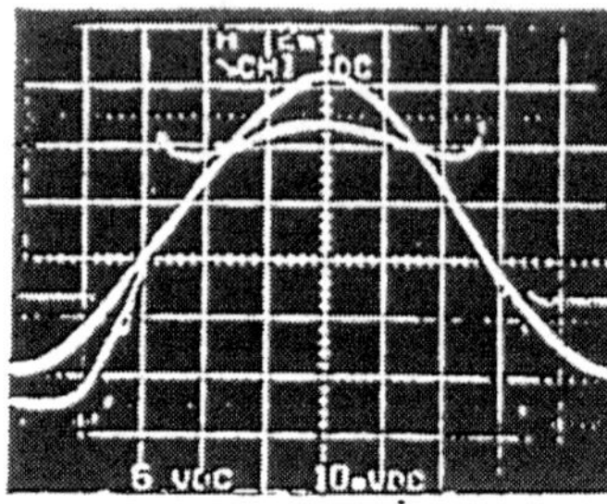

Fig.5 V_P in direct measurement at 50Hz. I_D=0.35A.

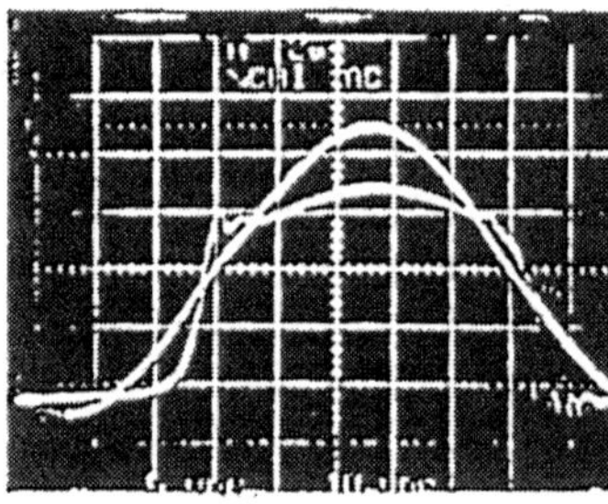

Fig.6 V_P in probe measurement at 50Hz. I_D=0.35A, I_P=3mA.

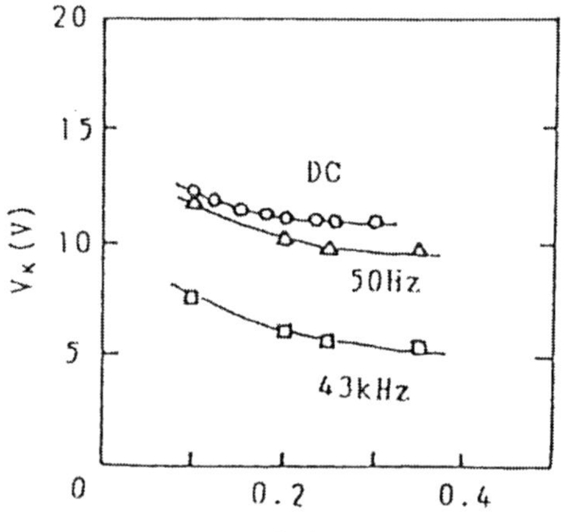

Fig.7 V_K-I_D in effective value at each operation frequency

Variable Luminous Color Discharge Lamp

Ryohei ITATANI[A], Makoto KUBO[A], Masaharu AONO[B], Masatoshi RYOKO[C]

[A] Department of Electronics Engineering, Kyoto University
Yosidahonmachi, Sakyo-ku, Kyoto 606, Japan
[B] Department of Electrical Engineering, Ehime University
3 Bunkyocho, Matsuyama, Ehime 790, Japan
[C] Dept. of Electrical Engineering, Akashi College of Technology
679-3 Nishioka, Uozumicyo, Akashi 674, Japan

The luminous color of the gaseous discharge lamps depends on the distribution of the emission spectra of the gas or the metal vapor filled in. However, in the case of dc discharge of rare gas, the negative glow and the positive column of a lamp emit the light of different color. The negative glow emits the light from high excitation levels of the atoms and positive column emits from the low excitation levels. It results from the variation of the electron energy in the discharge region of a lamp. The high energy electrons in the vicinity of the cathode are necessary to maintain the discharge. The energy of electrons in the positive column is usually low compared to that in the negative glow, but it can be raised without preventing the discharge. Thus it is possible to control the luminous color of a lamp if the electron energy in the positive column can be changed.

We have developed the method that can actively control the electron energy distribution in the positive column by using electronic pulse discharge. Selecting the gas and metal vapor and using this method, it is possible to realize the variable luminous color discharge lamp and fluorescent lamp.

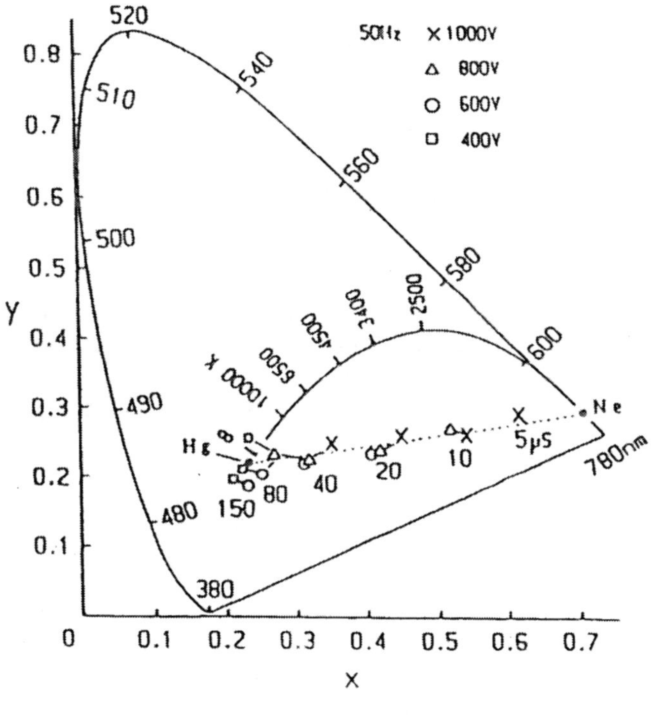

Fig.1 CIE chromaticity diagram for luminous color in the Hg-Ne lamp

For example, by applying the square wave pulse voltage with variable width to the lamp filled with mercury vapor and neon gas, luminous color of the positive column can be changed from the color close to red emitted by neon to blue of mercury. Figure 1 shows CIE chromaticity diagram of the experimental results. When the applied voltage on the lamp containing mercury vapor and neon at 10 torr is varied from 1,000 V to 400 V and duration of pulses is changed from 5 μsec to 150 μsec with a repetition frequency of 50 Hz, the luminous color changes from red {(0.545, 0.262) by CIE chromaticity coordinates} to blue (0.217, 0.198).

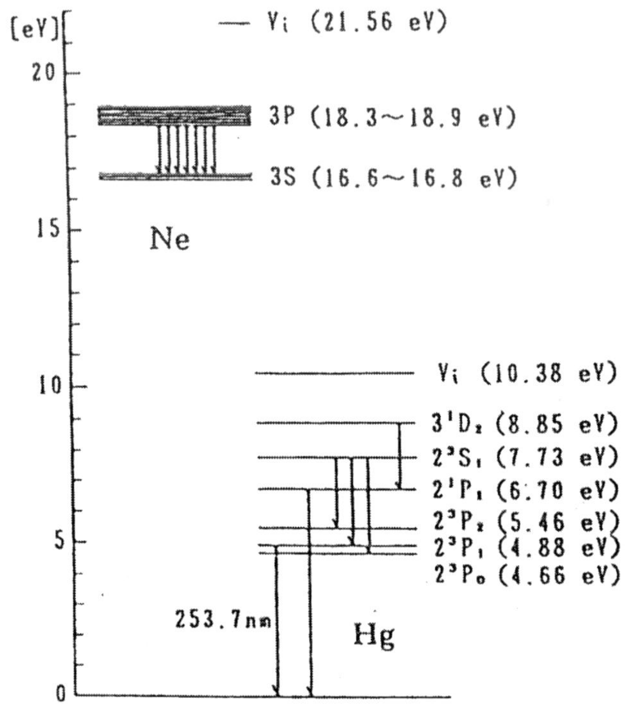

Fig.2 Excitation energy level of Hg and Ne

The energy levels of neon and mercury are shown in Fig.2. The excitation voltages of neon is higher than those of mercury by 8-10 eV. It is well known that, in pulse discharges, the electron energy is fairly high in the current rise period and becomes low after a steady state is reached[1]. When the mercury-neon lamp is discharged by the pulse with a long duration, the electron energy is low in average and it emits blue light of mercury as seen from Fig. 2. But when it is discharged by the pulse with a short duration, the electron energy is high and it emits red light of neon.

We have developed several types of variable luminous color discharge lamps using a precise control of the pulse wave form and/or additional fluorescent materials[2].

References

(1) B.T.Barnes : Phys. Rev. 86 (1952), 352.
(2) M.Aono,H.Kurokawa,M.Kubo,R.Itatani: J.Light and Vis. Env. 13 (1989),1.

A LOW PRESSURE MERCURY LAMP WITH ENHANCED UV-C RADIATION FLUX

E. Arnold, K.-J. Dietz, G. Gnass, S. Reber, W. Schwarz
Heraeus Noblelight GmbH, Heraeusstraße 12-14, D-W-6450 Hanau /Germany

Low pressure mercury lamps are widely used in the field of water degermination. This purpose requires high intense UV-C radiation at the normally used strongest mercury intercombination line at 253.7 nm. The efficieny of conventional lamps can reach 40 - 50 %, but the total radiation flux is limited by the mercury vapor pressure. The temperature of the quartz envelope for optimal operation should not exceed 40 - 50 °C, setting upper limits to the maximal electrical input power at fixed geometry. Larger lamp diameters or active cooling systems are required to further increase the efficacy of low pressure mercury lamps.
Instead of using pure mercury, we have doped the lamp with an amalgam. This causes a substantially reduction in mercury vapor pressure inside the lamp /1/, consequently the power consumption of the lamp can be increased.

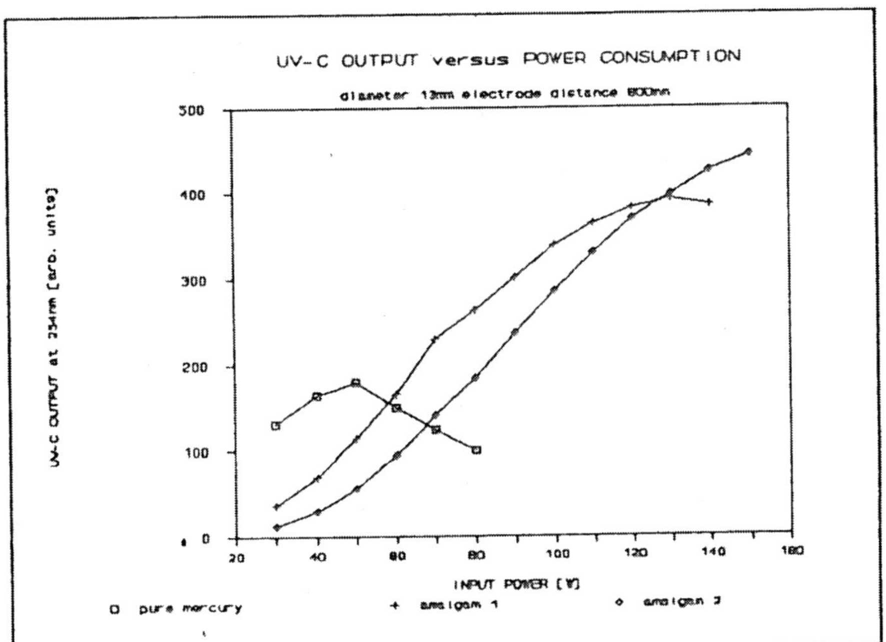

Figure 1: UV-C output as a function of electrical input power for a lamp with 13 mm inner diameter and 800 mm electrode distance. Ambient temperature 20 °C.

The new lamp has a 2 - 3 times higher UV-C-radiation flux at 253.7 nm compared to a conventional lamp with equal geometry, whereas the efficiency is only slightly smaller (30 %).

The mercury vapor pressure can be tuned to its optimal value for different applications by changing only the mercury content of the alloy /2/. This opens a wide field for new designs of small, compact and more efficient reactors for water degermination.

/1/ G. Frank, Z. Naturforschung 26a, 150-153 (1971)

/2/ B. R. Coles et al., J. Less-Common Metals 5, 41 (1963)

IMPROVEMENT OF COMPACT FLUORESCENT LAMPS BY FILLING DIFFERENT MIXTURE GASES

Chen Dahua, *Ma Guoqin, *Yang Leishun & Cai Zuquan

The Institute for Electric Light Sources, Fudan University, Shanghai, China

*Substar Fluorescent Lamp Factory, Boai Count, Henan Province, China

Abstract

Further improvement of compact fluorescent lamps is recently a very interesting subject for the research work. In our paper we attempt to find a new approach without the changing the conventional technical process to improve the Characteristics of compact fluorescent lamps by filling different mixture gases.

According to the theoretical model for ϕ38MM fluorescent lamps given by Dr. Waymonth and the three revisions for the compact fluorescent lamps by Prof. Zu Zholong et. al.. We present our viewpoints and calculation to analyse the changing for energy losses of particle elastic collision, ionization and radiation in a gaseous discharge.

The results of our analysis are as follows :

(1) When the nitrogen component of mixture gases is increasing in ϕ10MM discharge tube the energy losses of elestic collision are enhanced from 11.5ev to 59.3ev.

(2) When the nitrogen component of mixture gases is increasing in ϕ10MM discharge tube the transport rate μ_i of mercury positive ions Hg^+ is enhanced from 1.10 to 5.41, but the ionization losses are much strongly enhanced from 1.92 10 ev to 1.668 10 ev.

(3) When the electron temperature T_e is increasing , at first, the probability F of electron exciting H_g is growed up faster than the probability F_1 of electron exciting Hg atom to produce the radiation of 253.7 nm. When the electron temperature T_e is over the threshold value, that is to say, the nitrogen component in mixture gases is higher than 40%, the 185.0 nm radiation is greatly incresed and 253.7 nm radiation is reduced in positive column of gaseous discharge.

In the paper we also discribe our experimental set. The photometric and electric parameters of compact fluorescent lamps are given in the table when the filling mixture gases are changed.

The conclusion of our experiments is that the power and total flux is increased when the neon component of mixture gases is increased, but the efficacy of the compact fluorescent lamps is reduced and lamp voltage is enhanced to be up 5-30%. If the krypton is filled, the power and total flux of compact fluorescent lamps will be reduced to be down 5-10%, and the lamp voltage is also reduced too.

EXPERIMENTAL AND THEORETICAL STUDY OF THE NEGATIVE GLOW AND FARADAY DARK SPACE REGIONS OF A FLUORESCENT LAMP

S.Coe *, J.A.Stocks †, A.J.Tambini †

* THORN EMI Central Research Laboratories, Hayes, Middlesex, U.K.

† GE Thorn Lamps Limited, Advanced Technology Department,

Melton Road, Leicester LE4 7PD, U.K.

Although not responsible for significant light production in a fluorescent lamp, the negative glow and faraday dark space effectively control the behaviour of the rest of the lamp and are therefore of significant interest. The faraday dark space can be considered as a compensation region between different plasma conditions in the negative glow and the positive column. The positive column has an approximately maxwellian distribution of plasma electrons; the negative glow has an overall higher electron density including some high energy beam electrons from the cathode. This study is intended to improve our understanding of the physics of these regions. This extends the understanding achieved by Waymouth [1].

A moveable langmuir probe was used to measure longitudinal maps of plasma potential, electron density and electron temperature through the negative glow and faraday dark space regions of a fluoresent type discharge. Probe biasing, data collection and analysis are all controlled by a PC based system. The probe is kept red hot by electron bombardment, then swept through its characteristic in 40ms to avoid contamination effects. Program software allows the probe voltage ramping scheme to be adjusted to suit the widely different plasma conditions along the axis of the discharge.

The negative glow and faraday dark space regions have been modelled by dividing the electrons into two groups: monoenergetic beam electrons from the cathode sheath and plasma electrons generated by ionising collisions within the discharge. The main physics contained in the model includes:

(i) Excitation and ionisation of the mercury and noble gas by both beam and plasma electrons.

(ii) 2-D ambipolar diffusion of the plasma electrons and ions.

(iii) A self-consistently calculated, axially varying plasma electron temperature.

For particular discharge conditions the model is able to compute the electron density, electron temperature and plasma potential axial profiles as well as the magnitude of the ion flux returning to the cathode. This is a more complete model than those of Ingold [2] and Coe [3].

Figure 1 shows a comparison between the predicted (solid line) and measured (data points) electron density profile for a fluorescent type discharge. Pressure was 1 torr, discharge current 190mA. The electron density is seen to reach a maximum in the negative glow, then fall off to a shallow minimum in the faraday dark space before reaching a constant value in the positive column. The model accurately reproduces the measured length of the negative glow, and also provides a good estimate of the peak electron density.

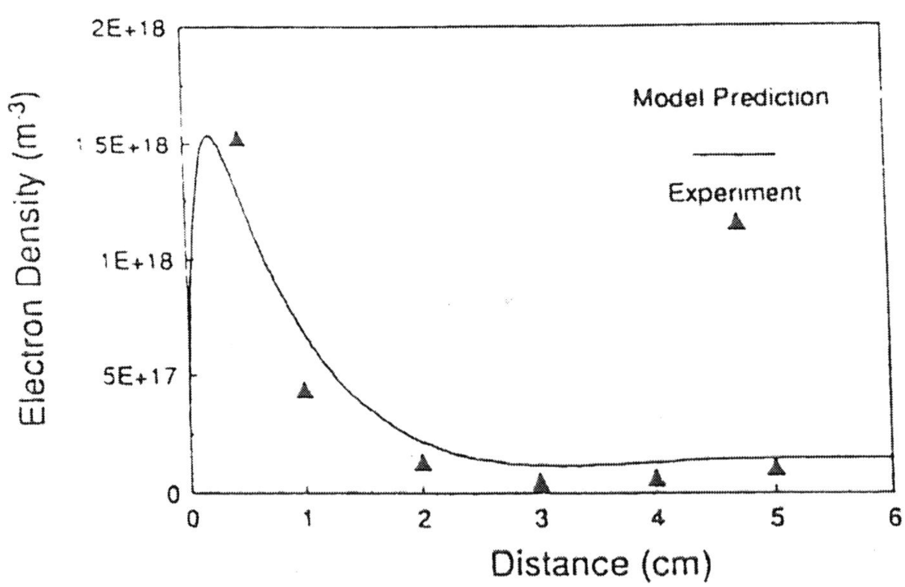

Figure 1. Measured and calculated electron density profiles.

Joint model/experimental results will also be presented for longitudinal maps of plasma and floating potential, and electron temperature.

Also, electron energy distribution functions calculated from measured probe characteristics are presented which show the presence of high energy "beam" electrons near the cathode. Figure 2 shows such a result measured 0.5 cm from the cathode. Evidence of beam electrons can be seen from the presence of a secondary peak at approximately 13 V, corresponding to the full energy of the cathode fall.

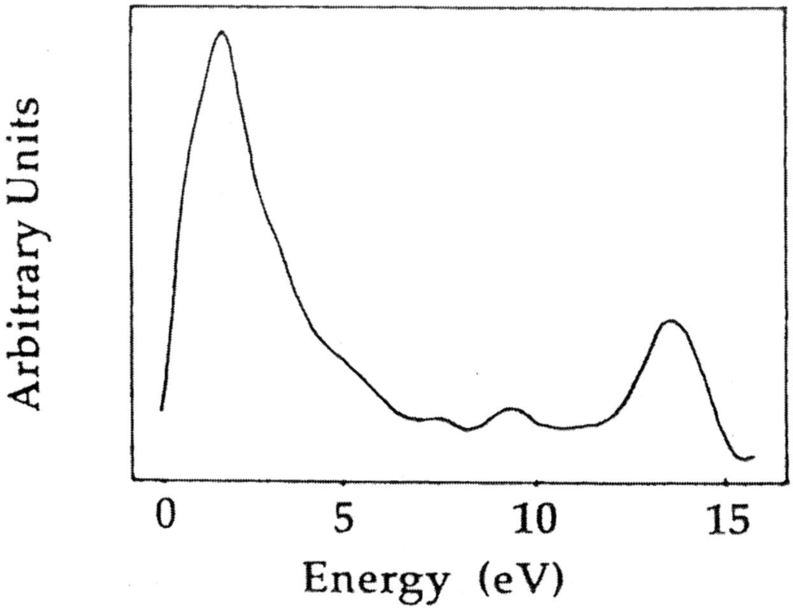

Figure 2. EEDF at 0.5cm from cathode, showing presence of beam electrons.

ACKNOWLEDGEMENT

The authors wish to thank G.G.Lister and D.O.Wharmby for many helpful discussions throughout this study.

REFERENCES

[1] J.F.Waymouth, J Appl. Phys. 30, 1404, 1959

[2] J.H.Ingold, Phys Rev A 43, 3093, 1991

[3] S.E.Coe and G.G.Lister, To be published J. Appl. Phys.

MINIATURISATION IN ELECTRONIC COMPACT FLUORESCENT LAMPS

J.J. de Groot
Philips Lighting B.V., Terneuzen, The Netherlands

Self-ballasted Compact Fluorescent Lamps (CFL) were introduced in Europe around 1980 as replacement for incandescent lamps. In these 10 years of existence they have found widespread application, with a variety of types, usually based on a magnetic type of ballast.

Around 1986 more expensive electronic types appeared on the market with their advantages of higher efficacy, lower weight, faster and flicker-free ignition. These electronic types still were based on the same discharge tubes, as used for the non-integrated lamp types, with separate magnetic ballast.

Recent developments are based on an integral system optimization of discharge tube, electronic circuit and housing together, leading to a more compact CFL, approaching the dimensions of the incandescent lamp.
Basic developments are:
* discharge tube: at high-frequency operation there is more freedom in the choice of the discharge tube voltage, and therefore in arc length, as compared with 50Hz-ballasts; this leads to very short as well as long discharge tubes.
* electronic circuit: using Surface Mounting Technology with small component dimensions, a more compact circuit can be realized
* housing: optimising the connection between discharge tube and electronic circuit, together with a narrow-closing plastic lamp shell, leads to a compact system.

Recent developments are described with examples of the PLC*Electronic (PLC*E) as well as the recently developed SLS*Electronic (SLS*E) lamp range.

The PLC*E lamp is an electronic integrated CFL, with four parallel narrow-diameter discharge tubes. The tubes are interconnected, to pass the discharge, by so-called bridges in the glass. In base-up position the mercury vapour pressure is controlled by the cold "heads" of the discharge tube.
The luminous flux range covers 400-1500 lm, corresponding to the luminous flux of 40-100 W GLS lamps, see table 1.
Fig 1 shows the development for the 400 lm type PLC*E lamp from 1987 till 1992, in comparison with the 40 W incandescent lamp.

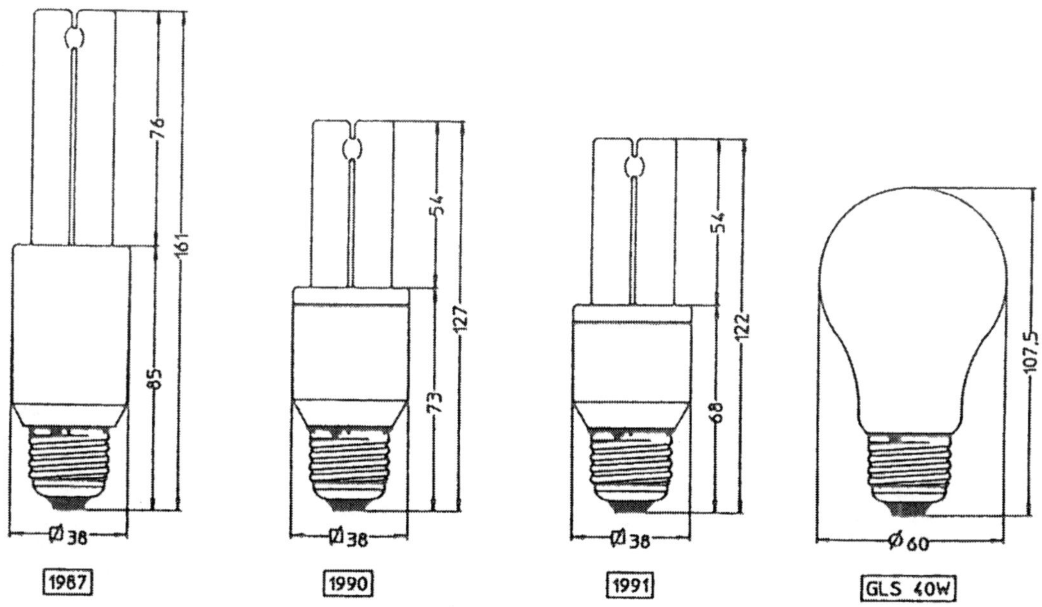

Fig 1. Development of 400 lm PLC*E lamps during recent years, compared with incandescent lamp (dimensions in mm).

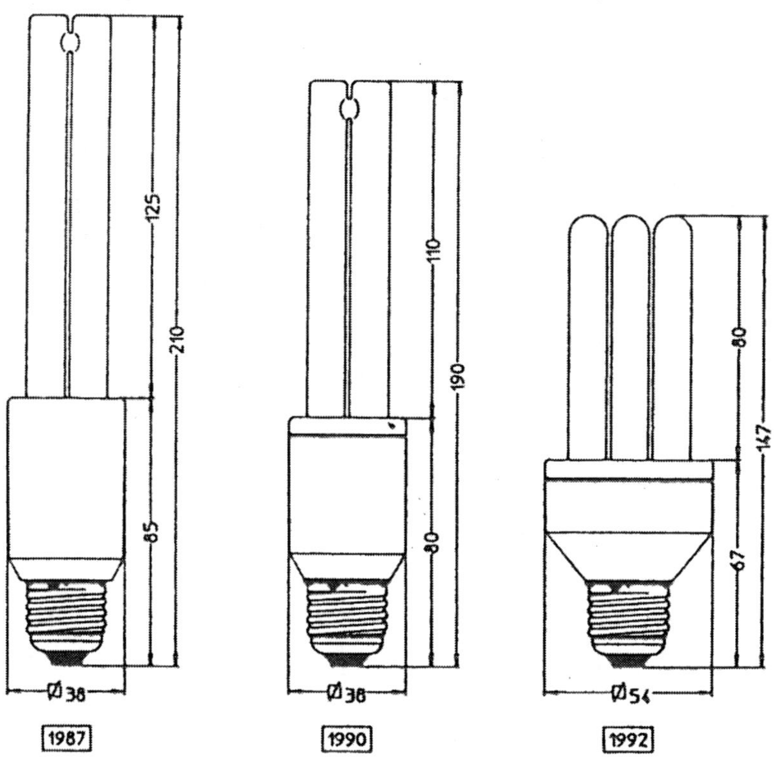

Fig 2. Development of 20W/1200 lm lamps (PLC*E and SLS*E) during recent years.

LASER DIAGNOSTICS OF ELECTRODE OPERATION IN LOW-PRESSURE Hg-Ar LAMPS

M.Duncan, R.Devonshire and J.A.Whitby

High Temperature Science Laboratories
Chemistry Department
Sheffield University
Sheffield S3 7HF, UK

At the last Symposium (LS:5), held in York, Woodward and Schulman reported a new laser technique for non-intrusive in situ electrode diagnostics in fluorescent lamps [Ref.1]. The technique is referred to either as Photoemission Optogalvanic Spectroscopy (POGS) or as Plasma-Enhanced Photoemission (PEP). Despite the mature state of development of the fluorescent lamp there remains a need for more direct information about electrode operation and the factors that affect it. This need also exists for the region of plasma around the electrodes. We have extended the POGS studies of Woodward and Schulman [Refs.1 & 2] and will report a range of investigations on both a.c. and d.c. operated Hg-Ar low-pressure lamps, either without or with phosphor coatings, and on special devices, using the technique.

The POGS experiments were performed on 20W and 40W T12 lamps and on 18W T8 lamps. Electrode configuration was either an open wound coil or a linear, "stick", variety; the latter being the most convenient to investigate. To allow the laser excitation beam to impinge on well defined areas of the electrodes the phosphor coating in the electrode region was omitted from the manufacturing process, or was removed immediately prior to measurements being taken. The laser excitation originated from a single longitudinal mode, Q-switched, Nd:YAG laser operated at 20Hz, or from a tunable dye laser pumped by the Nd:YAG laser. In the POGS experiments radiation falling on the cathode surface causes photoemission of electrons which are amplified in the cathode fall to create an optogalvanic signal. This POGS signal in the lamp current, which is generated at a known, variable, point in the mains cycle in a.c. operated lamps, is separated and captured by appropriate electronics and either recorded as a transient (changes are significantly more rapid (100x) than OGS signals from the bulk plasma) or analysed to determine the magnitude of the initial drop in lamp current (the signals are large compared to OGS signals).

Our initial experiments established that the technique was capable of giving a reproducible POGS profile, over periods of an hour or more, for an electrode operated under stable conditions; repeated sweeps, at intervals of several minutes, of the electrode along its length by a thin sheet of laser radiation in a plane normal to the electrode's axis gave signals that were identical. The POGS profile could be related to the temperature profile of the electrode [Ref. 3]. A fatigue effect was detected when a fixed area of the electrode was continuously excited for long periods, however, the POGS profile of the same electrode was unaffected by the prolonged, localised, exposure to the laser radiation. Dramatic effects in the POGS profile were observed when a lamp is switched on, off, and on again, and also when a separate high power laser pulse is used to ablate selected regions of the emissive coating of the electrode [Ref. 3].

The technique may also be used to obtain quantitative information about the electrode's surface. A study of the variation of the POGS signal from a fixed point on the electrode with the electrode's heater current led to a value of the zero field thermionic emission (ZFTE) work function at that point of 2.3 eV, in good agreement with measurements based on a different technique [Ref. 4].

Large variations in the magnitude of the POGS signal during the mains cycle in a.c. operated are observed; the maximum occurs during the cathode cycle. The behaviour is consistent with theoretical models of the growth and decay of the cathode fall during the cathode half-cycle [Ref. 5].

References:

1. D.R.Woodward and M.B.Schulman, "Electrode Diagnostics Using Plasma-Enhanced Photoemission", Proceedings Fifth International Symposium on the Science & Technology of Light Sources, York, September 1989 (Sheffield: University of Sheffield, 1989), Abs. 82L, p263-264.

2. M.B.Schulman and D.R.Woodward, "Plasma-Enhanced Photoemission as a Discharge Lamp Diagnostic", Appl.Phys.Lett., **55**(1989), p1618-1620.

3. M.Duncan and R.Devonshire, "Optogalvanic Spectroscopy as a Diagnostic of Operating Fluorescent Lamp Electrodes", in Optogalvanic Spectroscopy, eds. R.S.Stewart and J.E.Lawler (Bristol: Institute of Physics, 1991), 257-262

4. H.J.W.Schenkelaars, J.J.Heuvelmans, A.v.d.Kooi and J.Schlejen, "Zero Field Thermionic Emission in HF Operated Fluorescent Lamps", Ref. 1, Abs. 79P, p257-258.

5. J.F.Waymouth, "Electric Discharge Lamps", The MIT Press, Cambridge, 1971.

THE AXIAL ELECTRIC FIELD OF COMPACT FLUORESCENT LAMPS

D.Y.Fang
Institute for Electric Light Sources
Fudan University, Shanghai, P.R.China

1.INTRODUCTION

In order to design the compact fluorescent lamps, it is essential to know the value of the axial electric field of the low pressure Ar + Hg discharge with small tube diameter. Based on Lama et al's model[1], this paper presents a preliminary study of electric field strength for compact fluorescent lamps.

2.THEORETICAL ANALYSIS

The electric field E can be calculated from the energy balance equation which balances the rate of energy extracted from the electric field against the rate of energy loss by various processes:

$$jE = P_r + P_{el} + (\text{other losses}) \qquad (1)$$

where P_r is the radiation loss, P_{el} the elastic scattering loss, j the average current density:

$$j = e\mu_e E N_e \qquad (2)$$

here μ_e is the electron mobility, N_e the electron density. For the conventional fluorescent lamps, Lama et al had deduced an analytical formula to calculate P_r and P_{el}. The calculated values showed good agreement with experiments. In the case of compact fluorescent lamps with small tube diameter, because of the rise of electron temperature, P_r and P_{el} should be calculated by using new electron temperature. The de-excitation rate coefficients and ionization rate coefficient are calculated by using the cross section of Rockwood, Morgan and Vriens[3-5].

With the decrease of the discharge tube radius, the wall loss of the column will play an important role. Thus, for small diameter compact fluorescent lamps the wall loss in equation (1) should be additionally considered. According to Waymouth[2], the wall loss power P_i is:

$$P_i = 18.5(2.4/R)^2 \mu_i k T_e \quad (3)$$

where μ_i is the mobility of ions, T_e the electron temperature, R the inner tube radius, k the Boltzmann constant.
Because of the higher wall load of compact fluorescent lamps, the saturated Hg vapor pressure is much higher than that of conventional fluorescent lamps with larger diameter. Besides the contribution of Ar to μ_e, therefore, the effect of Hg vapor pressure on μ_e should be considered. Thus, the expression of μ_e according to Chen[6] is as follows:

$$\mu_e = \frac{1.148\times 10^9}{p_a \cdot T_e^{1.55}} \cdot \left\{ 1 - 1.7612\cdot 10^9 \frac{p_m}{p_a} \cdot \int \frac{x^{0.9}\exp(-x^2)dx}{x^{3.5}T_e^{1.75} + 9.0825\cdot 10^8 \cdot p_m/p_a} \right\} \quad (4)$$

where p_a and p_m are the reduced Ar and Hg pressure, respectively, $x = (mv^2/2kT)^{1/2}$.
From equation (1), using the method similar to that of lama et al and additionally considering the wall loss, the dependence of μ_e on T_e and Hg vapor pressure and the variation of rate coefficients with T_e, the axial electric field E for small diameter Ar+Hg discharge tube can be obtained:

$$E^2 = \frac{(C_4/R)(N_0 \cdot N_a)^{1/2}}{1 + C_3 R N_e N_0} + \frac{C_5 N_a^2}{[\ln(C_0 R^2 N_a N_0)]^{3/2}} + \frac{C_6}{R^2 \ln(C_0 R^2 N_a N_0)} \quad (5)$$

where N_a is argon density, N_0 mercury density, R the inner tube radius and C_0, C_3, C_4, C_5, C_6 are constants.

3. RESULTS AND DISCUSSION

For $R = 0.007$ m, $p_a = 3$ Torr and cold spot temperature $T_c = 43$ °C, the calculated constants in equation (5) are as follows: $C_0 = 1.76\cdot 10^{-36}$ m^4, $C_3 = 7.16\cdot 10^{-37}$ m^5, $C_4 = 4.98\cdot 10^{-20}$ $V^2 \cdot m^2$, $C_5 = 1.03\cdot 10^{-41}$ $V^2\cdot m^4$, $C_6 = 1.02$ V^2.
For $R = 0.007$ m, $p_a = 3$ Torr and $T_c = 43$ °C, the calculated E as a function of the current are plotted in Fig.1. For comparison, the experimental data[7] are shown as well. It shows that our work fits the experiment results quite well.

4. CONCLUSIONS

It is then concluded that for compact fluorescent lamps with

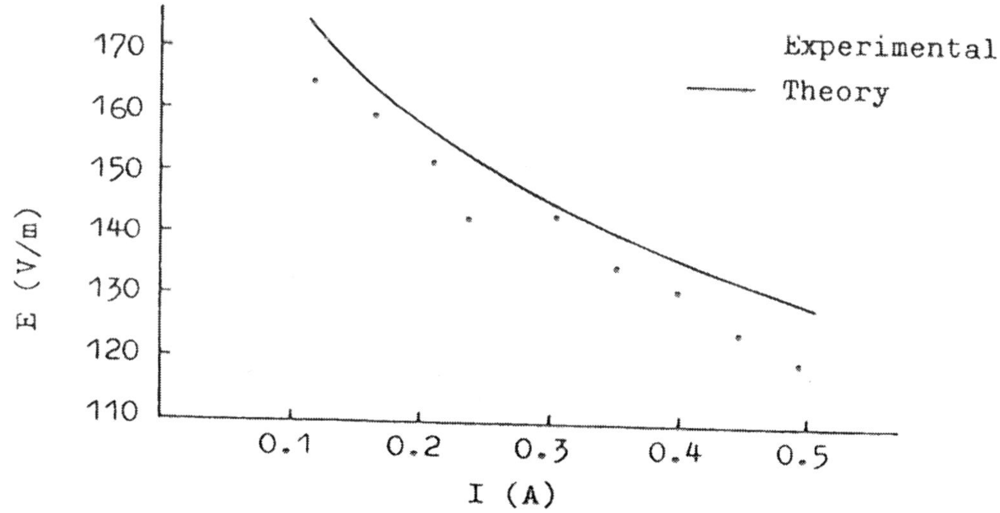

Fig.1: Axial electric field vs current for low pressure Ar+Hg discharge with R=0.007m, p_a = 3 Torr and T_c =43 °C.

smaller discharge tube diameter, the axial electric field strength E calculated according to our work are in good agreement with the experimental results.

5.References

[1] W.L.Lama et al, ApplyOpt., 21, 1806, 1982

[2] J.Waymouth, Electric Discharge Lamps, The M.I.T.Press 1971, Chapt.5

[3] S.D.Rockwood, Phys. Rev., A8, 2348, 1973

[4] W.L.Morgan and L.Vriens, J.Appl.Phys., 51, 5300, 1980

[5] L.Vriens, J.Appl.Phys., 44, 3980, 1973

[6] Z.F.Chen and H.G.Jones, J.Phys. D: Appl.Phys., 17, 337, 1984

[7] S.L.Zhu and B.H.Zhang, Proc. 3rd International Symposium on the Science and Technology of Light Sources Toulouse, France, April 1983

CATHODE-PLASMA INTERFACE IN THE FLUORESCENT LAMP

John Ingold
GE Lighting, Cleveland, OH 44112 USA

Introduction. Computer simulation of the axial temperature distribution of a fluorescent lamp cathode was described at LS-5 [1]. The temperature distribution was obtained by solving the time-dependent, one-dimensional heat balance equation for the stick cathode. Calculated temperature distributions are in good qualitative agreement with those measured by optical pyrometry, and calculated hot-to-cold resistance ratios are in good agreement with those determined from hot-to-cold resistance measurements. The model accurately predicts hot spot formation and behavior in agreement with observation.

Interaction of the cathode with the plasma of the NG is not understood as well. It is assumed that electrons are emitted from the cathode by two mechanisms: thermionic emission and ion ejection. Electron current density J_e leaving the cathode is related to the ion current density J_p arriving at the cathode by the equation $J_e(z) = J_R(z) + \gamma J_p(z)$, where γ is Townsend's second ionization coefficient and J_R is thermionic emission current density. Some of the ions produced in the NG find their way back to the cathode, being accelerated through the CF and acquiring the same energy as electrons going in the opposite direction. Impinging on the cathode, these ions give up their kinetic energy eV_k plus recombination energy $e(V_i - \phi)$ to the cathode surface. Consequently, the cathode surface is heated by the amount $J_p(V_k + V_i - \phi)$. The ratio of ion current to electron current at the cathode was found by Waymouth [2] to follow an assumption made by Druyvesteyn and Penning [3]. Using an ingenious experimental technique, Waymouth showed the relation to be approximately of the form

$$\frac{I_p}{I_e} = \alpha(V_k - V_i); \quad V_k > V_i, \tag{1}$$

where the constant α depends on gas type. The principal interaction between the cathode and the plasma of the NG is embodied in Eq. 1, which empirically relates ion current at the cathode to electron current and CF. This paper is concerned with fundamental derivation of a relation to replace Eq. 1, taking into account the most important physical processes operating in the NG. By observation, the cathode in an operating fluorescent lamp is engulfed by a glowing discharge, the NG. Line radiation characteristic of mercury and that of the buffer gas is normally observed emanating from the NG near the cathode. Visual observation shows a hot spot on the cathode surface with no discernible cathode dark space, a negative glow which extends 2-3 cm, and a Faraday dark space (FDS) of roughly the same length before the beginning of the positive column. The NG described thusly has been studied experimentally [2] [3] [4] [5], providing the following qualitative picture.

Model. Electrons emitted from the hot spot are accelerated through a very thin space with a high electric field, called the cathode sheath, into the NG of the lamp discharge. The magnitude of the potential drop V_k in the CF is somewhat larger than the ionization potential of mercury, being usually 12-23 V. Each electron entering the NG from the cathode supposedly has energy equal to V_k. In the first approximation, these energetic electrons lose momentum, but not energy, in elastic collisions with mercury and argon atoms. This process results in a spherical shell velocity distribution [6] for the energetic electrons; momentum is randonized, but energy is not. In the NG, these electrons have constant energy in the amount of eV_k and density which decays exponentially with distance away from the cathode, according to observation [4]. Theoretical Langmuir probe characteristics based on this picture are compared with those measured [4] in Fig. 1.

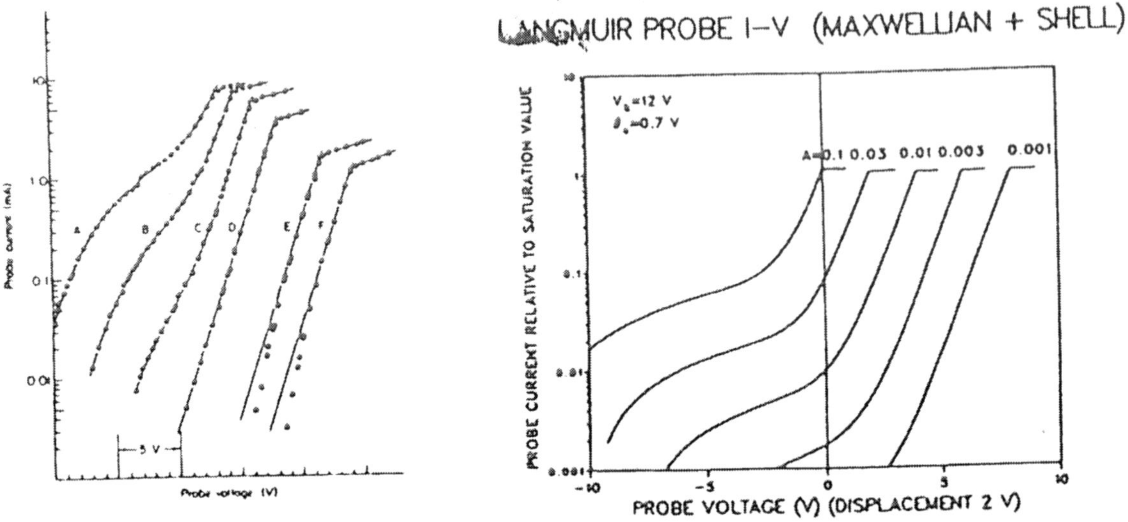

Figure 1. Theoretical Langmuir probe characteristics on the right are compared with those measured on the left. Measured curves A-F in the left hand graph, taken at distances 0.25 cm to 4.25 cm from cathode, are displaced vertically to separate them. Theoretical curves in the right hand graph simulate moving away from cathode by decreasing beam electron density, of which the value relative to plasma electron density is denoted by the parameter A. Note qualitatively similar shapes in the two graphs.

CF thickness is small due to the high ion current density impinging on the cathode; consequently, energetic electrons are injected into the NG with an average energy equal to the CF potential drop. In this paper, these energetic electrons are called *beam* electrons, to distinguish them from low energy electrons, called *plasma* electrons, which are produced by ionization of the gases in the discharge tube by beam electrons. Energy losses due to processes other than inelastic collisions between beam electrons and atoms can be neglected. Having undergone an inelastic collision, however, a beam electron loses a large fraction of its energy, and is therefore converted from the beam population n_b to the plasma population n_p.

In the PC of a typical mercury plus rare gas discharge, the electron temperature is about 12,000 K and the radially averaged electron density about 2×10^{11} cm^{-3} [7], giving a Debye length of about 2×10^{-3} cm, which is much smaller than the tube radius of about 1.8 cm. Therefore, it can be assumed that the PC of this discharge is controlled by ambipolar diffusion. According to measurement [4], electron temperature in the NG is about the same as that in the PC, and electron density somewhat higher, so that it is reasonable to assume that the NG is also controlled by ambipolar diffusion.

Based on this qualitative picture of the hot cathode negative glow, a model for the cathode-plasma interface in the fluorescent lamp has been developed [8]. A key result of this model is the ratio of ion current density to that of electrons at the cathode-plasma interface. This relation is

$$\frac{J_i}{Je} \approx \frac{f_{ib}\alpha_b}{\alpha_b + k_p}, \tag{2}$$

where f_{ib} is the ratio of ionization collision frequency to total inelastic collision frequency of beam electrons, α_b^2 is the ratio of inelastic collision frequency to diffusion coefficient of beam electrons, and $k_p^2 \approx \Lambda^{-2}$, where Λ is the characteristic diffusion length of the discharge tube. Comparison between this expression and measurement [2] is shown in Fig. 2.

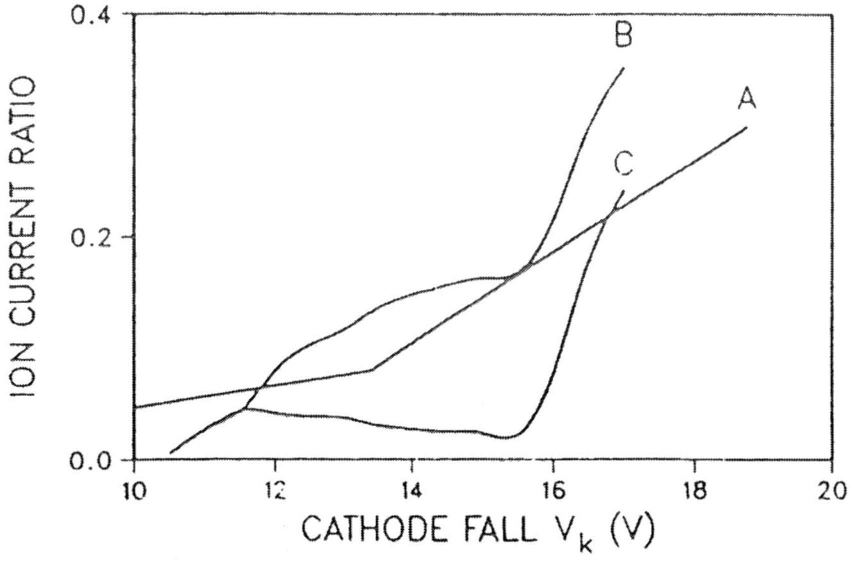

Figure 2. Comparison between measured [2] (Curve A) and calculated (Curve C) current ratio for a gas mixture of 0.9 Torr Ne, 0.1 Torr Ar, and 1 mTorr Hg (no Ar-Hg Penning effect). Curve B same as Curve C, except it is assumed that 30% of Ar excitation produces ion pairs by Ar-Hg Penning effect. Curves B and C calculated according to Eq. 2.

Results. It is found that the negative glow can be maintained by beam electrons without ionization by plasma electrons. The model predicts combined length of the negative glow and Faraday dark space in good agreement with measurement. In addition, the model closely predicts the shape of measured axial distribution of plasma electron density. The model also predicts ion current to the cathode in qualitative agreement with observation. This ion current is much larger than the radial ambipolar diffusion current density at the insulating wall of the discharge. The model also predicts two reversals in the axial electric field—one in the negative glow near the cathode where axial current is supported by ambipolar diffusion, and one at the end of the Faraday dark space near the positive column where axial current can no longer be supported by ambipolar diffusion.

References

[1] A. K. Bhattacharya, J. H. Ingold, T. F. Soules, LS-5, York, England, 10-14 Sep 1989. Paper 69:P.

[2] J. F. Waymouth, *Electric Discharge Lamps* (MIT Press, Cambridge, MA, 1971) Chap. 4.

[3] M. J. Druyvesteyn and F. M. Penning, Rev. Mod. Phys. 12, 87 (1940)

[4] J. F. Waymouth, J. Appl. Phys. 30, 1404 (1959)

[5] R. C. Wamsley, J. E. Lawler, J. H. Ingold, L. Bigio and V. D. Roberts, Appl. Phys. Lett. 57, 2416 (1990); R. C. Wamsley, T. R. O'Brian, K. Mitsuhashi, and J. E. Lawler, *ibid.*, **59**, 2947 (1991)

[6] J. H. Ingold, Phys. Rev. A 40, 3855 (1989)

[7] W. Verweij, Philips Res. Rpts. Suppl. 2 (1961)

[8] J. H. Ingold, Phys. Rev. A 43, 3093 (1991)

A NOVEL LOW WATTAGE FLUORESCENT LAMP

Li Xianli
Najing Telecommunications Technical Research Institute
3 Biao Ying. Najing, People's Republic of China

Abstract:

The performance and structure of a novel low wattage FL lamps are presented. These FL lamps work with low current and high luminous efficiency. In order to extend lifetime, Dual-filaments electrode are used.

Introduction

The luminous efficiency of existing low wattage FL lamps, no matter it is straight type (4w or 6w) or compact type (5w), are reduced seriously because the voltage of tube is too low. At the same time, the energy loss of the inductive ballasts of these FL lamps increased greatly. Therefore the total luminous efficiency of these lamps is much lower than other convetional FL lamps and they are not widely used. In fact the lumen of these low wattage FL lamps is equivalent to 15w to 25w incandescent lamps. If the performances of these low wattage FL lamps are impreoved, they can be used widely in various places where high illuminance is not required. But these problems have not been fully studied yet.

Parameters and performances

The main feature of the novel low wattage FL lamps is its low lamps current, which is much less than corresponding existing products. So the voltage of its tube can be disigned higher and the luminous efficiency of these tubes can be improved greatly, the energy loss, weight and size of the inductive ballasts matched these FL tubes are reduced respectively. With the result that the total luminous efficiency of the lamp system can be two times of the corresponding existing products. The total weight of lamps can be reduced about half and the cost can be reduced by 1/3.

lamp type / parameters and performances	existing low wattage compact FL lamps (5w)	new low wattage compact FL lamps (4w)
power voltage (V)	220	220
power frequency (HZ)	50	50
power of tube (W)	5	4
current of tube (A)	0.180	0.055
voltage of tube (V)	35	100
lumen · (Lm)	200	250
loss of inductive ballast(w)	5	1.5
weight of inductive ballast (g)	400	150
efficiency of tube (Lm/w)	40.0	62.5
efficiency of system (Lm/w)	20.0	45.5

Table 1 - Comparison of the parameters and performances.

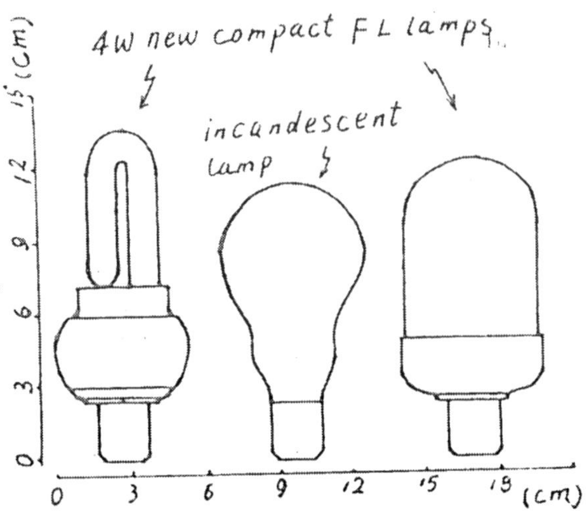

Figure 1 - Practical size of 4w new compact FL lamps with inductive ballast.

Table 1 illustrates the comparison of parameters and performances for new low wattage compact FL lamp (4w) with the existing compact FL lamp (5w).

Dual-filaments elestrode structure

In order to meet the demand for preheating filament, the diameter of filament of the low current FL lamps have to be very thin. But the temperature of the hot spot on the very thin filament will be sure to increase quickly during operation. It leads to very short lifetime of tube.

A special designed dual-filaments electrode structure had been applied to the new low wattage FL lamps shown in Figure 2. One of the dual-filamets is used as preheating filameht and it is preheated by current at the moment of start. The other is non-preheating filament. Two ends of the non-preheating filament are short-circuited.

It is worthy to point out, that a main feature of this filament electrode is automatic transfer of the hot spot from preheating filament to non-preheaing filament after some time the tube

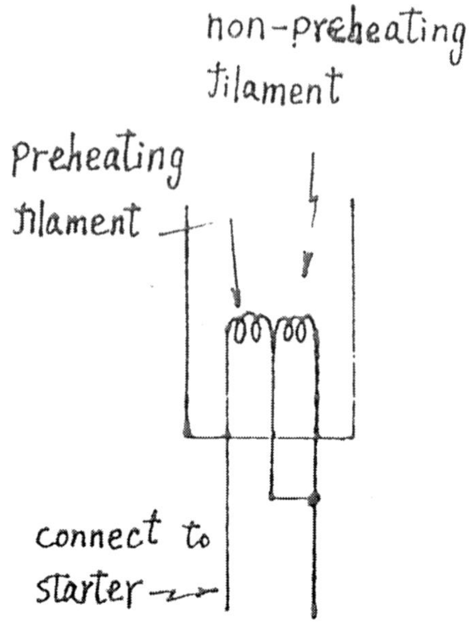

Figure 2 - Electrode structure with the non-preheating filament.

had operated. Therefore, it is made a new operating model with double-hot spots or three-hot spots on two filaments. The phenomenon of multi-hot spots can be seen easily at the later time of life. So the filament temperature is reduced greatly and the lifetime of tube is extended effectively. For example, the filament diameter of tube of the 4w new compact FL lamps is 0.018mm, when traditional single filament structur is used, the lifetime of tube can not reach 1000 hours . While above double-filaments structure is used, it is more than 5000 hours.

The principle of the hot spot tranfer between preheating filament and non-preheating filament is based on the discharge system has to keep the lowst energy condition. Analysis and experiments have shown that this dual-filament structure could also be applied to most other fluorescent lamps. Nearly every type fluorescent lamp can use this filament structure to extend the lifetime of lamps. It is especial useful for the high power FL lamps.

As is known, since the tube current is lower, the inside diameter of tube would be reduced and the pressure of augon gas would be increased.

Reference

1. Xianli Li 1989 High Efficiency FL Lamps With Low Power. Symposium on the Science and Technology of the Compact FL Lamps. Nanjing.
2. Waymouth, J.F. 1971. Electric Discharge Lamps. Cambridge, MA: MIT Press.
3. Misono, k. 1990. Cathode Fall Voltage of Low-Current Fluorescent Lamps. J of the IES 20 (no. 2) 108.

GENERAL SCALING LAW FOR THE LUMINOUS FLUX OF FLUORESCENT TUBES

F.A.S. Ligthart
Philips Lighting B.V., Eindhoven, The Netherlands

From experimental data, a general scaling law relating wall loading and luminous flux at optimum mercury pressure of fluorescent lamps was found. Within the experimental limits, the relation is independent of the noble gas mixture (Ar and Ar-Ne mixtures at optimum pressure were used), the operating frequency (50 Hz and 40 kHz) and the tube diameter (7.8, 10.3 and 14.5 mm were used). The tubes were provided with the three-band phosphors $(Ba,Eu)MgAl_{11}O_{17}$ or BAM for the blue emission, $(Ce,Tb)MgAl_{11}O_{19}$ or CAT for the green emission and $(Y,Eu)_2O_3$ for the orange-red emission. Thanks to the excellent maintenance behaviour of these phosphors at the high wall loadings occurring in the small diameter tubes, the luminous data are not distorted by maintenance effects.

Fig.1 shows the luminous flux (lumen/cm) versus the discharge power (Watt/cm), both per cm of discharge column. The data include 50 Hz operation of 400 Pa Hg-Ar lamps at different diameters and high-frequency operation of 533 Pa Hg-Ne75%/Ar25% at 10 mm diameter. More data are available; they were left out for clarity. All measurements were done at optimum mercury pressure. The measurements show the well-known saturation effect of the luminous flux at the higher wall loadings, the saturation being stronger for the smaller diameters.

Our scaling model implies the following: the reduced quantities Φ/O and P/O are introduced, Φ/O represents the optimum luminous flux of a tube divided by its area and P/O represents the power consumption of the discharge column divided by its area. It appears (see Fig.2) that between these quantities, Φ/O and P/O, **one simple relation** approximately holds for all 400 and 533 Pascal Mercury/Argon and Mercury/Argon-Neon discharges, both for 50 Hz and for high-frequency operation, and for diameters in the range from 7.8 to 14.5 mm.

Fig.2 shows that, at a wall loading of 50 mW/cm^2, the maximum attainable luminous flux is about 5.5 lm/cm^2 (implying a column efficacy of 110 lm/W) while, at a wall loading of 100 mW/cm^2, the maximum attainable luminous flux is about 10 lm/cm^2 (implying a column efficacy of 100 lm/W).

Fig. 1

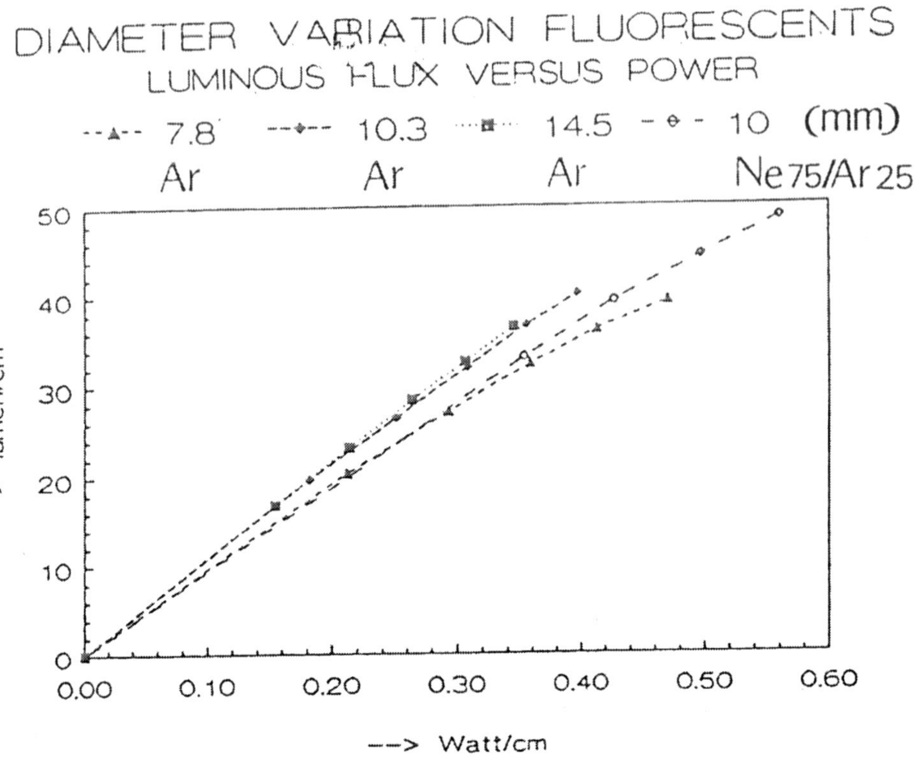
DIAMETER VARIATION FLUORESCENTS
LUMINOUS FLUX VERSUS POWER
diameter 7.8 10.3 14.5 10 (mm)
Ar Ar Ar Ne75/Ar25
--> lumen/cm
--> Watt/cm
0 10 20 30 40 50
0.00 0.10 0.20 0.30 0.40 0.50 0.60

Fig. 2

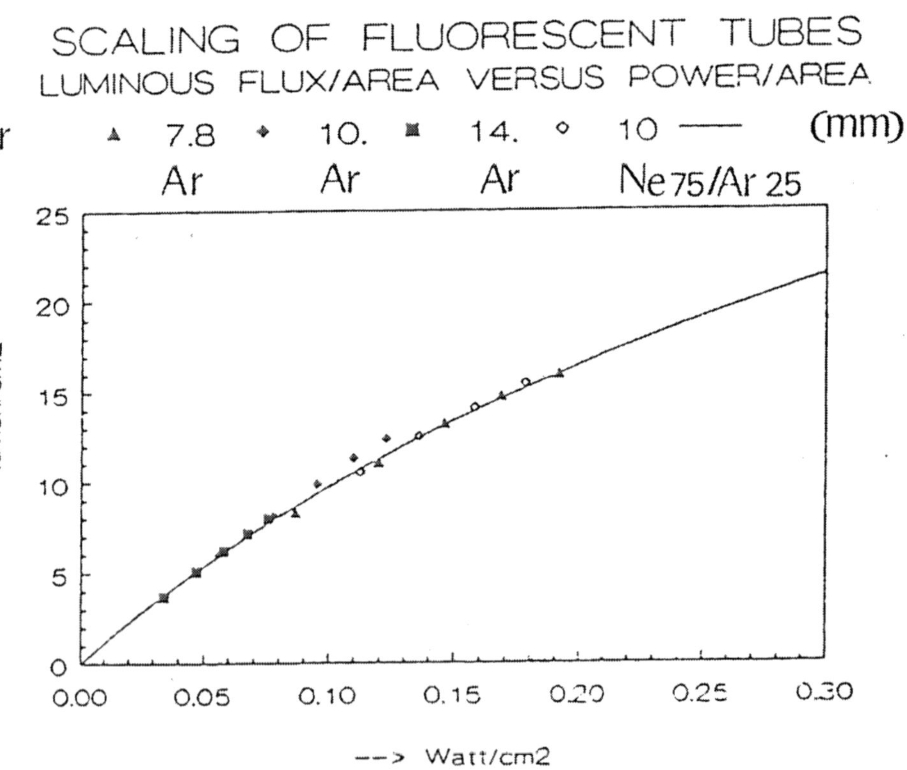
SCALING OF FLUORESCENT TUBES
LUMINOUS FLUX/AREA VERSUS POWER/AREA
diameter 7.8 10. 14. 10 (mm)
Ar Ar Ar Ne75/Ar25
--> lumen/cm2
--> Watt/cm2
0 5 10 15 20 25
0.00 0.05 0.10 0.15 0.20 0.25 0.30

FLUORESCENT LAMP ELECTRODE DIAGNOSTICS BY LASER INDUCED FLUORESCENCE

Philip E. Moskowitz
GTE Electrical Products
Danvers, MA

This paper describes a laser induced fluorescence (LIF) experiment to measure the flux of free barium from the electrode of an operating fluorescent lamp. Much of the activity in this field in the past few years[1,2,3] has taken advantage of the laser as a nonintrusive diagnostic tool. The work described here uses the continuous wave dye laser's extremely narrow (<1 MHz) spectral bandwidth and tunability to measure the radial density profile of Ba(I) in the vicinity of the coil electrode as a function of various lamp parameters. As such, it is an application of the technique described in ref(3). However, in the present work, the spectral profile of the Ba(I) resonance at each spatial point was recorded so that the local temperature could be ascertained as well. The flux has been calculated using these data, and estimates made as to the relative amount of barium being evaporated from the coil. Details of the experimental effort follow.

The Ba-Sr-Ca emissive coating is lost from the electrode primarily by evaporation from the hottest portion of the coil: the "hot spot". During lamp life, the coating is depleted as the hot spot migrates along the coil. The relative density and absolute temperature of atomic barium in the vicinity of the hot spot were recorded as a function of: lamp current phase angle, distance from the coil, fill pressure, and ballast type. Data were taken on F40T12 lamps with the phosphor cleared away from the end. The laser beam was parallel to the axis of the coil, and the resulting laser induced fluorescence was collected at a right angle to the coil using a lens, a 0.22m spectrometer, and a photomultiplier. Barium excitation was achieved from the ground state to the 6^1P_1 level at 553nm. This scheme did not provide for an alternate decay channel, hence both excitation and detection were at the same wavelength. This coincidence placed an upper limit on the signal-to-noise ratio due to laser scattering. Therefore, densities could not be measured farther than 5mm from the coil. Data were recorded by scanning the laser frequency over the resonance. A typical spectrum is shown in Figure 1. The lineshape is a Voight profile. However, under these experimental conditions of temperature, pressure, and laser power. Doppler broadening dominates. Consequently, data were fit to a Gaussian.

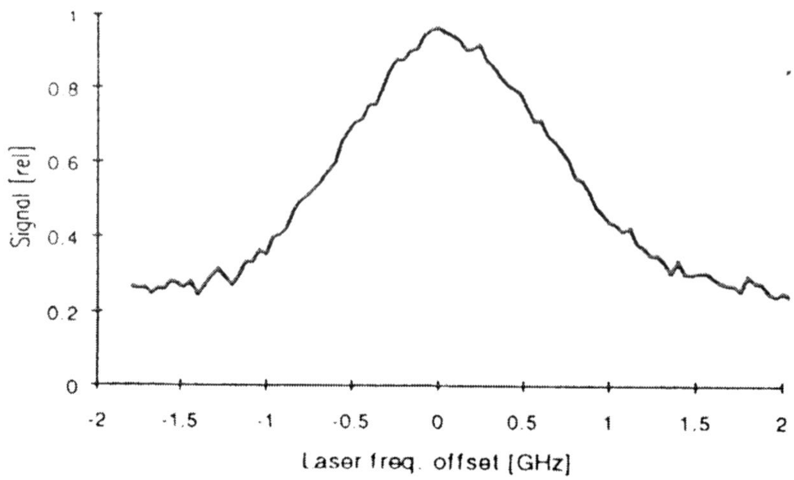

Figure 1

The relative density (n) is proportional to the area under the lineshape profile, while the temperature (T) is proportional to the square of the linewidth.

Figure 2 shows the Ba(I) density in arbitrary units as a function of current phase angle (0 = start of anode phase), acquired on the 60Hz inductive ballast. It is in qualitative agree-

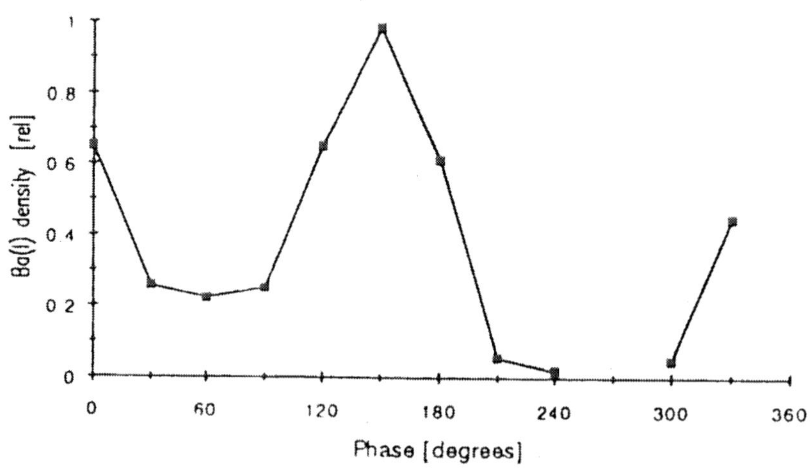

Figure 2

ment with previous work,[1] and shows that the bulk of the barium emission occurs during the anode phase.

Figure 3 shows the relative density distribution for radial distances of 1 to 5 mm from

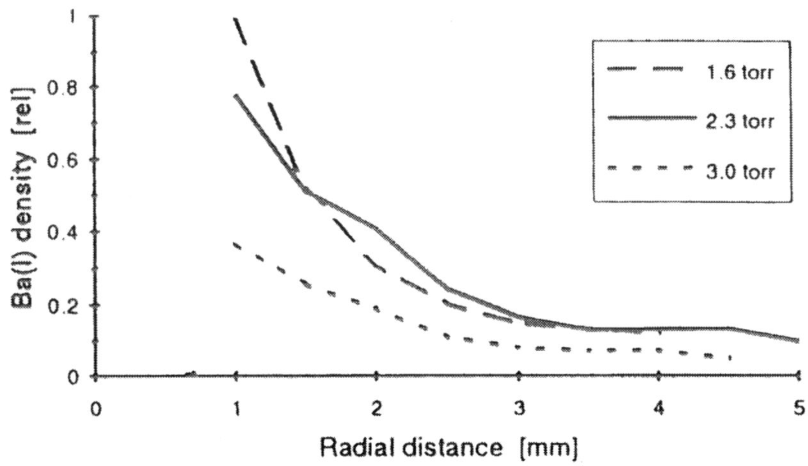

Figure 3

the hot spot, recorded in 0.5mm increments. This distribution was recorded for 3 lamps with 100% argon fill at pressures of 1.6, 2.3, and 3.0 Torr operating with a 60Hz "reference" inductive ballast. To acquire the spatial density profile, only the lamp was moved; the laser beam and detection optics remained fixed. Axial density distributions (parallel to the central axis of the lamp) yielded distributions similar to Figure 3. While the decrease in barium pressure ($p = nk_BT$) with distance fits an exponential form as stated in ref(3), the decay is more rapid here. This could be explained by a different buffer gas type or pressure, or a different ballast. Details were not given in ref(3).

For purposes of estimating barium loss rates, a flux can be calculated by assuming spherical symmetry with the center located at the hot spot. The flux will be equal to the local gradient of the barium pressure times the diffusion coefficient, divided by the local temperature (k_BT). Since no absolute density measurements were made, all experimentally derived fluxes can be related to one "reference flux": the 2.3 Torr lamp on the reference 60Hz ballast. A qualitative analysis using the data in Figure 3 (and the temperature data) would lead one to conclude that the most rapid depletion of the barium will occur in the 1.6 Torr case. Temperature distributions and data as a function of the other variables mentioned above will be presented.

[1]A. K. Bhattacharya, J. Appl. Phys. 65, 4595 (1989)
[2]A. K. Bhattacharya, J. Appl. Phys. 65, 4603 (1989)
[3]R. Hilbig, U. Chittka, *Proceedings of the 5th Int'l Symp. on the Science and Technology of Light Sources*, York, edited by R. Devonshire and T. Meads (University of Sheffield, Sheffield, 1989)

VUV RADIATION OF THE RARE-GAS DISCHARGE FLUORESCENT LAMP AT PULSED DISCHARGE

Takehiko Sakurai, Takeo Saikatsu and Osamu Myodo
Mitsubishi Electric Corporation
Consumer Products Laboratory, 2-14-40, Ofuna, Kamakura 247, Japan

Recently, the rare-gas discharge fluorescent lamps, using rare-gas (xenon etc.) as discharge gas instead of mercury vapor, have been used for some industrial applications in place of conventional fluorescent lamps because of their merits------instantaneous start-up time and largely independent of temperature-induced fluctuation. But, the luminance and efficiency of the rare-gas discharge lamps are smaller than them of the conventional fluorescent lamps.

At the former experiments, we have discovered that the luminance and the efficiency of the rare-gas discharge fluorescent lamps were increased by means of the operation of a pulse mode and the optimum selection of a gas filling pressure. Then, we tried to observe the waveforms of luminous emittance at pulsed discharge.

From the waveforms of the luminous emittance, we also discovered that the large emission which exceeded the emission at energization-period of pulsed discharge happened at die-period when the gas filling pressure was high. (Fig. 1)

To investigate this discharge in detail, we observed the waveforms of vapor-ultraviolet radiation of xenon atom. As a result, at low gas pressure area, only the resonance radiation from xenon atom (147nm) could be observed. On the other hand, at high pressure, the radiation from xenon's excimer (172nm) which did not occur at low gas pressure area also could be seen. At high pressure, this radiation happened strongly at die-period of pulsed discharge, as same as xenon atom's resonance radiation. (Fig. 2)

Consequently, at the rare-gas discharge fluorescent lamps with high pressure xenon, it is thought that the increace of luminance is mainly caused by the radiation from xenon's excimer.

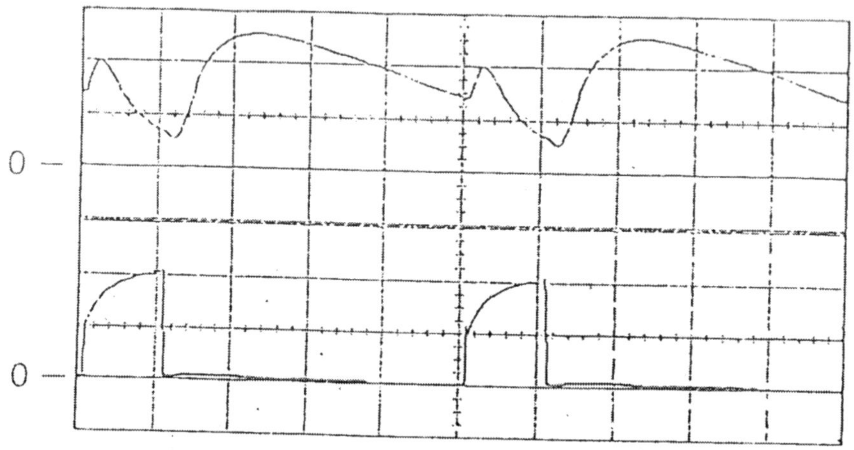

Fig. 1 Waveform of emission of the blue lamp
Waveform of emission(UP) and lamp current(DOWN)
(3333Pa, Xe100%, 10kHz, 20 μs/div)

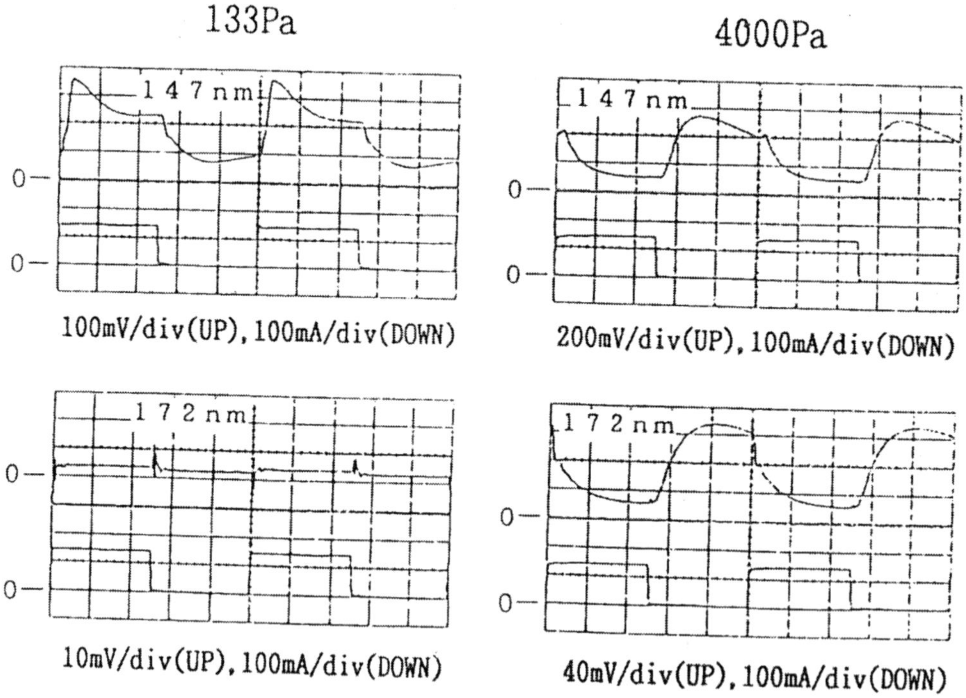

Fig. 2 Waveforms of VUV radiation of 2 different wavelength
VUV radiation(UP) and lamp current(DOWN)
(Xe100%, 10kHz, 20 μs/div)

CIRCUIT PARAMETERS MEASUREMENT OF HIGH FREQUENCY ELECTRIC DISCHARGE LAMPS AND ELECTRONIC BALLASTS

WANG Liancai, JIA Gengxin, YANG Lixi
Zhengzhou Institute of Technology, P.R.China
GUO Jianmin
Academy of Railway Sciences, P.R.China

Electric discharge lamps, in the high frequency working condition, have some advantages, such as easy starting, small voluem for the ballast devices and low power loss. Therefore, many light source researchers are interested in the lamp. But the lamp is a nonlinear circuit element and is working under high frequency, which make it difficult to measure the circuit parameters. The measurement of the circuit parameters and the efficiency of the electronic ballast is also an urgent problem to be solved.

On the basis of analysing the present measuring methods, and according to the nonlinear circuit theory, samples of the lamp voltage $v(t)$ and the current $i(t)$ are taken with our new device, which is made by some intergrated electronic elements, through Fourier's transformation:

$$v(t)=V_0+\sum_{k=1}^{\infty}V_k\sin(k\omega t+\Phi_k),\quad i(t)=I_0+\sum_{k=1}^{\infty}I_k\sin(k\omega t+\phi_k)$$

According to the definition of the effective value of the voltage and the current,

$$V=\sqrt{V_0^2+V_1^2+V_2^2+\cdots}\qquad I=\sqrt{I_0^2+I_1^2+I_2^2+\cdots}$$

according to the definition of the average power:

$$P=\frac{1}{T}\int_0^T p\,dt=\frac{1}{T}\int_0^T v(t)\cdot i(t)\,dt$$

Some mathematical treats (square, sum and root) have been done with the electronic elements, circuit parameter measurement of high frequency electric discharge lamps and electronic ballast has been achieved through some circuit transformations. If the new measuring device combines the PC, it will become automatic.

The paper will recount the principle, the realization channel and the use of the new measuring device.

REFERENCES

(1) WANG Liancai, YANG Lixi, QIN Kejing, JIA Gengxin, "Circuit Model of Discharge Lamps". The Fifth Internat. Symp. on Sci. Technol. of Light Sources. York, England. 10—14 Sept. 1989.

(2) Leon O.Chua, "Introduction to Nonlinear Network Theory", McGraw-Hill Book Company, New York, 1969.

(3) QIU Guanyuan, "Circuit Theory".

(4) C.A.Desor and E.S.Koh, "Basic Circuit Theory".

(5) SHEN Lei, "Principle and Application of COMS Intergrated Circuit".

(6) KONG Youlin, "Intergrated Operational Amplifier and Its Application".

FREQUENCY PROPERTY OF LOW PRESSURE ELECTRIC DISCHARGE LAMPS

WANG Liancai, YANG Lixi, JIA Gengxin
Zhengzhou Institute of Technology, P.R.China

YANG Laishun
YAXING Fluorescent Lamps Factory, P.R.China

Frequency property of the low pressure electric discharge lamp shows the relationship between the lamp circuit parameters and the frequency of source. Taking the lamp for a circuit element, and studying its circuit property in different frequency band is the key solution to the matching of the ballast and the lamp.

Drived by the large power and wide frequency band sine source, the waves of the lamp voltage and current, and the dynamic state $v-i$ characteristic are studied with some new instruments, while the electric power of the lamp is kept constant. On the bases of the achievements in the lamp researches by S.T.Henderson, W.Elenbaas and others, we found some new experimental phenomena, and discovered the characteristic frequency range of the lamp. These results will provide reliable theoretical bases for the research of the electrodeless discharge lamp and the high frequency discharge lamp, as well as some key parameters for the development of the electronic ballast and an important experimental support for the new circuit model of the lamp —— current controlled memristor.

REFERENCES

(1) WANG Liancai, YANG Lixi, QIN Kejing, JIA Gengxin. "Circuit Model of Gas Discharge Lamps". The Fifth Internat.Symp. on Sci. Technol. of Light Sources. York, England. 1989.
(2) Leon O.Chua, "Introduction to Nonlinear Network Theory".
(3) Waymouth, "Electric Discharge Lamps". The M.I.T. Press.
(4) CAI Zuquan, et al. "Principle and Application of Light-Source Electric Appliances".

STUDIES OF THE VUV XENON RESONANCE RADIATION SOURCES

Shaolong Zhu, Wei Zhou, Li Sun
Institute for Electric Light Sources, Fudan University, Shanghai, China

Vacuum ultraviolet(VUV) radiation sources are current interest in many research fields and in some modern industries as well. The synchrotron radiation is certainly the most powerful VUV source, but it is expensive to build and operate. Besides, it is not always available for most potential users. It is well known that the xenon resonance radiation is in the VUV range (λ=147.0nm) and has been used as VUV sources for many years. However, its radiation characteristics are not fully understood and there are no quantitative data available up to now. This paper presents studies of the VUV xenon resonance radiation sources, both theoretically and experimentally, which finally leads to the optimization of the VUV output.

The discharge plasma is enclosed in a glass tube of diameter 12mm and length 40mm. At one end of the tube, a MgF_2 single crystal disk of thickness 0.8mm is vacuum-tight sealed as a VUV output window. In the middle of the tube inserted is a pair of tungsten probes of diameter 0.4mm and length 2mm with seperation of 2mm, which serves as measurement of electron temperatures and densities. The whole tube is evacuated to 10^{-3}Pa and then after baked thoroughly it is filled a He-Xe mixture with various ratio and pressure.

The plasma is excited by r-f electrodeless discharge (~120MHz, 1W) and its radiation is detected by a VUV detect system, which consists of a VM-505 VUV monochrometer, sodium salicylate fluorescent convertor and a GDB-413 photo-multiplier. The detect system has been absolutely calibrated beforehand. So that the VUV (λ=147.0nm) axial radiances at various He-Xe mixtures and pressures are obtained. At the same time the electron temperatures and densities on the axis are measured by the double probes.

In order to model the discharge, a simplified xenon four atomic level model is introduced, i.e. ground, ionized and two excited levels. The two excited levels are virtual, which consist of a real radiation level and its nearby metastable, i.e. the first excited level consists of $^3P_1$ and $^3P_0$, and the second of $^1P_1$ and $^3P_2$. The other real higher levels are much less populated, therefore we neglect them for simplification. Because of its much higher excitation and ionization energies, the helium is only served as a buffer. A set of rate equations of the four level system are set up. The rate coefficients are derived from the counterparts of the real energy states, which are calculated from the corresponding collision cross-section. The effective life time of the resonance state is calculated by Holstein equation.

The rate equations are solved numerically and the particle densities at excited levels are obtained, from which the axial radiances of the discharge plasma are calculated by the radiation transfer equation. The results are compared with the experiments. It seems in reasonable agreement, though the theoretical values are systematically little smaller than experimental.

The results also show, both theoretically and experimentally that the He-Xe discharge plasma is a more efficient VUV (147.0nm) source than pure xenon. At about filling pressure 400Pa and 2.5% Xe, the axial radiance of the He-Xe discharge plasma reaches the maximum, about $10W/Sr.m^2$. The total VUV (147.0nm) photon output is about 5.9×10^{14}. This VUV monochromatic source could be applied to molecular photo-excitation, photo-ionization and so on.

SEMICONDUCTOR LIGHT SOURCES

F. Beleznay

Research Institute for Technical Physics of HAS
1325 Budapest 1., P.O.Box 76.

Special optical properties such as foton emission and absorption were with us from the very begining of semiconductor physics. Forbidden band and the existence of two kinds of charge carriers of controllable concentration and their injection in a current carrying p—n junction are the most basic properties of any semiconductor material. During injection these charge carriers are recombined and photons of the energy of the band gap may generated, hence a forward biased p—n junction is a very simple, natural light source. This application has been realized in the early days of semiconductor research. Even the title of one of the early International Conference in Semiconductor Physics in Garmisch in 1956 was different as today, being called "Semiconductors and Phosphors". It was the general belief that these simple diode light sources made from materials of appropriate band gaps (such as ZnS, phosphor of early fluorescent lamps) may replace other incandescent light sources. Although phosphors are made from different materials today, ZnS still exists in electroluminescent panels for diaply applications.

The history of semiconductor light sources can be classified into 3 different eras.

1. The dawn of modern physics of semiconductor light sources

This is the time when an unified understanding was evolving that some semiconductors, called phosphors were a special calss of semiconductors with efficient recombination. This process was not complete until the discovery and understanding direct and indirect bands in semiconductors, details of pair recombination and the different forms of charge injection over the fifties. The evolution of this understanding led to the development of the all solid state laser which were independently achieved by three groups in 1962.

2. Extensions and sophistication

One might note the twenty year time span from this initial discovery in the early sixties to the general semiconductor laser manufacturing technology that was evolving in the early 1980s. During this time the knowledge of some materials and studies of defects in semiconductors were closely coupled with device development. This has been particularly true of III—V materials, such as GaP and GaAs. Early work on light emitting diodes (LEDs) led to the studies of isoelectronic traps, such as the Zn–O and N luminescence centers in GaP and GaAsP. The effort to improve the reliability of GaAs/GaAlAs lasers in the 1970s brought such topics as recombination–enhanced defect motion and dislocation climb into prominence with the semiconductor physics community.

At the same time perhaps one of the most important and far–reaching device breakthroughs was the recognition that epitaxial films of different semiconductors are the most useful for device technology and performance. Most semiconductor laser today is made from these heterostructure layers. All this knowledge resulted in extremaly compact, high power and long lifetime semiconductor lasers based on GaAs.

Devices made of these III–V compounds work in the near IR region. The II–VI semiconductors however offer a wider range of band gap energies from ¡0.02 eV to ¿4. eV. In spite of a considerable effort, the limitations in the microscopic and macroscopic material properties of II–VI's have led to a poor control of the electrical properties of II–VI semiconductors. ZnS, ZnSe, CdS and CdSe potentially offer devics in the visible spectral region. The resurgence in II–VI materials in the early 1980s has occured due to the hope that the current semiconductor technology developed for Si and GaAs when applied to II–Vi materials, will result in better and novel devices from these semiconductors. These transformations and transfer of our knowledge characterizes semiconductor research in recent years.

3. Present

The development of new semiconductor lasers established a new era of telecommunication. Fiber optics is replacing microwaves and cables, and is a basic technology for one of the biggest business ever, the total telecommunication. These new lasers are based on InP, another III–V compound.

With the development of laser diodes and sophisticated epitaxial technologies lasers and LEDs of different wavelengths are being developed. Red emitting visible lasers are replacing other light sources in different applications. After 30 years of intensive research green– and blue–emitting diodes are also developed. Last summer, 3M announced the first room–temperature green diode lasers, made of ZnSe, since that GaN and ZnTe/ZnS diodes have been developed, and their low threshold current could be a major advantage in developing long–lived lasers. AlGaN/GaN LEDs are on their way to market. All these novel devices are made of thin layers, sometimes superlattices of different semiconductors using the technology developed for longer wavelength commercial lasers.

With this record of history as our guide, the natural question to ask is, "What's next?". Most probably, LEDs and lasers in the visible region will find many new applications and may replace some of the other light sources in another twenty years!

Session 3

Chairman: R. Bergman

HT Chemistry + Materials

31:I **G. Eriksson**
Integrated thermochemical databases

32:P **T. R. Brumleve, M. Fields, S. A. Mucklejohn**
Preparation, vapour pressure and thermochemical parameters of thulium(III)iodide

33:P **W. van Erk, P. Cobben, P. Bennema**
Wall blackening in metal iodide lamps seen from a crystal growth point of view

34:P **O. Horacsek, Cs. Tóth, K. Horacsek**
Void growth and its relation with the bubble dispersion in doped tungsten wires

35:P **D. M. Jenkins, J. S. Ogden**
Matrix isolation studies on high temperature lamp chemistry

36:P **S. A. Mucklejohn, A. T. Dinsdale**
Modelling multicomponent metal halide systems

37:P **S. A. Mucklejohn, D. L. Trindell, R. Devonshire**
An assessment of the vapour pressures and thermochemical parameters for the rare earth tribromides

38:P **T. D. Russel**
Silica and tungsten corrosion in discharge lamps

39:P **J. Szöllősi, I. Somogyi**
Strengthening possibility of glass bulbs

40:L **G. M. Fordsdyke**
The reactions of vitreous silica in discharge lamp environments

41:L **Z. Boksay**
Inhomogeinity as an inherent property of silicate glasses

42:L **G. Zilberstein, H. J. Kim**
Creep and fracture behavior of NS tungsten in oxygen-doped atmospheres

Integrated Thermochemical Databases

Gunnar Eriksson
Lehrstuhl für Theoretische Hüttenkunde
RWTH Aachen
Germany

A review of the fully Integrated Thermochemical Database systems which are accessible to the public will be given. An Integrated Thermochemical Database (ITD) is defined as a system which offers extensive thermochemical databanks and powerful software, and which can perform equilibrium thermodynamic computations typically associated with complex chemical equilibria and phase diagram calculations in multicomponent multiphase systems. Private databases will not be considered. Also, the review does not include databases or software dealing with kinetics or heat and mass transfer.

The role of a scientific group who supports and maintains an ITD is to couple proven state-of-the-art software with self-consistent critically assessed thermodynamic data, and to make the programs and data readily accessible to the scientific community.

The principal accessible ITDs include: CSIRO, Australia, F*A*C*T, Canada, MTDATA, U.K., THERDAS, Germany. THERMO-CALC, Sweden, and THERMODATA, France. Some of these databases are on-line and some can be leased or purchased for private use. All centres offer some form of professional service. In principle, on-line ITDs can be accessed from any point in the world, although legal contracts between databases and communication costs may limit the access.

The European ITDs are members of SGTE (Scientific Group Thermodata Europe), a legally constituted consortium of databases whose membership is composed of the following laboratories: National Physical Laboratory (NPL), Teddington, U.K.; Harwell

Laboratory, U.K.; LTPCM, Grenoble, France; THERMODATA, Grenoble, France; Institut de Recherches de la Siderurgie Francaise (IR-SID), Metz, France, RWTH Aachen, Germany; Royal Institute of Technology, Stockholm, Sweden. The members of SGTE tend to share the same thermochemical data although most of the software programs are independently supported.

The most commonly used software programs are those which minimize the total Gibbs energy of a system to predict complex chemical equilibria. The following may be mentioned: ChemSage at THERDAS and F*A*C*T, MULTIPHASE at MTDATA, MELANGE at THERMODATA, and POLY3 at THERMO-CALC.

PREPARATION, VAPOUR PRESSURE AND THERMOCHEMICAL PARAMETERS OF THULIUM(III) IODIDE

T.R.Brumleve
APL Engineered Materials Inc., 2401 North Willow Road, Urbana, IL 61801, U.S.A.
&
M.Fields and S.A.Mucklejohn
GE Thorn Lamps Limited, Advanced Technology Department, Melton Road, Leicester LE4 7PD, U.K.

As part of our continuing work on the preparation, analysis, vapour pressure and determination of thermochemical parameters of metal halides employed in high-pressure discharge lamps (1-3), we have studied the synthesis and properties of thulium(III) iodide.

High purity metal halides with low concentrations of oxygen-, carbon- and hydrogen-containing species are essential for use in discharge lamps and for the determination of physicochemical properties. Modelling the behaviour of multicomponent metal halide systems used in lamps (4) requires reliable thermochemical parameters for the solid and liquid phases of the pure components. Similarly, accurate parameters are required for the gases to predict the gas phase composition above the condensed phases.

Thulium(III) iodide was prepared by direct combination of thulium metal (99.9+ per cent, Research Chemicals, Phoenix, Arizona) and resublimed iodine (reagent ACS). The details of the synthetic procedure have been described previously for iron(II) iodide in reference (1). The large, pale yellow platelets of anhydrous TmI_3 were handled and packaged in an argon-filled drybox.

The product was analyzed for total Tm content by EDTA titration (result: 30.73±0.06 mass per cent Tm; theory: 30.73), and for total iodide by argentometric titration (result: 68.98±0.30 mass per cent I^-; theory: 69.27). The product was also analyzed for trace H_2O and OH^- by coulometric titration (result: 11±5 ppm by mass H_2O plus OH^-). Analysis of 68 trace elements by inductively coupled plasma emission spectroscopy yielded total impurities of less than 100 ppm by mass.

The origins of the thermochemical parameters available for the TmI_3 system are typical of the situation for many of the rare earth trihalides. Early compilations (5-7) gave estimated values for various parameters and issued [illegible] for further experimental work to be carried out. This work was long delayed in most cases partly a reflection of the limited use of these materials in industrial applications but mainly because high purity samples are difficult and expensive to prepare, the materials react with silica at temperatures above approximately 1000 K, are high melting and are relatively involatile. Reliable vapour pressure measurements for many of the rare earth trihalides cover very restricted temperature ranges. The {monomer + dimer} equilibria have been studied for only a few of the series.

The most comprehensive account of the vapour pressure of thulium(III) iodide is the elegant work by Dettingmeijer and Dielis (8). By a combination of Knudsen effusion with mass spectrometric detection and mass loss monitoring, they determined the vapour pressure above the solid from 874 to 1032 K. They were able to confirm the presence of monomers and dimers in the vapour and to show that the mole fraction of the dimer increased with increasing temperature from x(dimer) = 0.21 at 880 K to 0.43 at 1010 K.

The only experimental values for the molar heat capacity of TmI_3 are those obtained from differential scanning calorimetry by Gardner and Preston (9). The melting temperature (1288 K) is too high for measurements to be made on the liquid phase using this technique.

There are no experimental values reported in the literature for the enthalpy of formation of TmI_3<s>, the enthalpy of fusion nor the molecular constants (Tm-I bond length and vibrational frequencies) for TmI_3<g>. Estimated values for the vibrational frequencies and bond lengths are provided by Lynch (10) and Myers and Graves (11). TmI_3<g>, in common with the other gaseous rare earth trihalides, is thought to have D_{3h} symmetry, *i.e.* a planar structure. We have calculated the molar heat capacity and entropy of TmI_3<g> over the range 298.15 to 6000 K from the equations given in the reference (12). The Tm-I bond length and the vibrational frequencies were taken from reference (11). We have used energy levels for Tm^{3+}<g> taken from reference (13) to represent the electronic energy levels of TmI_3<g>. Only states with energies less than 50000 cm^{-1} were included.

Estimated values for the enthalpy of formation at 298.15 K vary from -669 $kJ.mol^{-1}$ (5) to -567 $kJ.mol^{-1}$ (14). The most widely quoted (15-17) value is -602 $kJ.mol^{-1}$. References (7) and (18) both cite -573 $kJ.mol^{-1}$ whilst reference (19) uses -582 $kJ.mol^{-1}$.

We have used the vapour pressure measurements from reference (10) together with our own values to derive a comprehensive set of thermochemical parameters TmI_3<s>, TmI_3<g> and Tm_2I_6<g>. These are supplemented by estimated values for TmI_3<l>. The parameters are presented in the format adopted by SGTE (20).

REFERENCES

(1) S.A.Mucklejohn, N.W.O'Brien and T.R.Brumleve, *J. Phys. Chem.*, 89, 2409 (1985).

(2) T.R.Brumleve, S.A.Mucklejohn and N.W.O'Brien, *J. Chem. Thermodynamics*, 21, 1193 (1989).

(3) T.R.Brumleve, S.A.Mucklejohn and N.W.O'Brien, *J. Chem. Thermodynamics*, 24, in the press (1992).

(4) S.A.Mucklejohn and A.T.Dinsdale, *6th Int. Symp. Science & Tech. Light Sources*, Budapest, Hungary (1992).

(5) L.Brewer, L.A.Bromley, P.W.Gilles and N.L.Lofgren, *The Chemistry and Metallurgy of Miscellaneous Materials*, editor: L.L.Quill, McGraw Hill, p.76 (1950).

(6) L.Brewer, *The Chemistry and Metallurgy of Miscellaneous Materials*, editor: L.L.Quill, McGraw Hill, p.193 (1950).

(7) *Iodide Metals and Metal Iodides*, R.F.Rolsten, John Wiley & Son (1961).

(8) J.H.Dettingmeijer and H.R.Dielis, *J. Less Common Metals*, 139, 331 (1988).

(9) P.J.Gardner and S.R.Preston, *Thermochimica Acta*, 175, 129 (1991).

(10) D.A.Lynch, Personal communication.

(11) C.E.Myers and D.T.Graves, *J. Chem. Eng. Data*, 22, 436 (1977).

(12) *JANAF Thermochemical Tables*, 3rd edition, parts I and II, M.W.Chase, C.A.Davies, J.R.Downey, D.J.Frurip, R.A.McDonald and A.N.Syverud, *J. Phys. Chem. Ref. Data*, 14, Supplement No.1 (1985).

(13) *Atomic Energy Levels - The Rare-Earth Elements*, W.C.Martin, R.Zalubas and L.Hagan, U.S. Department of Commerce, *NSRDS-NBS 60* (1978).

(14) C.Hirayama, J.F.Rome and F.E.Camp, *J. Chem. Eng. Data*, 20, 1 (1975).

(15) *Selected Values of Chemical Thermodynamic Properties*, R.H.Schumm, D.D.Wagman, S.Bailey, W.H.Evans and V.B.Parker, U.S. Department of Commerce, *NBS Technical Note 270-7* (1973).

(16) *The NBS tables of chemical thermodynamic properties*, D.D.Wagman, W.H.Evans, V.B.Parker, R.H.Schumm, I.Halow, S.M.Bailey, K.L.Churney and R.L.Nuttal, *J. Phys. Chem. Ref. Data*, 11, Supplement No.2 (1982).

(17) Y.C.Kim and J.Oishi, *J. Less Common Metals*, 65, 199 (1979).

(18) L.G.Lishenko, T.S.Nazarova, Yu.I.Polyakov and A.A.Rosen, *Zh. Neorg. Khim.*, 18, 921 (1973).

(19) D.W.Smith, *J. Chem. Ed.*, 63, 228 (1986).

(20) I.Ansara and B.Sundman, *Computer Handling and Dissemination of Data*, editor: P.F.Glasser, CODATA (1987).

WALL BLACKENING IN METAL IODIDE LAMPS SEEN FROM A CRYSTAL GROWTH POINT OF VIEW

W. van Erk, P. Cobben* and P. Bennema*

Philips Lighting BV, p.o. box 80020, 5600 JM Eindhoven, The Netherlands
* R.I.M. Laboratory of Solid State Chemistry, Faculty of Sciences, University of Nijmegen, Toernooiveld, 6525 ED Nijmegen, The Netherlands

Introduction

In high-pressure gas discharge lamps with a filling containing rare-earth iodides or scandium iodide these iodides are so reactive towards oxygen that the pressure of tungsten oxide iodides will be very low. In contrast to halogen incandescent lamps therefore in these gas discharge lamps no regenerative tungsten cycle is possible. There is only transport of tungsten from the hot electrode tips to the wall, leading to a continuous decrease of the lumen maintenance.
As can be seen from table 1, for the lumen maintenance not only the amount of transported tungsten is important.

Table 1

Lamp	Filling	lumen maintenance (%)	tungsten mass (mg)
A	Hg (+ Tm metal as oxygen getter)	60	0.45
B	Hg + TmJ_3/NaJ/TlJ	90	0.40

The way tungsten is deposited is important as well: on the inner side of the wall of the discharge vessel of A a very fine deposit of tungsten is present. In the metal halide lamp B, however, tungsten is deposited as relatively large particles. This causes only a small reduction of the light transmission of the wall.
The particles have well defined facet faces, which means that blackening has very much to do with the growth of tungsten crystals on the wall.
Our contribution will be a first exploration of how crystal growth theories can help us in understanding more about wall blackening.

Nucleation

When the lamp is switched on for the first time the wall is still virginal. The first step in the growth process will be nucleation followed by growth of the stable nuclei.
In general the nucleation rate is strongly dependent on the surface energy, the temperature and the supersaturation. The supersaturation ß, the driving force of the growth process, is the chemical potential difference between a so-called growth unit in the vapour phase and a growth unit in the crystalline phase, divided by the thermal energy kT. The supersaturation ß is a dimensionless number. For growth from the vapour phase it is equal to:

$$\beta = \ln \frac{P}{P_e}$$

where p is the actual pressure of tungsten in the gas phase
p_e is the equilibrium pressure of tungsten.

ß can be obtained from thermodynamical calculations assuming local thermal equilibrium.

- In the mercury lamp (A) there will only be physical transport of tungsten and in that case the supersaturation is high: a typical value is ß = 70. This is due to the very low value of p_e.
- In the metal iodide lamp (B) also the tungsten iodides WJ and WJ_2 are formed at the electrode tip. Calculations show that in this type of lamps ß is about 20, a value much lower than for the mercury lamp. This is because of the much higher values of p_e.

The consequence of this is that in mercury lamps the nucleation rate is high, resulting in many and small nuclei, whereas in metal iodide lamps the nucleation rate will be much smaller.

The nucleation rate is exponentially dependent on the value of $ß^2T^3$, where T is the wall temperature. Calculations show that in order to obtain a low nucleation rate (low $ß^2T^3$) the electrode tip temperature should be low, and the wall temperature should be high.

Growth

After stable nuclei have been formed they will grow with a rate which is proportional to the tungsten pressure near the wall. Therefore the higher the electrode temperature the higher the growth rate. During lamp life the supersaturation at the wall will decrease due to competition between growth and nucleation. Therefore, it may be expected that nucleation is only important in the first period of lamp life.

Conclusions

- In metal iodide lamps the effect of wall blackening on the lumen maintenance depends not only on the amount of tungsten which is deposited but also on the morphology of the deposit.
- Wall blackening can be considered as a crystal growth process and crystal growth theories should therefore be included in the modelling of the complicated process of wall blackening.
- With respect to mercury lamps without emitter, metal iodide lamps owe their good long-term behaviour to the fortunate circumstance that the supersaturation is so low that the tungsten is deposited as relatively large particles.

VOID GROWTH AND ITS RELATION WITH THE BUBBLE DISPERSION IN DOPED TUNGSTEN WIRES

O. Horacsek, Cs. Tóth* and K. Horacsek
Research Institute for Technical Physics of the Hungarian Academy of Sciences, Budapest IV. Fóti ut 56.
* TUNGSRAM Co, Budapest IV. Váci ut 77.

To produce long life incandescent lamp of high luminous efficacy it is necessary that the coiled filament retains its original shape over a long period of time. This demand can be met by an appropriate microstructure having high stability even at high operating temperatures and under drastic shocks and vibrations. To satisfy the desired requirements, bubble strengthened tungsten is applied extensively as filament material ensuring its excellent creep strength by a dispersion of potassium-filled bubbles.

Although the purpose of the doping treatment is to produce a suitable bubble dispersion of high stability, this requirement cannot be perfectly fulfilled. Several investigations have shown that void growth can occur even in high quality lamp filaments (1-3). The aim of this paper is to outline a growth mechanism suggesting that the sensitivity to void formation depends largely on the features of the bubble structure of the filament.

RESULTS AND DISCUSSION

The mechanism of void growth in creep tested wires, and lamp filaments was studied by standard metallographic methods. To obtain information about the arrangement of the bubbles at low magnification on a "macroscopic" scale, the sectioned and polished filaments were over-etched. The results :

a) Void formation was observed not only on the grain boundaries but also within the grain interiors.

b) Void growth was accompanied by bubble coarsening. In the neighbourhood of large voids, bubble denuded regions extending more than 2 μm were observed. (Fig.1)

While the critical radius of an empty void for growth, r_c, is given by the expression

$$r_c = \frac{2\gamma}{\sigma} \qquad (1)$$

where γ is the surface energy of the void and σ is the applied stress, a gas-filled and equilibrium sized bubble of radius r can grow if (4)

$$r > \frac{2}{3\sqrt{3}}\,\frac{2\gamma}{\sigma} \qquad (2)$$

A simple estimation with a surface energy of γ = 2300 J/m and σ = 10 MPa, suggests that at the relatively low stress levels resulting from the own weight of the coils, the bubbles in the lamp filaments having radii of the order of 0.01 µm, are too small to act as pre-existing void nuclei. Growth of these subcritical potassium bubbles is possible only at some preferred sites on the grain boundaries where a local stress concentration associated with grain boundary sliding lowers the critical size for void growth (5). This mechanism works, however, only at the grain boundaries and, hence, can not account for the observed void formation within the grain interiors.

Fig.1 shows that void growth inside the grains was accompanied by a coarsening of the initially linearly arranged bubble network. Since bulk diffusion of the insoluble potassium is highly improbable, bodily migration of the bubbles and their subsequent coalescence seems to be the only mechanism by which the bubbles within the grains can coarsen. As a result of collision and coalescence, bubbles can become large enough to satisfy relation (2) and, hence, to act as nuclei for void growth.

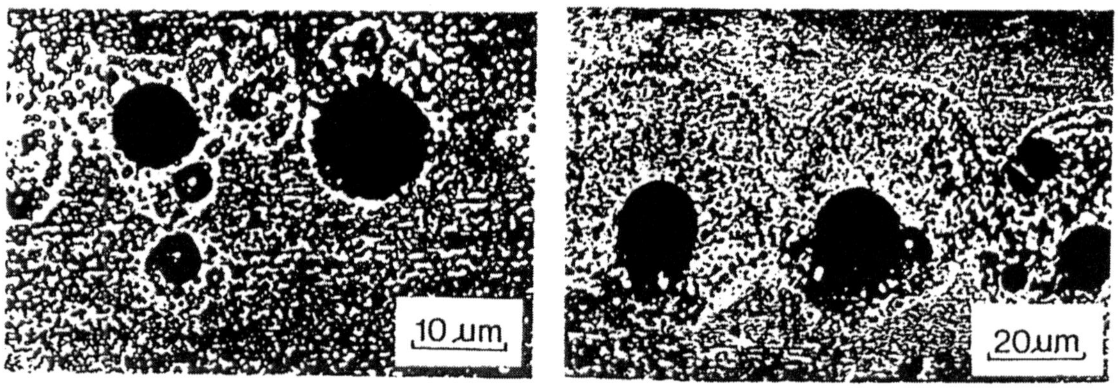

Fig.1 Intragranular voids in bubble strengthened tungsten wire

The suggested void formation mechanism is a result of several interconnected processes. In the first stage void nucleation takes place by coalescence of migrating subcritical bubbles. Once the critical size is attained continuous vacancy flux is directed to the growing void. Because of this vacancy flux, however, each bubble in the vicinity of the growing void experiences a force which leads to diffusive drift in the opposite direction. This counterflow of the bubbles results in the development of a bubble denuded region around the growing void. Since initially the interbubble spacing in an unstressed doped wire is everywhere less than 0.3 µm, the development of relatively large bubble denuded regions gives an indirect evidence for bubble migration. From the measured extents of the bubble-free regions and the burning time, we obtained for the rate of bubble migration values of the order of 10 [illegible] µm/sec.

When in the densely spaced bubble network the small, subcritical bubbles move away from the growing void, they will necessarily collide and coalesce. In this way practically immobile large bubbles are formed around the growing void. If these coalesced bubbles attain the critical size, they also will grow by vacancy absorption. As a result of this additional vacancy flux (and conterflow of bubbles), a second ring of coalesced bubbles forms. At the same time, the central void progressively grows, overtakes and annexes the first ring of the enlarged immobile bubbles. During further growth the second ring of immobile bubbles will also be annexed by the central void and this process repeatedly continues: new rings of coalesced immobile bubbles are formed and then occupied by the large void.

Since the coalescence of subcritical bubbles plays a decisive role in the intragranular void formation, it depends largely on the features of the bubble network. Two kinds of coalescence must be taken into consideration. In situ coalescence can occur when in some places in the filament the bubbles are too densely spaced and a bubble that is expanding overlaps and merges with its neighbours. Although this mechanism probably works occasionally in bubble strengthened tungsten, the observed disordering of the initially linearly arranged bubbles and the development of bubble denuded zones refer to the fact that bodily migration is the most important cause of coalescence.

Obviously, in this case the most effective way to prevent void formation is to reduce the mobility of the subcritical bubbles. Probably both the size and the shape of the bubbles has an effect on their mobility. Although there is a rather wide interval of sizes in which the mobility decreases with increasing bubble diameter, the presence of a very high internal pressure in the smallest bubbles reduces the surface diffusion coefficient and, hence, their mobility (6). Another effect which acts to prevent bubble migration is the faceted surface of the bubbles. If the surface of the potassium bubbles is free of contaminations and facets can develop on their surface, it can be expected that their mobility drops, because in this case the rate is no longer determined by the diffusion of atoms over the bubble surface, but by the rate at which ledges can nucleate on the flat facets.

(1) Day J.H. (1973), Lighting Res. Techn. 5, 36

(2) Horacsek O. and Horacsek K. (1977), Z. Metallkd. 67, 264

(3) Garbe S. and Hanloh S. (1983), Philips J. Res. 38, 248

(4) Hyam E.D. and Sumner (1962), In: Proc. Conf Radiation Damage in Solids. Intern. Atomic Energy Comm., Vienna Vol. 1, p. 323

(5) Briant C.L. and Walter J.L (1988 Acta Metall. 36, 2503

(6) Mikhlin E. Ya. (1979), Phys Stat. Sol. (a) 56, 763

MATRIX ISOLATION STUDIES ON HIGH TEMPERATURE LAMP CHEMISTRY

D.M. Jenkins and J.S. Ogden
Department of Chemistry, University of Southampton, U.K.

This poster describes the application of matrix isolation infra-red spectroscopy (1) to the study of chemical reactions in high temperature lamps. A particular emphasis has been placed on the study of tungsten corrosion as this remains a major cause of lamp failure. A knowledge of the role and stability of the transporting species is central to the future development of these devices.

The matrix isolation technique allows unstable or reactive species to be trapped and studied in an inert environment. The technique yields high quality spectra and has previously been used to fingerprint a number of tungsten halide and oxide halide molecules (2,3,4). Under favourable conditions, isotopic fine structure can be resolved, which aids spectral assignment.

Once the matrix isolated species has been identified, the experimental vibrational frequencies, together with other molecular parameters, can be used to calculate the corresponding thermodynamic parameters of the molecule (5). In this study the matrix isolation technique has been used to probe the high temperature region near a metal filament in contact with a reactive gas mixture.

Figures 1a, 1b and 1c show simplified, schematic diagrams of the apparatus used in this study. In a typical experiment, the gas flow into the apparatus is regulated using a PTFE needle valve [A]. The vapour in the region surrounding the heated filament [B] is sampled by means of a 3 mm aperture [D] in the bulb wall. The vapour then enters the matrix isolation apparatus [E] where it is mixed with a thousand-fold excess of an inert gas [F], typically Ar or N_2.

The vapour is condensed onto a CsI window [H], maintained at ca. 12 K by a closed cycle liquid helium refrigerator [I]. This cryotip unit is supported inside a vacuum shroud [J] maintained at ca. 10^{-9} atm using a diffusion pump backed by a rotary pump [L]. The cryotip unit can be rotated on an O-ring within the vacuum shroud, allowing deposition of the matrix and spectroscopic observation to be made at right angles to one another (**Figure 1c**). Infra-red spectra are recorded on a Perkin Elmer 983G spectrometer interfaced to a PE 3600 data station.

Experiments have been performed using both tungsten and molybdenum filaments. WO_2Cl_2 is the principal species observed when O_2/Cl_2 gas mixtures are passed over a tungsten filament. Similarly, WO_2Br_2 is generated when O_2/Br_2 gas mixtures are used. **Figure 2a** shows a typical low resolution spectrum of the $Mo/O_2/Br_2$ system. The most intense absorption bands may be assigned to MoO_2Br_2 **(3)**. This assignment is supported by the observation of molybdenum isotope fine structure under high resolution (**Figure 3b**).

An analogous series of experiments, passing pure halogens (Br_2 or Cl_2) over a tungsten filament generates vapour phase species not previously identified within lamp-like environments. The matrix isolation infra-red spectra indicate the presence of new binary tungsten halide molecules. The poster will discuss the identity of these new species and will provide estimates of their thermodynamic functions.

Acknowledgements

The financial support of the SERC and GE Thorn Lamps Ltd is gratefully acknowledged.

References

1 See for example, Advances in Spectroscopy Vol.17 (1989), Spectroscopy of Matrix Isolated Species, Editors: R.J.H. Clarke and R.E. Hester, John Wiley & Sons, Chichester.

2 A.K. Brisdon, E.G. Hope, W. Levason and J.S. Ogden: J. Chem. Soc. Dalt. Trans. 313 (1989).

3 W. Levason, R. Narayanaswamy, J.S. Ogden, A.J. Rest and J.W. Turff: J. Chem. Soc. Dalt. Trans. 2009 (1982).

4 W. Levason. R. Narayanaswamy, J.S. Ogden, A.J. Rest and J.W. Turff: J. Chem. Soc. Dalt. Trans. 2501 (1981).

5 See for example, Molecular Spectra and Molecular Structure,II, Infra-red and Raman Spectra of Polyatomic Molecules, G.H. Herzberg, Van Nostrand, New Jersey (1966).

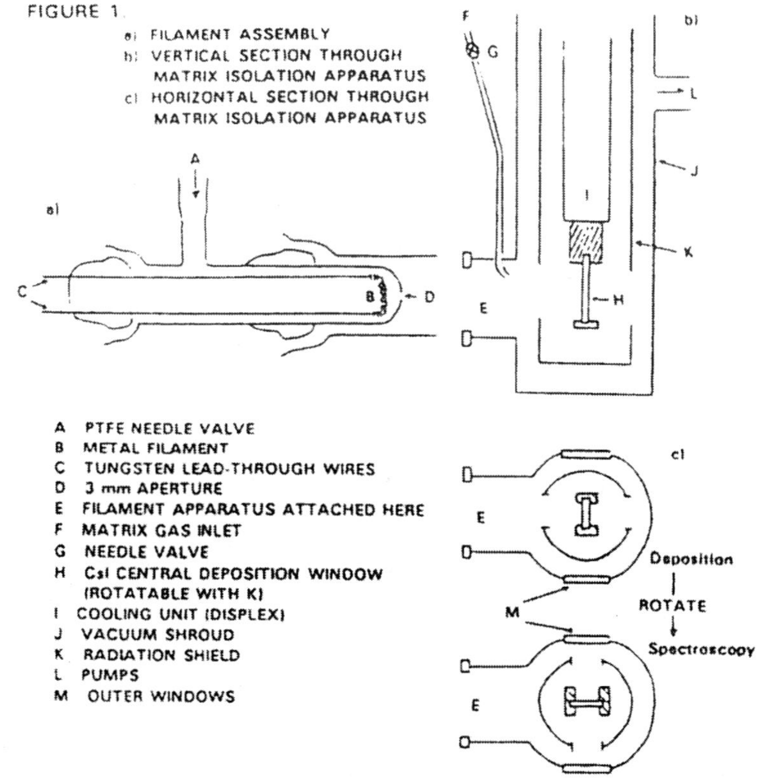

FIGURE 1.
a) FILAMENT ASSEMBLY
b) VERTICAL SECTION THROUGH MATRIX ISOLATION APPARATUS
c) HORIZONTAL SECTION THROUGH MATRIX ISOLATION APPARATUS

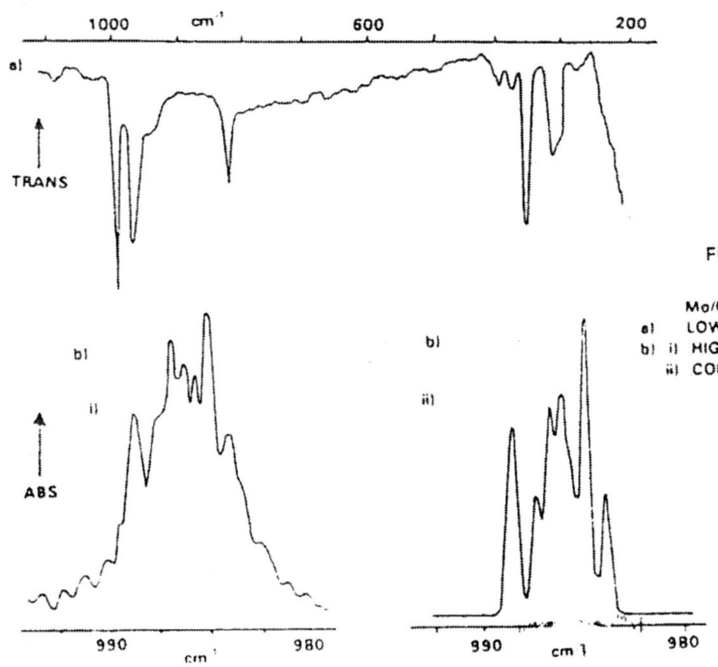

FIGURE 2.

$Mo/O_2/Br_2$ SYSTEM
a) LOW RESOLUTION SPECTRUM
b) i) HIGH RESOLUTION SPECTRUM
ii) COMPUTED SPECTRUM

MODELLING MULTICOMPONENT METAL HALIDE SYSTEMS

S.A.Mucklejohn
GE Thorn Lamps Limited, Advanced Technology Department,
Melton Road, Leicester LE4 7PD, U.K.
&
A.T.Dinsdale
National Physical Laboratory, Division of Materials Metrology,
Teddington, Middlesex TW11 0LW, U.K.

Metal halide systems can be employed in high-intensity discharge lamps to produce white light sources with high efficiencies and good colour rendering properties. Some applications, however, require the spectral output to be restricted to a specific range. Since many metal halide lamps need complex mixtures to deliver the required spectral output, an understanding of the chemistry of the dose system over a large temperature range is essential.

A comprehensive study of the properties of a metal halide system over the range of lamp cool-spot temperatures demands many careful, well designed and time consuming experiments. Very few metal halide mixtures have been subjected to the detailed study and analysis which permits the determination of the partial pressures and gas phase composition in the system which form the basis for predicting the spectral characteristics of the lamp. Phase diagrams, however, are reported for many binary metal halide systems and these contain vital information which ultimately allows the gas phase composition to be calculated. Complete thermodynamic descriptions for the binary metal halide systems can be derived from the phase diagram and any experimental thermodynamic data. The behaviour of the condensed phases in the multicomponent systems can then be predicted from these critically assessed thermodynamic data and used to calculate the equilibrium gas phase composition for a given temperature, composition and pressure.

Three compilations (1, 2, 3 to 6) have proved to be extremely valuable in the search for phase diagrams of binary metal halide systems. These can be supplemented by searching the original literature for material published in the past four years.

Thermodynamic data for the pure components, together with the binary phase diagram information can be assessed using thermochemical databanks in conjunction with routines developed by Lukas *et al.* (7, 8). We have used one such databank (9) to model several binary and multicomponent metal halide systems. A detailed review of thermochemical databanks has been published recently (10).

Several models are available to represent binary and multicomponent systems although it is difficult to assess which is the most appropriate unless reliable experimental information is available for guidance. We have used the sublattice model to represent the {InI+MI} (M=Na, K, Rb) systems, the experimental values were taken from reference (11).

The phase diagram for the {KI+TlI} system is reported in reference (12) but is restricted to the composition range x(TlI)=1.000 to 0.679. The limited composition range made the evaluation of the system difficult but satisfactory agreement between the calculated phase diagram and the experimental values was eventually obtained.

The thermodynamic parameters obtained from assessing the {InI+KI} and {KI+TlI} systems were combined, together with parameters derived for the {InI+TlI} system, to predict the behaviour of the {InI+KI+TlI} system. The sublattice model was used to represent the data for this system and to calculate the partial pressures in the gas phase from 600 to 1200 K.

Although more effort is required to refine the methods outlined here, the phase equilibria calculations provide a sound basis for representing data for metal halide systems of the type used in high-intensity discharge lamps. To derive full benefit from these modelling studies they must be complemented by diagnostic measurements on lamps.

REFERENCES

(1) *Phase Diagrams for Ceramists*, E.M.Levin, C.R.Robbins and H.F.McMurdie, The American Ceramic Society Inc., Ohio (1964) and Supplements 1969, 1975, 1981, 1983, 1987, 1989.

(2) *Phase Equilibria in Binary Halides*, V.I.Posypaiko and E.A.Alekseeva, IFI/Plenum Data Company, New York (1987).

(3) G.J.Janz, G.L.Gardner, U.Krebbs and R.P.T.Tomkins, *J. Phys. Chem. Ref. Data*, 3, 1 (1974).

(4) G.J.Janz, R.P.T.Tomkins, C.B.Allen, J.R.Downey Jr., G.L.Gardner, U.Krebbs and S.K.Singer, *J. Phys. Chem. Ref. Data*, 4, 871 (1975).

(5) G.J.Janz, R.P.T.Tomkins, C.B.Allen, J.R.Downey Jr and S.K.Singer, *J. Phys. Chem. Ref. Data*, 6, 409 (1977).

(6) G.J.Janz, R.P.T.Tomkins and C.B.Allen, *J. Phys. Chem. Ref. Data*, 8, 125 (1979).

(7) H.L.Lukas, E.Th.Henig and B.Zimmermann, *CALPHAD*, 1, 225 (1977).

(8) H.L.Lukas, J.Weiss and E.Th.Henig, *CALPHAD*, 6, 229 (1982).

(9) R.H.Davies, A.T.Dinsdale, T.G.Chart, T.I.Barry, M.H.Rand, *Proc. 6th Int. Conf. High Temperatures, Chemistry of Inorganic Materials*, Gaithersburg, Maryland, U.S.A. (1989).

(10) C.W.Bale and G.Eriksson, *Can. Met. Quart.*, 29, 105 (1990).

(11) P.I.Fedorov, N.S.Malova and V.N.Zhikharev, *Russ. J. Inorg. Chem.*, 17, 286 (1972).

(12) I.I.Il'yasov, G.G.Shchemelava and A.G.Bergman, *Russ. J. Inorg. Chem.*, 6, 355 (1961).

AN ASSESSMENT OF THE VAPOUR PRESSURES AND THERMOCHEMICAL PARAMETERS FOR THE RARE EARTH TRIBROMIDES

S.A.Mucklejohn
GE Thorn Lamps Limited, Advanced Technology Department,
Melton Road, Leicester LE4 7PD, U.K.
&
D.L.Trindell and R.Devonshire
High Temperature Science Laboratories, Department of Chemistry,
The University of Sheffield, Brook Hill, Sheffield S3 7HF, U.K.

The rare earth halides are attractive components for doses in high-intensity discharge lamps because they offer the opportunity when combined with other metal halides to design light sources with high efficacies, good colour rendition and colour temperatures in the range 3000 to 7000 K. The properties of many of the rare earth halides, however, are poorly characterised. The bromides have received even less attention in the scientific literature than the chlorides and iodides. Comprehensive thermochemical parameters for the rare earth tribromides are not available in any of the standard compilations of thermodynamic functions. Pankratz (1) offers the most detailed collection of thermochemical parameters for the rare earth halides published to date (March 1992).

Thermochemical parameters are required for the calculation of the chemical composition in the lamp to assist in the understanding and hence prediction of tungsten corrosion, silica corrosion and spectral output. The data are also needed for predicting the behaviour of doses by modelling multicomponent metal halide systems (2).

There are very few experimental studies reported in the literature for the rare earth tribromides, for example the vibrational frequencies have been measured for $NdBr_3$<g> (3, 4), $GdBr_3$<g> (4) and $LuBr_3$<g> (4) but not for the others in the series. We have used the estimated frequencies in reference (5), which assumes the molecules to have planar (D_{3h}) structure, where experimental values are not available. The rare earth trihalides are notoriously difficult to prepare as high purity materials (6,7) and react at moderate temperatures with silica. This, together with their low volatility, accounts for the paucity of experimentation reported for the tribromides.

Vapour pressures have been reported above the liquid phases (8) and the solid phases (8, 9), values for the latter cover a very restricted temperature range. Enthalpies of formation have been determined experimentally (10) for $REBr_3$<s> (RE = La, Pr, Nd, Sm,

Gd, Dy, Er) and we have used these together with the estimated values for the remainder of the series from reference (11).

We have critically assessed all experimental values found in the literature for the rare earth tribromides. These have been supplemented by estimated values to derive a complete set of thermochemical parameters for all the series.

To improve the parameters selected here, reliable experimental values are needed for the vibrational frequencies of the gaseous molecules and for the molar heat capacities of the solid and liquid phases.

The parameters have been selected such that they predict vapour pressures in the range 900 to 1500 K close to those reported from experimental studies.Thermochemical parameters are given in the format adopted by SGTE (12). Listings in the JANAF Tables style (13) can be derived from this format. The following parameters are listed: enthalpy of formation for the solid at 298.15 K; entropy of the solid at 298.15 K; transition temperatures and enthalpies of transition where appropriate; melting temperature; enthalpy of fusion at the melting temperature; enthalpy of formation of the gas at 298.15 K; entropy of the gas at 298.15 K; coefficients for polynomials describing the temperature variation of the molar heat capacities of all phases using equation [1].

$$C_p(T)/J.K^{-1}.mol^{-1} = A+B(T/K)+C(T/K)^2+D(T/K)^{-2} \qquad [1]$$

ACKNOWLEDGEMENTS

DLT thanks the SERC and GE Thorn Lamps Limited for financial support.

REFERENCES

(1) L.B.Pankratz, *Thermodynamic Properties of Halides*, United States Department of the Interior, National Bureau of Mines Bulletin 674 (1984).

(2) S.A.Mucklejohn and A.T.Dinsdale, *6th Int. Symp. Science & Tech. Light Sources*, Budapest, Hungary (1992).

(3) J.C.Wells, J.B.Gruber and M.Lewis, *Chem. Phys.*, 24, 391 (1977).

(4) N.S.Loktyushina and A.A.Mal'tsev, *Russ. J. Phys. Chem.*, 58, 1602 (1984).

(5) C.E.Myers and D.T.Graves, *J. Chem. Eng. Ref. Data*, 22, 436 (1977).

(6) F.L.Carter and J.F.Murray, *Mat Res. Bul.*, 7, 519 (1972).

(7) J.D.Corbett, *Inorg. Nucl. Chem. Letters*, 8, 337 (1972).

(8) A.Makhmadurodov, M.Temurova and A.Sharipov, *Izv. Akad. Tadxh. S.S.R., Otd. Fiz. Mat., Geol. Nauk.*, 39 (1989).

(9) E.Shimazaki and K.Niwa, *Z. Anorg. Allgem. Chem.*, 314, 21 (1962).

(10) C.Hurtgen, D.Brown and J.Fuger, *J. Chem. Soc. Dalton Trans.*, 70 (1980).

(11) Y.C.Kim and J.Oishi, *J.Less-Common Metals*, 65, 199 (1979).

(12) I.Ansara, B.Sundman, *Computer Handling and Dissemination of Data*, editor P.F.Glasser, CODATA (1987).

(13) *JANAF Thermochemical Tables*, 3rd edition, parts I and II, M.W.Chase, C.A.Davies, J.R.Downey, D.J.Frurip, R.A.McDonald and A.N.Syverud, *J. Phys. Chem. Ref. Data*, 14, Supplement No.1 (1985).

SILICA AND TUNGSTEN CORROSION IN DISCHARGE LAMPS

Timothy D. Russell
Discharge Engineering Department
Lighting Technology Division
GE Lighting
Cleveland, OH 44112

Introduction

The development of lighting systems using high-intensity, metal halide discharge light sources has been attracting much attention in the last few years due to their long life and highly efficient light emission. However, the push to achieve ever brighter, more efficient, longer lived light sources is severely stressing the physical and chemical properties of normally used discharge tube materials.

This work will concentrate on chemical problems internal to the arc tube, mainly on the consequences to arc tube performance due to the chemical attack of the hot silica wall by scandium iodide, and the transport of tungsten due to high iodine partial pressures in the arc that result from various metal-loss processes that occur in an arc tube. It will draw heavily on data obtained during the development of the Discharge Forward Lighting (DFL) Headlamp System. The same chemical problems, however, are also present in all discharge lamps.

Silica Wall Corrosion

The DFL arc tube is made from GE 214A fused quartz. It is dosed with NaI and ScI_3. The external arc tube hot spot temperature is 890° C.

Operation of these lamps in either a continuous or a controlled on-off cycled mode results in both wall darkening and in the appearance of an opaque, etched area on the quartz wall. This etched area has been analysed by SEM-EDS and X-Ray diffraction to consist of crystals of Scandium Silicate embedded in the fused quartz wall. We can estimate from the size of the crytals and the extent of the wall coverage that there is 40-50 micrograms of scandium silicate present, or that approximately 50% of the scandium originally added to the arc tube as scandium iodide has reacted with the quartz wall. DCP emission spectrometry analysis for water-soluble scandium that was dissolved from the arc tube was in good agreement that 50% of the added ScI_3 had reacted with the silica wall.

Although the thermodynamics of this reaction are not rigorously known, experimental work [1,2,3] would suggest that much less than 50% of the ScI_3 should have been reacted.

Although it is not certain what is the cause of the extent or the rate of the observed ScI_3/SiO_2 reaction, the consequences of the reaction are better known. At 1000 hours, calculations predict that the total amount of Na in the vapor phase, from NaI, Na_2I_2, and $NaScI_4$, is essentially unchanged from that at 0 hours. The total amount of scandium has decreased to about 50% of the 0 hour amount. About 1 bar of "free-iodine" (from HgI_2), is also present after 1000 hours.

The effect of the excess iodine at 1000 hours on the dissociation of NaI and ScI_3 to Na and Sc can be calculated. While at 0 hours virtually all of the NaI is dissociated in the 3000-5000 K region of the discharge, at 1000 hours the effect of the large amount of excess iodine is to suppress this dissociation so that only about 60% of the NaI is dissociated. This suppression of the metal iodide dissociation to atomic species due to excess iodine is one aspect of the well-known arc constriction caused by mercuric iodide (excess iodine).

The effects of these two processes, the contraction of the 3000-5000 K region of the arc and the suppression of the NaI and ScI_3 dissociation due to excess iodine, can be seen in the lumen output and color temperature of the arc tube. The color shift is roughly along the black-body curve towards higher CCT, lower CCX and CCY. This shift is in the direction of reduced emission from both scandium and sodium, even though only the vapor phase concentration of scandium species has been reduced, with the vapor phase concentration of sodium species remaining unchanged. The color shift is not in the direction of a high-pressure sodium lamp but rather in the direction of a high pressure mercury lamp. The loss of scandium due to its reaction with silica has resulted in a loss of both metals from the discharge; scandium due to loss of ScI_3 and formation of involatile $Sc_2Si_2O_7$, and sodium, plus additional scandium, lost due to the suppression of the dissociation of the halides in the 3000-5000 K region of the discharge. The biggest shift in color occurs in the first 100 hours, indicating the reaction is quite rapid initially, then slows as either some equilibrium condition is approached or else the hot zones of the silica arc tube walls are being passivated by a layer of scandium silicate crystals that result from the reaction.

The figure below shows the lumen maintenance curves of the DFL arc tube. Whether the arc tubes are tested by being cycled off-and-on or burned continuously, during the first 100 hours of burning there is an approximate 10% lumen drop. Although some of this initial drop in lumens is undoubtedly due to the halides being repositioned to the coolest regions of the arc tube, most of this loss is thought to be due to the ScI_3/SiO_2 wall reaction. Attempts to model the consequences of this reaction on lumens have resulted in an expected 7% loss in lumens due to the change in vapor phase composition and arc constriction [4].

Tungsten Transport

Yet another consequence of the ScI_3 reaction is the wall darkening that results due to tungsten being transported from the electrodes to the silica walls. This is the result of the excess iodine generated in the ScI_3 reaction. The partial pressures of the tungsten species responsible for the transport can be calculated from the thermodynamic data from Dittmer, et.al.[5], and are similar to previous calculations of Fischer [6]. At 0 hours, before any ScI_3 has reacted, the only atomic iodine in the discharge in the vicinity of the electrodes is the result of the dissociation of the sodium and scandium iodides that are vaporized from the molten halide pool which resides at the silica walls at a temperature of 1100 K. There is sufficient halide vaporized that the partial pressure of atomic iodine at the electrodes is approximately 0.03 bar.

At 1000 hours, after the ScI_3 reaction has happened, the HgI_2, which is fully vaporized at the coolest spot wall temperature, produces a partial pressure of atomic iodine of approximately 2 bar at the electrodes.

Although, as stated by Fischer [6], a rigorous calculation of the tungsten transport rate in discharge lamps would require detailed modelling of the arc plasma and temperature profiles of the electrodes, by making some simple assumptions, we can arrive at theoretical lumen maintenance curves for the DFL lamp at the two limiting cases: The first being that the 0 hour conditions persist throughout life, i.e, that there is no metal loss due to scandium reaction or due to sodium diffusing out of the silica arc tube; and second that the conditions that were described above as existing at 1000 hours are assumed to have existed from 0 hours and throughout life.

The tungsten that is solubilized into the gas phase at the electrode surface diffuses away from the electrode through a quiescent boundary layer or Langmuir sheath. At the edge of the Langmuir sheath, the tungsten species are convected to the arc tube walls, where the solubility of tungsten is quite low, as virtually all the iodine has recombined with sodium, scandium, and mercury. At the walls, the tungsten deposits, causing the observed wall blackening, loss of transmissivity, and a drop in lumen maintenance.

Included in the Figure is a plot of the calculated lumen maintenance curves that result. A significant point from these calculations is that the temperature of the electrode tips is sufficiently low that tungsten evaporation alone cannot explain the observed wall darkening (W Only curve). Also, the conditions that exist in the lamp at 0 hours, before any metal loss, cannot explain the loss of lumens due to wall darkening (WI_x - No Reaction curve). An arc tube designed to operate with the stated electrode temperatures and metal halide atmosphere, as given, would be expected to perform with excellent maintained lumens, at least if burned continuously.

The results of the calculation for the limiting case of a large amount of excess iodine existing at 0 hours predicts too much tungsten transport and lumen loss (WI_x - Reaction curve). Part of the discrepancy between the calculated and experimental results could obviously be due to over-simplified assumptions, particularly in the region of the electrode. Detailed modelling of this region is actually needed to improve the calculation. However, the largest single reason for the discrepancy is probably due to the uncertainty in the thermodynamic data on the tungsten iodides. A 10% error in the heat of formation of WI_2 would lead to a prediction of 85% lumen maintenance for the continuously burning arc tube after 1500 hours. The experimental value is 77%. As has been previously discussed by Hildenbrand[7] with the thermodynamic data of the tungsten fluorides, the error in the heats of formation could easily be 10% or greater. More experimental work is needed to verify the validity of the available thermodynamic data.

Regardless of the uncertainty of available thermodynamic data or the use of oversimplified assumptions, it is clear from the figure that without the large amount of excess iodine due to the ScI_3/SiO_2 reaction, that the walls should remain clear of tungsten deposits, at least in a continuously burning, non-cycling arc tube.

References

[1] J.E. Spencer, *ScI_3 Wall Reactions in Metal Halide Lamps*, G.E. Report 80-LRL-1764, June 24, 1980

[2] E. Fischer and R. Kaiser, in *High Temperature Lamp Chemistry*, (E.G. Zubler, editor), p. 218, The Electrochemical Society Symposium Series **Vol. 85-2**, Pennington, NJ (1985)

[3] R.K. Datta, unpublished results

[4] J. Dakin, private commumication

[5] G. Dittmer, U. Niemann, and E. Schnedler, in *High Temperature Lamp Chemistry*, (E.G. Zubler, editor), p. 185, The Electrochemical Society Symposium Series **Vol. 85-2**, Pennington, NJ (1985)

[6] E. Fischer, *ibid.*, p. 47

[7] D.L. Hildenbrand, *ibid.*, p. 172

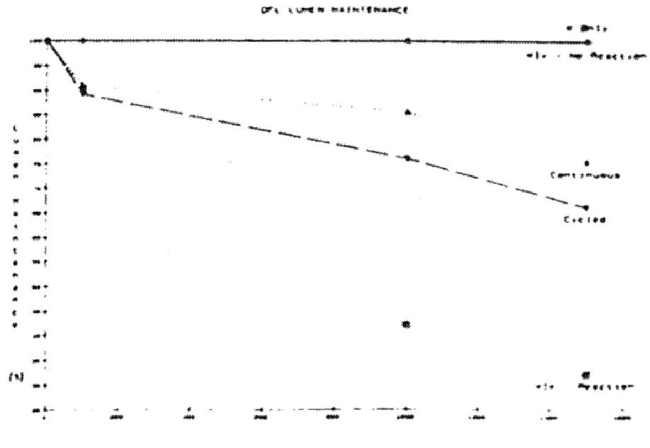

I. Szöllősi - I. Somogyi:

Strengthening possibility of glass bulbs

The theoretical strength of glass is better by 3 order of magnitude than the practical one.

In practice the causes of fragility of glass are: non ideal structure of glass, gas content, in-homogeneity.

- The glass contains a large quantity of pressed gas which stretches the structure even at ambient temperature.
- Influence of expanding gas at 1100°C, in vacuum furnace, caused the volume extension of glass up to six time, and it becomes hollow, like the sponge (photo 1)
- Alkali-glasses have a big inhomogeneity, there are areas of 10-100 μm in diameter riched in SiO_2, others riched in Na_2O, the alkali-riched parts of glass treated by heavy acids can dissolve, making craters (photo 2)
- The parts containing high Na_2O have high thermal expansion coefficient, and influence of the changing in temperature come into being cracks
- The water settling in the cracks both chemically (solution) and physically (heat developing) enlarge the micro-cracks, becoming starting point for total cracks
- The Na_2O content of glass surface increases at finishing, decreasing the strength. The higher Na_2O content of glass surface belongs to lower strength
- According to EDS measurements at the breaking surface of normal bulbs the Na_2O content is the highest at the edges and the lowest at the middle
- At secondary strong heating (e.g. heating of Ribbon bulbs

for bonding with lead glass tube) the Na_2O content of glass surface increases further and decreases the strength. The surrounding acidic air (H_2O+CO_2) induce the accumulation of Na_2O content in the surface. For better demonstration of phenomena, water and stronger acids than carbonic acid were vaporized - meanwhile heating - into the bulbs. Table 1.)

Table 1.

Sample	dissolved Na_2O mg/10 bulbs
Untreated bulbs	0.15
Heated bulbs (closed)	0.18
Heated bulbs (Atmosphere of water)	0.225
Heated bulbs (Atm. of Hcl)	0.39
Heated bulbs (Atm. of HF)	2.10

Considering the strength of finished lamp 100%, the strength of bulbs are 172,5%. The aim is to find a process, which prevents the accumulation of Na_2O in the glass surface during production; and so improves the strength of bulbs and would be eliminated the round cracks observable occasionally under the lamp-head.

Photo 1 Heated glass in vacuum furnace at 1100°C (15x)

Photo 2 Glass surface treated [illegible]

THE REACTIONS OF VITREOUS SILICA IN DISCHARGE LAMP ENVIRONMENTS

G. M. Forsdyke
GE Thorn Lamps Limited, Advanced Technology Department,
Melton Road, Leicester LE4 7PD, U.K.

High Intensity Discharge (HID) lamps are usually composed of an outer envelope and an inner arc tube which is made of vitreous silica. The arc tube is known to undergo a number of reactions within this lamp environment, the extent of which are not clearly understood. The reactions and the reaction rates have a great effect on the acheived (and rated) life and the long term colour rendition of the lamp. This literature review was undertaken to obtain a clearer view of how silica reacts in a lamp environment and to what extent the life of an arc tube, and therefore the lamp life, are affected.

The devitrification of silica and crack propagation (delayed stress fracture) could be considered to be purely physical phenomena. However the presence of impurites that can occur within the lamp environment have been shown to enhance and cause these phenomena.

Devitrification has been shown (1) to occur in the presence of tungsten(VI) oxide where the oxide acts as a nucleation site for the formation of crystobalite from vapour phase silicon monoxide. Other species including other oxides have also been shown to accelerate the devitrification of silica.

Crack propagation has been shown to occur and be promoted by the presence of water (2) which is a known lamp contaminant. The rate of chemical reaction has been shown to be increased by the presence of strained edge shared silica tetrahedra as is created by the dehydroxylation of the silica or the formation of a lamp shape. The overall reaction scheme can be represented by the reaction [1], however the actual mechanism proceeds via three discrete steps. These are:

1) bonding of the water molecule to the silica via hydrogen
2) simultaneous proton and electron transfer to rupture one of the bridging oxygen atoms of the silica surface
3) bond rupture between the hydrogen atom and the rest of the water molecule to give two hydroxyl groups.

$$2H_2O\langle g,l\rangle \ + \ SiO_2\langle s\rangle \ = \ Si(OH)_4\langle s\rangle \qquad [1]$$

Although the hydroxyl groups are actually spread across the two edges of the crack tip. It has also been shown that there is a steric effect in the crack propagation mechanism and molecules with a molecular diameter of greater than 0.5 nm will not contribute to crack propagation.

The reaction of silica with rare earth metal trihalides (rare earth being defined as including ${}_{58}Ce$ to ${}_{71}Lu$ plus ${}_{21}Sc$, ${}_{39}Y$ and ${}_{57}La$) have been investigated by a number of workers (3,4,5) and all agree that the following steps occur.

$$MX_3<g> \;=\; M<g,c> \;+\; 3X<g> \qquad [2]$$

$$xM<g> \;+\; nSiO_2<c> \;=\; M_xO_y.mSiO_2<c> \;+\; SiO<g> \qquad [3]$$

However there is dispute as to the full extent of the reactions of the rare earth metal trihalides and the nature of the products. Some authors propose that the following reactions takes place,

$$SiO_2<s> \;+\; SiX_2<g> \;=\; 2SiOX<g> \qquad [4]$$

$$SiO_2<s> \;+\; SiX_4<g> \;=\; 2SiOX_2<g> \qquad [5]$$

others however have observed reaction [6]. These differences and in particular the work on scandium triiodide will be discussed fully.

$$4MX_3<g> \;+\; 7SiO_2<s> \;=\; 2M_2Si_2O_7<s> \;+\; 3SiX_4<g> \qquad [6]$$

REFERENCES

(1) T. Ekstrom, R. J. D. Tilley; J. Cryst. Growth, 38, 197 (1977)

(2) T. A. Michalske, B. C. Bunker; J. Appl. Phys., 56, 2686 (1984)

(3) D. M. Speros, R. K. Datta, D. A. Lynch; J. Electrochem. Soc. 128.4, 778, (1981)

(4) W. van Erk, T. Rietveld; High Temperature Lamp Chemistry I, E. Zubler editor, The Electrochemical Society Symposium Series, Pennington NJ, 85.2, 57 (1985)

(5) E. Fischer, R. Kaiser; High Temperature Lamp Chemistry I, E. Zubler editor, The Electrochemical Society Symposium Series, Pennington NJ, 85.2, 218 (1985)

Inhomogenity as an inherent property of silicate glasses

Z. Boksay
Eötvös University Budapest, Hungary

In the lighting industry a great amount of glass is used. Its transparency, high electrical resistance, variability in the composition according to the requirements are highly appreciated. The technical application of the glass is admittedly more advanced, in general, than the current knowledge on the structural background. However, at a certain stage of the developement the insuffiency of the scientific basis may be an impeding factor.

Observations considering the faulty surface and the fragility of glasses led to an assumption that the glass was not as homogeneous as usually regarded. Thus a method has been elaborated to detect the supposed uneven distribution of the components. Omitting the details from here the essentials of the method are described as follows. Breaking the glass into small grains, fractioning the grains by sieving, and analysis a number of grains from each fractions. Six different types of glasses have been investigated in this way so far and none has been found to be homogeneous.

From the analytical data the distribution density of the composition is calculated and plotted against one or two variables. Usually these are the mole fraction, x, of a given component, the ratio of two components and/or the grain mass. The figures presented as examples refer to a lithia soda silica glass. The three dimensional Fig. 1

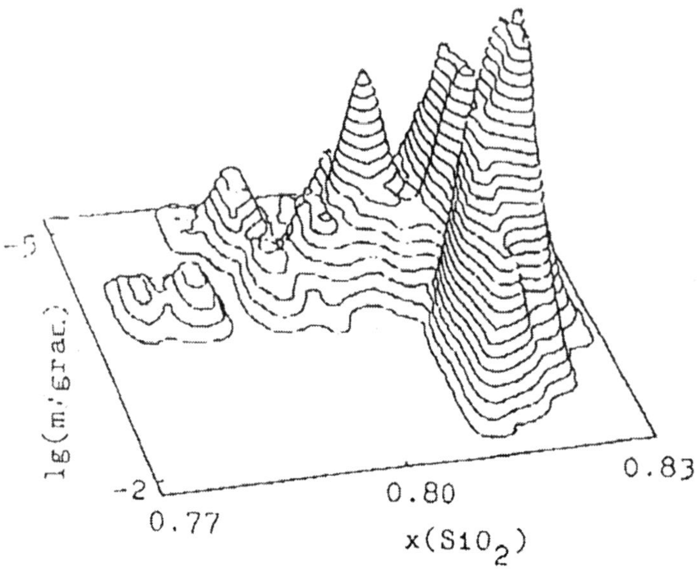

Fig. 1

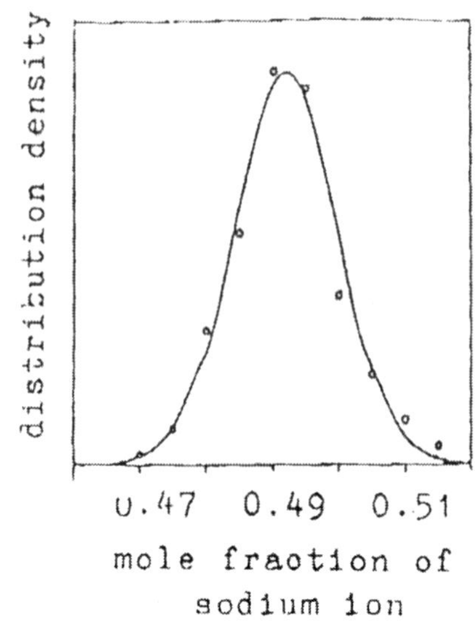

Fig. 2

indicates that the grains may be classified as of higher and of lower silica content, resp. No grain was found with high mass and low mole fraction of silica. Fig. 2 shows that the mixing of the lithium and sodium ions is extended to a narrow interval and approximatelly follows the Gaussian law (solid line).

Thus both the segregation and the association of the components have been observed. This fact leads to some consequencies. When, due to a project or an idea, a component is, partly or totally, replaced by another one it is not irrelevant whether the affinities of the two components to the rest of them is similar or not. There are components which can remarkably infuence the glass properties, even in small amount. It would be useful to know which components they are attached to. It looks worth-while to supervise the powder method for determining the durability of the glass or, at least, to consider that the sieve fractions differ in composition, i.e. neither of them can represent the glass as a whole.

In the lighting industry a great amount of glass is used. Its transparency, high electrical resistance, variability in the composition according to the requirements are highly appreciated. The technical application of the glass is admittedly more advanced, in general, than the current knowledge on the structural background. However, at a certain stage of the developement the insuffiency of the scientific basis may be an impeding factor.

Observations considering the faulty surface and the fragility of glasses led to an assumption that the glass was not as homogeneous as usually regarded. Thus a method has been elaborated to detect the supposed uneven distribution of the components. Omitting the details from here the essentials of the method are described as follows. Breaking the glass into small grains, fractioning the grains by sieving, and analysis a number of grains from each fractions. Six different types of glasses have been investigated in this way so far and none has been found to be homogeneous.

From the analytical data the distribution density of the composition is calculated and plotted against one or two variables. Usually these are the mole fraction, x, of a given component, the ratio of two components and/or the grain mass. The figures presented as examples refer to a lithia soda silica glass. The three dimensional Fig. 1

CREEP AND FRACTURE BEHAVIOR OF NS TUNGSTEN IN OXYGEN-DOPED ATMOSPHERES

G. Zilberstein and H.J. Kim
GTE Laboratories Incorporated, 40 Sylvan Road, Waltham, MA 02254, U.S.A.

Oxygen and moisture are the most notorious contaminants in incandescent and W halogen lamps. Interactions between oxygen (air) and tungsten at elevated temperatures that could occur during lamp fabrication can result in permanent damage to tungsten, i.e., brittleness and loss of sag resistance. Excessive coil sag is a critical issue in evolving models of W halogen lamps having a small envelope size, a focused beam, and, in particular, the lamps with an IR-reflective coating. In the latter lamp type, confinement of a coil sag to a minimum is crucially important to maintain the effectiveness of the coating.

The objective of this work was to examine the strain rates, deformation, and fracture behavior of non-sag (NS) tungsten during high temperature creep in an oxygen-doped inert gas to gain a better understanding of degradation of sag performance of coils in lamps contaminated with residual air. The test samples were electropolished NS-W wires, 0.178 mm in diameter, from three major lamp wire producers (A, B, and C) because the differences in the manufacturing practices relating to purity, powder processing, ingot size, or thermomechanical procedures could affect the materials' microstructure and hence their creep properties and sag performance in lamps.

Steady-state creep was evaluated on fully recrystallized 10 cm long vertically mounted and uniaxially stressed (10–80 MPa) wire samples at 2800 K in 10^{-6} torr (or better) vacuum and 100 cc/min flow of UHP argon doped with 0.03 and 0.002% O_2. SEM was used for post-creep evaluation of the wires' deformation and failure modes. Microstructural analysis of wire samples recrystallized prior to the creep tests showed that interlocking grain structure developed in ~80 mm of the wire's length, which was consistent with the evaluated temperature distribution from grip to grip. Hence 80 mm was used as a measure of the gauge length for strain rate calculations. Grain aspect ratio (GAR) varied between ~18 and ~30. Wire A had the highest GAR and wire C the lowest. K-bubble structure in each material was quantitatively evaluated on fully recrystallized wires using intergranular fracture surfaces, Fig. 1. It shows noticeable differences in the mean K-bubble diameter and the breadth of diameters distribution among the three materials, with wire A having the finest and hence the most effective for restraining grain boundary movement, bubble structure, and wire C the least effective. While the largest bubble diameter in wire A does not exceed 60 nm, 11.4% of a total bubble area in wire C is occupied by K-bubbles with diameters over 80 nm.

Results of the creep measurements, Fig. 2, show that in vacuum and 0.002% O_2, wire A has the highest creep strength compared to B and C, whereas in 0.03%O_2, strain rates of all the wires are virtually the same and are ~1–1.5 orders of magnitude higher than in vacuum. Post-creep SEM analysis showed grain boundary (GB) cavitation and sliding to be a common deformation and failure mode for all the wires tested in vacuum. Wire A had the least number of cavities and virtually no failures, and wire C the most. All samples had failed during testing in 0.03% O_2. Time-to-rupture vs stress, Fig. 3, again demonstrates a superior creep strength of wire A. In this regime, wires A and B deformed almost exclusively through the bulk showing coarse slip-band structures and ductile transcrystalline fractures. It appeared that GB cavitation or sliding did not take any part in creep deformation of wires A and B. In contrast, wire C showed pronounced GB cavitation and sliding at all applied stresses. It is interesting that the rupture time is appreciably less sensitive to the stress level for wire C than A and B, apparently suggesting that oxidation (of GB surfaces), and not stress, was the process governing its cavitation and sliding resulting in diffusion-controlled creep. Conversely, a strong rupture time dependence of stress for wires A and B could imply a stress- rather than an oxygen-induced deformation process, and hence a dislocation creep mechanism. It is unclear why and how the deformation mode and mechanism of wires A and B have changed from a diffusion-controlled in vacuum to a dislocation-controlled in the presence of 0.03% O_2. Comparing only the observed deformation modes in 0.03% O_2, it appears, however, that the K-bubble structure was a determining factor in each wire's deformation behavior. A large fraction of oversize K-bubbles (>80 nm diameter) in wire C should provide enough "pre-existing flaws" on its GB, thus fixing GB cavitation as its major deformation mode [1]. The absence of oversize bubbles in wires A and B and the two orders of magnitude lower bubble density in the bulk than on GB, as it has been demonstrated on similar materials [2, 3] apparently made deformation through the bulk a more favorable process. Finally, the strain rates and a post-creep examination of wires A and C tested in 0.002% O_2 showed that wire A has maintained creep strength as high as was previously observed in vacuum, while wire C deformed again by GB cavitation and sliding independent of the applied stress.

The results reported here clearly demonstrate the crucial importance of the size and distribution parameters of a K-bubble structure in controlling creep deformation of NS-W wire in oxygen-doped atmospheres. The results also suggest that NS-W coils with a refined K-bubble structure (as in wire A) could tolerate low level oxygen impurities (to be quantitatively determined) in lamps without degradation of their sag performance.

REFERENCES

1. C.L. Briant and J.L. Walter, Acta Metall., 36, 9 (1988).
2. H. Warlimont *et al.*, Z. Metallk, 66, (1975).
3. P. Schade, Proc. Plansee Seminar, 24, (1976).

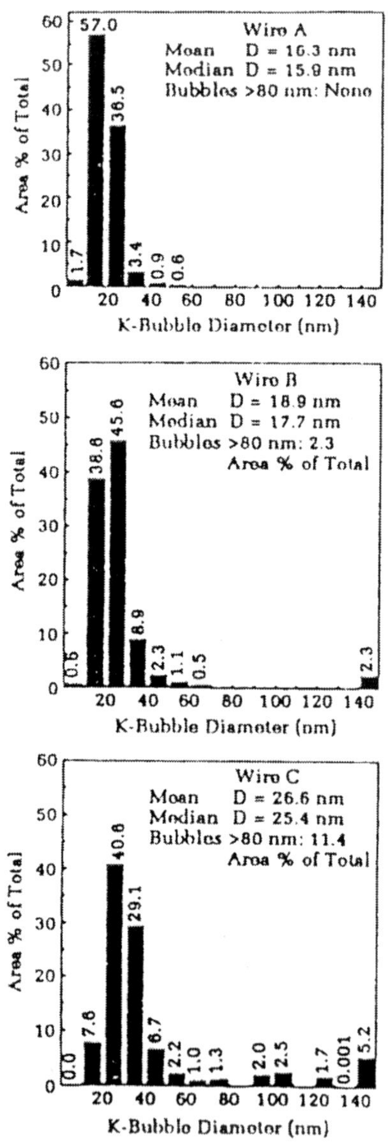

Fig. 1. K-bubble size distribution in A, B, and C NS-W wires.

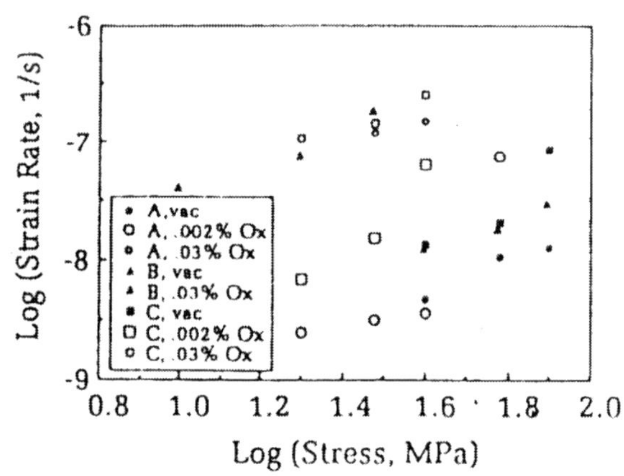

Fig. 2. Creep of A, B, and C NS-W wires in vacuum and Ar + 0.002% and 0.03% oxygen.

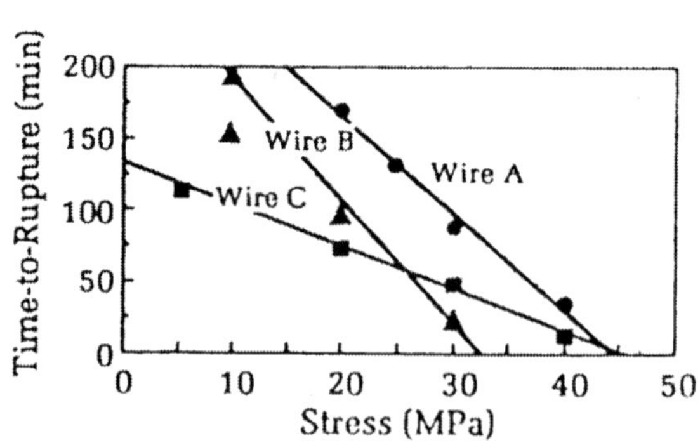

Fig. 3. Time-to-rupture during creep of A, B, and C NS-W wires in Ar + 0.03% oxygen.

Session 4

Chairman: S. Kamiya

Incandescent lamps

43:I **R. S. Bergman, T. G. Parham**
Applications of thin film reflecting coating technology to lamps

44:P **L. Bigio, B. Kubicki, P. J. Codella**
FTIR diagnostics of tungsten-halogen lamps: role of halogen conc., phosphorus, wall material and burning environment

45:P **É. Dancs**
Comparison of physical and life parameters of GLS lamps with and without phosphorus getter

46:P **W. Dieudonné, K. Schmitz, R. Stehling**
More efficiency for thermoforming processes

47:P **I. Gaál**
Radiation properties of coils

48:P **G. Hebbinghaus, G. Frank**
High quality infrared reflecting filters on lamps prepared by dip coating

49:P **L. Y. Lu, Y. Z. Gu**
Spectral irradiance standard lamp

50:P **A. Nagy, Cs. L. Tóth, I. Károlyi, A. Kálovics**
Effect of tungsten and molybdenum mandrel wire size variation on incandescent lamp performance

51:P **L. Nagy, L. Bigio**
Gas composition investigation in tungsten halogen lamps

52:P **J. M. Ranish, L. E. Hoegler**
A simple lead corrosion model for quartz halogen lamps

53:L **R. Kersten**
Cooking with light

54:L **M. Makhlouf, R. Devonshire, H. F. Boysan**
Non-equilibrium effects in the recalculating gas flows in incandescent lamps

APPLICATIONS OF THIN FILM REFLECTING COATING TECHNOLOGY TO LAMPS

Rolf S. Bergman and Thomas G. Parham
GE Lighting, Nela Park, Cleveland, OH, 44112

INTRODUCTION

The human eye is most sensitive to the peak of the spectral radiation output of the sun as filtered by the atmosphere. This corresponds to a blackbody radiator of about 5200°K whose peak radiation occurs near the maximum eye photopic sensitivity of 555 nm. Man-made incandescent light sources made with tungsten filament can only achieve temperatures of about 3000°K maximum.

Incandescent sources are inherently inefficient because the bulk of the radiated power occurs in the infrared region of the spectrum. Note from Figure 1 that most of the energy is radiated in the near infrared. The fraction of radiated energy in the visible region increases rapidly with temperature for an incandescent tungsten source as can be seen from Figure 2. The life of incandescent lamps, however, decreases at a rate about eight times faster than the increase in visible energy. Thus from a practical standpoint lamps with temperatures in excess of 3000 K are not in use for general lighting applications.

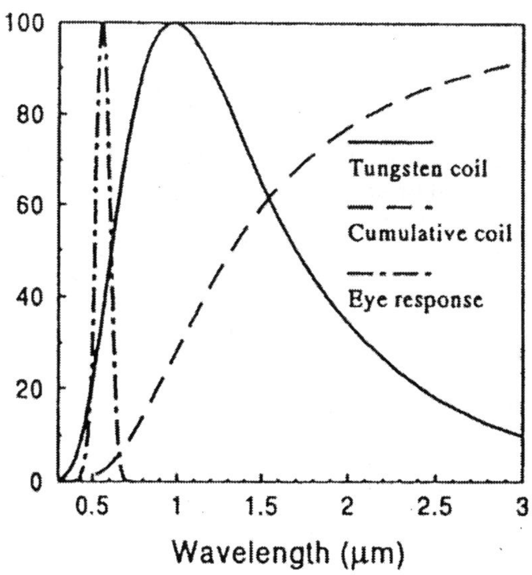

Figure 1. Spectral emittance of 2800°K tungsten coil and luminous efficacy (human eye response).

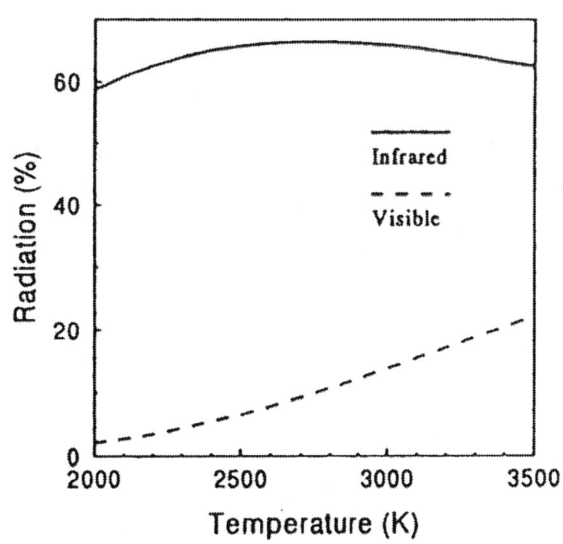

Figure 2. Emitted power for tungsten coiled coil.

The lighting industry has developed a large array of more efficient light sources, all based on gas discharges that radiate an appreciable portion of their energy in the visible spectrum. In spite of these more efficient sources the incandescent lamp still continues to be the dominant source used in homes. Thus it appears justifiable to invest significant resources to improving the energy efficiency of incandescent lamps, particularly if this could be done by recycling the wasted infrared radiation (IR) into usable light.

Lamp-makers have long been aware of the potential of improving the efficacy of incandescent sources by returning the IR to the filament. One of the earliest efforts is found from a patent issued in 1912, that uses prisms and mirrors to separate the IR from the visible and return the IR radiation(1). Obviously this method is cumbersome and not used. GE's effort focused on the development of multilayer dielectric films to reflect the IR and transmit the visible; this resulted in linear double-ended quartz products introduced in 1983 (2). Other systems such as the TiO2-Ag-TiO2 filter have also been developed and marketed (3). Köstlin and Frank (4) published a comprehensive review of lamp filter technologies through 1983. Bergman has reported on the system design requirements and constraints for making a low wattage compact quartz halogen lamp utilizing multilayer IR reflecting filters (5). The present review will discuss potential benefits and practical limitations of the use of IR reflecting, visible transmitting filters. It will focus on current materials and processes for coating multilayer dielectric interference films on lamps.

POTENTIAL BENEFITS/PRACTICAL LIMITATIONS OF IR FILTERS

Consider an incandescent or halogen lamp with an ideal filter that transmitted all visible radiation below 750 nm and reflected all the infrared radiation between 750 and 2600 nm to the filament where it was reabsorbed. What would be its efficacy? Assume the filament temperature is 2800°K. From Figure 1 we note that about 11 percent of the radiated power is below 750 nm. Another 13 percent is radiated above 2600 nm leaving 76 percent of the radiation to be returned to the filament. Further assume that non-radiative losses, i.e., gas and lead losses, account for 10 percent of the total power input. Thus 31.6 percent of the power needed to operate a non-IR lamp is needed in this case. Assuming constant lumen output in either case, this implies that the efficacy increase is a factor of 3.2 times that of the non-IR lamp, or about 57 LPW given a 2800°K non-IR lamp produces 18 LPW. For a 2900 and 3000°K source the same consideration yields efficacies 66 and 74 LPW, respectively, again assuming 10 percent non-radiative losses.

There are several practical limitations that prevent incandescent lamps with IR reflecting filters from achieving the very high efficacy improvements suggested above. First, there is the issue of the filter reflectivity. Presently there is no known way of making a filter with 100 percent reflectivity over a large bandwidth such as suggested above. The present manufacturing capability, using materials compatible with operation at 8000C on quartz limits the average reflectivity to about 80 percent where the upper reflective edge is about 2000 nm. Reasons for this will be discussed later. In addition the filter must have high transmission over the whole visible spectrum to give good quality white light. In particular absorption of radiation in the visible spectrum must be avoided.

Secondly, the lamp surface must be configured to return as much of the IR to the filament on one reflection. Optimization of the filament-reflector optics places severe restrictions on system configurations. Figure 3 present several choices. An ideal geometry would be a point or spherical source in a spherical envelope. Unfortunately feasible filaments are not point sources. Long cylindrical filaments placed on the axis of small diameter cylindrical envelopes or short cylindrical filament placed between the foci of elliptically shaped envelopes are practical geometries that results in good optical coupling. Furthermore the film-surface interface should produce specular reflected rays, i.e., diffuse radiation needs to be avoided as much as possible. Centering the filament throughout life, i.e., avoiding excessive sag, must also be considered since filament offset, due either to manufacturing or sag reduce the effective gain of the filter.

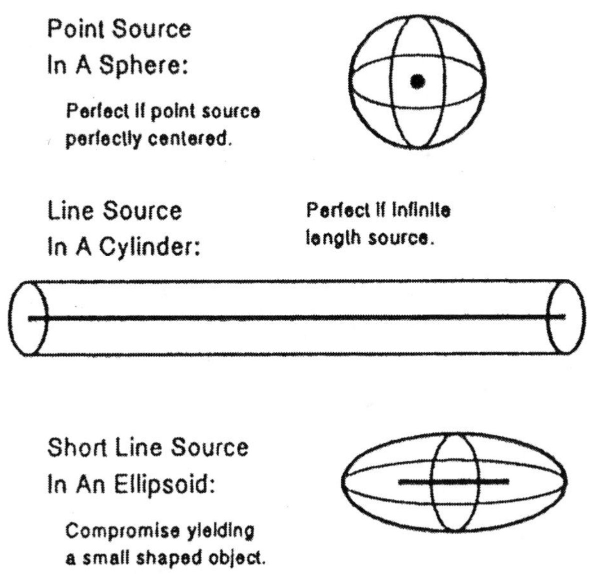

Figure 3. Geometric alternatives for IR reflectors/filaments.

Finally, while tungsten is desirable as a filament material because of its low vapor pressure, it has a rather low absorptivity for radiation in the infrared. Even with some increase in absorptivity due to coiling the absorptivity is only about 40 percent. Thus radiation reflected by the filter to the filament is likely to be scattered from the filament back toward the filter surface.

As described in Reference 5 by using a multi-pass ray-tracing model the fraction of radiation absorbed by the filament, f_a, in terms of the specular reflectivity, R, the geometrical gain factor, G, and the coil absorptivity, a, can be written as:

$$f_n = aGR/(1 - (1-a)GR). \qquad (1)$$

This relationship still assumes that the filament is centered on the optical axis of the source. If the filament is radially and/or axially offset from the optical axis the G factor must be reduced proportional to the offset times the original GR factor. Thus

$$G_K = G(1 - k + kGR) \qquad (2)$$

where k is the fractional offset of the filament normalized to its own diameter from the optical axis. Equation (2) is obtained assuming the radial offset produces an image which on second reflection returns to the actual source.

Figure 4 shows the expected efficacy gain based on Eqs.(1 & 2) as a function of the average weighted specular IR reflection over a bandwidth from 750 to 2000 nm. A constant coil absorption of 0.4 is assumed. Five curves are shown, that for no end loss (G=1), that where G=.9 which is typical of well made elliptical bulbs and cylinders with aspect ratios of eight to ten, and that where G=0.6 typical of cylinders with aspect ratios of one to two. For each of the last two cases the effect of radial coil offset by one-quarter its diameter is shown by the dashed lines. Also shown on Figure 4 are the operating points on three known IR products commercially available.

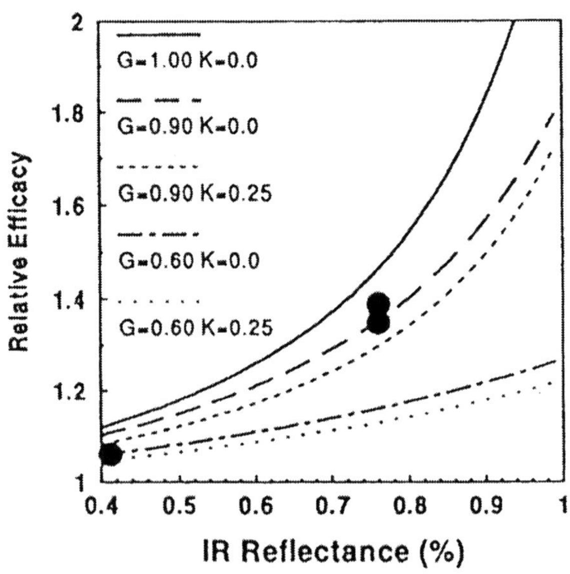

Figure 4. Effect of geometric losses and filament centering vs. reflectance 750-2000nm.

It is important to note that for elliptically designed envelopes the filament length should fill the space between the foci. If the filament is longer than the distance between the foci then radiation from the portions outside the focus are not effectively utilized, nor does reflected radiation reach these portions thus leading to potential filament temperature non-uniformity. On the other hand if the filament does not fill the whole space between the foci, some reflected radiation will be returned between the ends of the filament and the foci, which is equivalent to radial offset. Thus for elliptical filament tubes with IR filters, controlling axial positioning is as important as radial positioning.

MATERIAL CANDIDATES

Lamp operating conditions and dielectric stack design dictate the selection criteria for materials to fabricate an IR coating. These include:

- Thermal expansion coefficients of the material pairs matched to substrate
- Phase stability (morphology/scattering)
- Resistance to oxidation
- Resistance to reduction
- Environmental stability, particularly to moisture

The selection of the dielectric interference filter approach imposes the further constraint that refractive indices in a two material stack must be widely disparate. Examples of pairs of materials that may be appropriate for use in such IR reflecting films are listed in Table 2.

Table 2
MATERIALS SUITABLE FOR MULTILAYER DIELECTRIC INTERFERENCE FILTERS

High Index		Low Index	
ZnS	2.2	MgF	1.35
TiO_2	2.1-2.7	SiO_2	1.45
Ta_2O_5	2.18	Al_2O_3	1.85
Nb_2O_5	2.35		
Si_3N_4	2.0		
Y_2O_3	1.9		
ZrO_2	2.0-2.15		

Thermal stability requirements for coatings applied directly to halogen lamp envelopes (above 500°C) effectively eliminate the pair ZnS and MgF from consideration. However, these materials are widely used for visible reflecting (cold mirror) coatings on lamp reflectors.

An important requirement for an optical filter material is that it remain optically transparent and free of scattering at lamp operating temperatures for the life of the lamp--up to 5000 hrs. This may limit the use of some materials such as TiO_2 which has a tendency to become diffuse at temperatures above 750°C (6). However, because of its low cost and high refractive index, much work has been done to circumvent this material weakness. TiO_2/SiO_2 has been used for visible reflecting coatings (dichroic or "cold mirrors") on lamp reflectors in applications where temperatures are too high for ZnS/MgF or for ZnS/SiO_2.

The most commercially important material pairs for IR reflecting films on incandescent lamps have been Ta_2O_5/SiO_2, used since 1983 on General Electric high wattage quartz halogen lamps (7,8), and TiO_2/SiO_2, used by Toshiba and others on small low wattage halogen lamps (9-11). A 28 layer Nb_2O_5/SiO_2 coating has been described by Kuus (12), but has evidently not yet been commercialized. Figure 5 shows an example of a Ta_2O_5/SiO_2 coating that has been commercialized by GE.

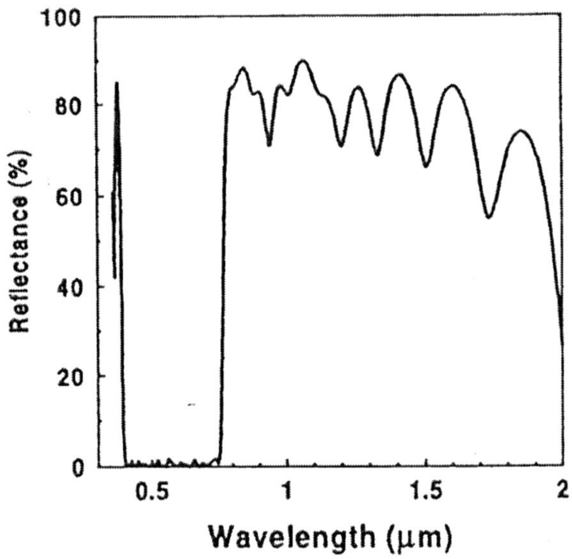

Figure 5. 46 layer Ta_2O_5/SiO_2 interference filter for infrared reflection.

The Si_3N_4/SiO_2 multilayer film process described by Verheijen, et al. (13) is a good example of the limitations of some material pairs when used for an IR reflecting coating. They described use of Si_3N_4/SiO_2 multilayer films for blue-reflecting, yellow-transmitting automotive headlamps, but they did not suggest its use in an IR reflecting film. Its use in an IR film design is certainly possible, but the lower refractive index of Si_3N_4 would require a significantly higher number of layers than an equivalent filter using TiO_2, Ta_2O_5, or Nb_2O_5.

If thin films were deposited with the properties we expect from bulk materials, the only other major requirement might be processing cost. Many deposition methods may be used to deposit the so-called same film resulting in the generic film with often different stoichiometries and with different properties. The process effects on film stress, density, moisture sensitivity, optical absorption, microstructure and optical scattering must be carefully controlled. Because film stress plays such an important role in the survivability of multilayer coatings, process effects on stress can be a significant factor. A general discussion of film deposition processes is necessary to demonstrate some of the interactions between process and material properties.

FILM DEPOSITION PROCESSES

Process Requirements

Several significantly different processes have been considered for fabrication of IR reflecting interference filters for lamp applications. These include:

- Physical vapor deposition (PVD)
  - evaporation
  - sputtering
  - hybrid processes
- Dipcoating or Spray coating
- Chemical vapor deposition (CVD)

The chemistry involved in deposition of IR reflecting films varies in complexity by several orders of magnitude depending upon which deposition process is selected. Excellent discussions of film deposition processes are available (14-16). Deposition processes are discussed here only relative to their impact on the chemistry and materials properties of IR reflecting films on lamps.

One of the most obvious requirements for optical coating processes is a high degree of precision in film thickness, which must be controlled to about ±4% depending upon the filter design. In addition, film thickness uniformity around the surface of a lamp must be of this same precision to avoid color variations.

Physical Vapor Deposition (PVD)

High precision optical coatings have been applied for many years primarily by physical vapor deposition processes including sputtering and evaporation. These processes involve the physical condensation of the material onto a substrate. They also may involve chemical reactions. An example of the latter is reactive evaporation, discovered in 1952 by Auwärter (17), where for example Si metal may be evaporated in an oxygen-containing atmosphere to form a silicon oxide film on a substrate. The physics of these processes is well established as are accurate optical thickness control methods. Thus these processes have represented the "state of the art" for several years for coating flat substrates. However, due to their largely line-of-sight coating limitations PVD processes have not been widely used for precision coatings on highly curved substrates.

The first commercially successful lamp application of a multilayer dielectric IR film was in 1981 when General Electric introduced a line of quartz halogen lamps using a high temperature Ta_2O_5/SiO_2 coating to achieve 30-40% efficiency gain (2). The coating, developed in a joint effort by GE and OCLI, represented an improvement upon the current "state of the art" evaporation processing by allowing tubing to be coated (7,8). The tubing was later processed into long cylindrical lamps. The evaporation process could not be economically adapted for coating the elliptical shapes that are most appropriate for such low wattage lamps. Another key disadvantage of this process was the high unit cost that limited applications in the highly cost-sensitive lighting field.

Historically, cathodic sputtering is the oldest vacuum process for depositing thin films, dating from 1852 (18). However, the economics of sputtering has prevented its application for films on lamps until recently. Significant progress has been made recently in several closely related hybrid PVD processes including plasma plating (19), DC magnetron sputtering (20), e-field assisted deposition, ion plating and ion-beam assisted deposition (21). All of these processes have in common the use of high energy fields and ions. These methods have the effect of producing more dense films with superior microstructure and resistance to moisture. In some cases these hybrid processes yield higher deposition rates and superior material utilization that can improve the economic viability of the approach.

Commercial lamp products using these new processes are only beginning to appear in the marketplace. Nevertheless, some of the processes may be able to meet the stringent cost and performance requirements for coating tubing and perhaps even those for small shaped halogen lamp envelopes.

Dip Coating

Film formation by dipcoating is a technology that has been used for many years in the glass industry. Schott Glass has used TiO_2/SiO_2 dip coatings for rear-view mirrors (1959) and anti-reflection coatings (1964) as well as for Pd/TiO_2 sun-shielding window glass (1969) (22). The technique found early application for IR reflecting coatings on lamps when Yoldas, then at Westinghouse, used dipcoating to deposit the inner and outer TiO_2 layers for a three-layer $TiO_2/Ag/TiO_2$ "metal" IR reflecting coating (23).

Dipcoating has several advantages including the low capital investment typically required and the simplicity with which mixed or doped materials may be deposited. A key disadvantage of dipcoating may be its apparent limitation to the coating of tubing rather than of shaped bulbs due to the non-linear variation of film thickness with withdrawal angle. The coated substrate must be heated to relatively high temperature

(>400°C) after each layer or pair of layers to convert the precursor polymeric structure into the desired oxide materials. This thermal treatment can impact the film microstructure and its usefulness in a multilayer optical coating (9). Uncracked films of 0.05-0.5 μm are readily achievable with dipcoating with multiple dips and high temperature cures required for multilayer films. Optical coatings of 16 layers have been reported in the literature (24,25).

Despite the advantages of dipcoating, it has not been applied to shaped lamps. This is probably due to the difficulty in controlling film thickness over a curved surface. The process has been successfully applied for the coating of tubing that has been fabricated into linear lamps. So far, sophisticated filters having a large number of layers (>20) have not been reported. This apparent layer number limitation, combined with the geometric analysis by Bergman (5) suggests that it will be difficult to achieve high lamp efficiencies in low wattage lamps fabricated from dipcoated tubing unless these problems can be surmounted.

<u>Chemical vapor deposition (CVD)</u>

Despite the fact that CVD has been in wide use for several decades, there has been only limited progress toward development of this process for deposition of multilayer optical coatings outside the lamp industry (26-28). In large part this stems from the complexity of the process, the chemistry and the deposition hardware that is required to create a viable CVD optical coating.

The use of thermal LPCVD for optical coatings in the lamp industry was first described by Verheijen, et al. (Philips) in 1987 (13). They described a single temperature process using TEOS to deposit SiO_2 and $SiH_2Cl_2+NH_3$ to deposit Si_3N_4 at 750°C in a conventional semiconductor tube furnace. The Philips process has been used to deposit blue-reflecting/yellow-transmitting filters on automotive lamps. This demonstrated that LPCVD was suitable for deposition of multilayer optical coatings directly on small, shaped halogen lamps.

The first descriptions of LPCVD for IR reflecting films on halogen lamps are not found in the published literature but instead are disclosed only in the patent literature. Kuus (Philips) described the fabrication of a 28 layer Nb_2O_5/SiO_2 IR reflecting coating that used niobium ethoxide to deposit Nb_2O_5 and coupled with SiO_2 from either SiH_4+O_2 or from TEOS (12). Parham et al. (GE) described the fabrication of a 24 layer Ta_2O_5/SiO_2 coating using tantalum ethoxide and DADBS (29).

Chemical vapor deposition involves a combination of gas-phase transport processes, gas-phase reactions, gas-surface interactions, surface diffusion, and surface reactions. Gas-phase intermediate reactions apparently are not unusual during the deposition of dielectrics. Such reaction complexities have been reported for the deposition of SiO_2 (30) and for TiO_2 (31-33). Unfortunately, the true reaction sequence is not fully understood for most useful LPCVD reactions. Simplified reaction pathways and kinetics are usually assumed so that empirical progress may be made in the absence of fundamental research. The number of competing CVD process variations and reagents is growing at a high rate. For the most part this growth is driven by the needs of the semiconductor industry for suitable dielectric processes at lower temperatures.

Other criteria enter into the choice of reaction chemistry because the choice of reagent can significantly affect optical properties and film microstructure. An extreme example is represented by the case of ZrO_2 films deposited via thermal CVD from zirconium acetylacetonate which may contain up to 15% carbon entrained in the film (34) while use of the tri-fluoracetylacetonate has yielded ZrO_2 films with <0.1% carbon. Similar extreme levels of carbon entrainment in films of Dy_2O_3 and Sc_2O_3 deposited from acetylacetonates have been observed by the authors. Use of alkoxides for film deposition generally leads to less severe contamination problems, but there remains a significant effect of precursor on the amount of carbon in the deposited film. Carbon entrainment can increase the optical absorption of a film significantly beyond the intrinsic properties of the film material. This extrinsic absorption can be caused directly by absorption of the carbon or indirectly by carbon-induced reduction of a metal oxide to a metal or substoichiometric oxide.

Finally, choice of the ideal chemical precursor for use in multilayer optical coatings must be tempered by its effect on film stress. This issue is discussed below.

FILM MATERIAL PROPERTIES

The material properties required for application of multilayer dielectric films to lamps were discussed earlier especially for optical and mechanical requirements. The effects of the deposition process on the following properties must be considered for lamp coatings.

Optical Absorption

Optical filters may be exposed to relatively high temperature environments that are chemically reducing. This may occur even during deposition, especially during reactive sputtering (35). In such an environment, many high-index materials including Ta_2O_5, Nb_2O_5 and TiO_2 may become oxygen deficient. For example, bulk or thin film TiO_2 may lose oxygen through the generation of oxygen vacancies (V_O) to form a sub-stoichiometric material best described as TiO_{2-x}. The change in stoichiometry increases both the optical absorption and the electrical conductivity. Demiryont (36) has shown that tantalum sub-oxide films may be deposited by ion beam sputtering of tantalum metal without adequate oxygen pressure in the vacuum chamber. The effect of sub-stoichiometry on optical properties may be severe. The Ta_2O_5 films produced by Demiryont were unacceptable for optical coatings due to wavelength-dependent optical absorption of up to 20% visible light. Similar sub-stoichiometric films have been reported for other high index films via PVD (37).

From the available literature on the defect structure of bulk TiO_2 and Ta_2O_5 (38,39) it is possible to estimate the stoichiometry of these oxides at lamp operating temperatures. The defect concentration is dependent on temperature and oxygen partial pressure (P_{O2}). Oxygen vacancy calculations show clearly that TiO_2 is much more susceptible to reduction than is Ta_2O_5. Experiments with single crystal rutile by Cronemeyer (40) showed that an oxygen weight loss of 0.1% could be produced by H_2 reduction at 800°C. This level of reduction increased the optical absorption sufficiently to produce a blue-black crystal.

In a clean inert system relatively low P_{O2} is needed to form an appreciable concentration of oxygen vacancies (V_O). However, the presence of relatively small amounts of H_2 or carbon (graphitic or hydrocarbon) in contact with these materials at lamp operating temperatures greatly enhances reduction from oxide to sub-oxide to produce unacceptable optical absorption. One example of this is apparent in the patent by Mellor et al. (41) where H_2 off-gassing from a hot quartz filament tube was thought to cause rapid darkening of a multilayer TiO_2/SiO_2 visible reflecting coating. It was found that the problem could be avoided by controlling the oxygen pressure in the lamp outer jacket. This was done through the use of BaO_2 and Cu to getter H_2 and maintain P_{O2} at a value that avoided TiO_2 darkening. Watanabe et al. (42) accomplished this through direct dosing of the outer jacket with a small amount of oxygen.

Phase Stability

As deposited, SiO_2 films are usually amorphous. In lamp applications they can be expected to remain non-crystalline over the life of a lamp. Most of the technologically interesting high refractive index materials are not stable as amorphous materials at lamp operating temperatures exceeding 550°C. These materials may be deposited either as amorphous or as crystalline films. The details of deposition generally significantly affect the crystalline form of the film. This is especially true for TiO_2 films where a wide variation in crystallinity and grain structure can be produced by process variations in all the deposition processes (37,43). Even when amorphous films are deposited, there may be subsequent grain growth or phase changes to a more thermodynamically stable polymorph. TiO_2 is sometimes deposited amorphous but it is also frequently deposited as a poorly crystalline material formed of extremely small crystallites having the anatase structure. Subsequent heat treatments, either intentional or through lamp operation, may result in an increase in optical scattering due to grain growth of the anatase phase at temperatures above 550°C. At about 650°C, the anatase to rutile transition occurs. This transition may occur slowly and may result in a mixed phase film due to an incomplete transition. Besides increased optical scattering, film cracking and spalling also may occur during these transitions.

Kawakatsu et al. (6,9) showed the importance of thermal treatments between layers during a dipcoating curing process as a means to produce low optical scattering in multilayer IR reflecting films on lamps. Further improvements have been made since then by taking advantage of the ease with which binary oxides may be formed with the dipcoating process. In one example, Yuge described the use of germania as a dopant to inhibit the anatase-rutile transition in TiO_2 (6). In a separate dipcoating example, Brock et al. showed that this transition could be inhibited by addition of several different dopants to TiO_2 including ZrO_2, HfO_2, Nb_2O_5, Ta_2O_5 (44). The inhibition likely occurs through formation of a solid solution of the dopant with TiO_2.

Whereas TiO_2 is often deposited as microcrystalline anatase, Ta_2O_5 and Nb_2O_5 are usually reported to be amorphous as deposited. Ta_2O_5 subsequently crystallizes at about 650°C while Nb_2O_5 has been reported to remain amorphous until as high as 1200°C (45). The crystallization of Ta_2O_5 in a multilayer stack has been shown to contribute to unacceptable film cracking and subsequent increase in optical scattering of the coating (29).

<u>Thin Film Stress</u>

The presence of stress in thin film coating/substrate systems is undesirable because it can limit the fabrication of high performance optical designs. Several reviews are available that consider the effects of stress and its management in multilayer systems (46,47), but there are evidently no comprehensive models that effectively address the issues most appropriate to lamps where very thick dielectric stacks consisting of many layers must withstand thermal cycles on brittle substrates.

The susceptibility of a film to failure, and the type of failure, is the result of a complex balance between film adhesion, the sign and magnitude of residual stresses, film thickness, fracture resistance of the substrate and interface, and the flaw distribution at or near the interfaces. When film stress exceeds either the strength of the film/substrate materials or their adhesive strengths, failure by one or more modes will occur. These failure modes have been discussed mostly for e-beam evaporated metal films on glass and ceramic substrates (48,49), but there are similarities in CVD films and in dielectric films deposited by other means on brittle substrates (50,51). One mechanism of stress relief occurs by cracking of the film to form a 2-dimensional pattern of islands that is usually called "mud cracking". Examples of this are shown in Figure 6 for Ta_2O_5/SiO_2 films deposited by different vapor deposition methods (52). These cracked films would be unacceptable for laser mirror applications where extremely high energy fluxes are intolerant of optical scattering. However, in many cases these films are adequate for lighting applications. The four most

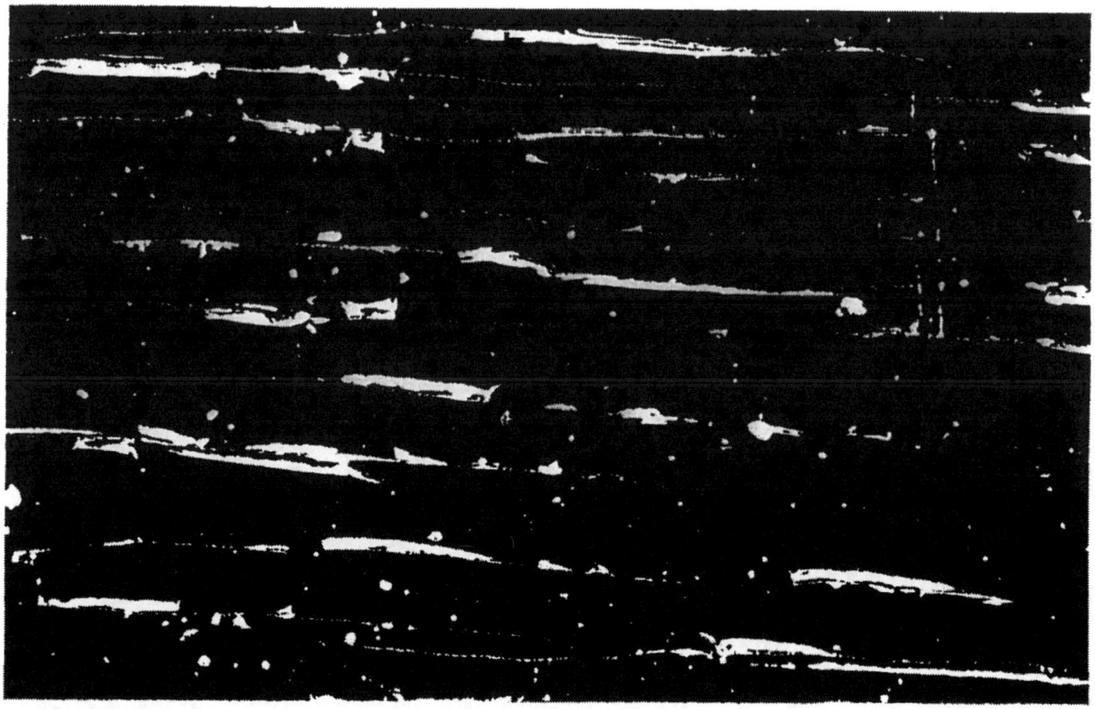

(a)

(b)

Figure 6. Crack patterns of annealed multi-layer Ta_2O_5/SiO_2 film; (a) PVD and (b) CVD. 1 cm = 140 μm.

likely failure mechanisms for films in residual tension are shown in Figures 7a through 7d (53).

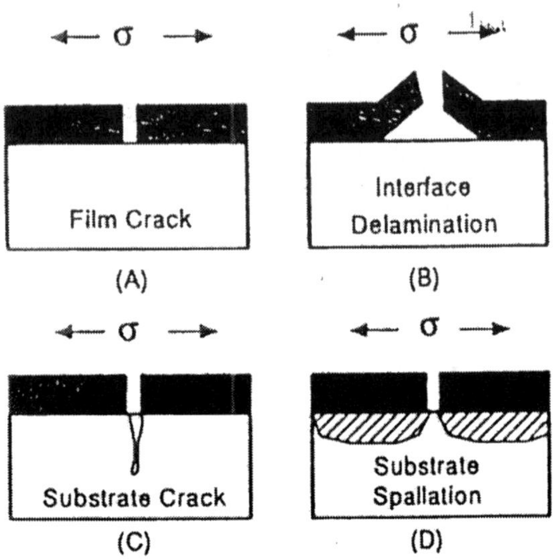

Figure 7. Mechanisms for failure in films.

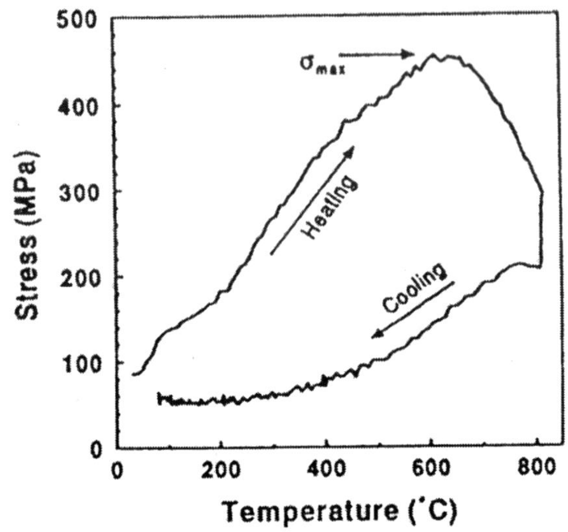

Figure 8. Stress relaxation with hysteresis

There are three sources of stress generation in a single layer thin film:

$$\sigma_{total} = \sigma_{intrinsic} + \sigma_{thermal} + \sigma_{extrinsic}$$

Intrinsic stress ($\sigma_{intrinsic}$) is a process dependent, microstructure-sensitive property caused by the modes of film growth and microstructure interactions (14). Although it is frequently the dominant stress it is relatively impossible to estimate from first principles. Thermal stress ($\sigma_{thermal}$) is caused by mismatch in thermal expansion between film and substrate and between individual film materials. Unlike intrinsic stress, thermal stress may be readily calculated by conventional analysis. Its contribution to total film stress can be made relatively clear. For example, the high thermal expansion of TiO_2 creates higher $\sigma_{thermal}$ on fused SiO_2 substrates than on glass substrates having greater thermal expansion. Finally, extrinsic stress ($\sigma_{extrinsic}$) is difficult to predict. It is caused by volume change upon crystallization of an amorphous phase such as TiO_2 and Ta_2O_5. It requires a detailed understanding of the microstructural changes occurring in the film over its thermal history. $\sigma_{extrinsic}$ may be anisotropic due to preferential crystallization of a thin film combined with anisotropy in the subsequent crystal structure.

Stress related problems occur in films of all materials deposited by all deposition technologies. For example, sol-gel SiO_2 dipcoatings have been reported to have tensile stresses as high as 44 MPa and compressive stresses as high as 56 MPa where the sign of the stress is dependent upon processing conditions (54,55). However, the most abundant work reported in the literature has been carried out in the microelectronics industry on SiO_2 films deposited by physical or chemical vapor deposition processes. Very little data is available on optically interesting materials such as TiO_2, Ta_2O_5, Nb_2O_5. The properties of thin SiO_2 films, as distinct from bulk material, can have a very significant effect in multilayer stacks. This can be seen clearly by observing the hysteresis in Figure 8. This figure shows stress as a function of temperature for a typical SiO_2 film deposited on a fused SiO_2 stress wafer by the authors. It closely resembles the stress vs. temperature behavior reported by Shintani et al. for SiO_2 films from SiH_4/O_2 (21,56). The room temperature stress in Figure 8 is much lower than the maximum observed stress due to the effects of ambient water vapor (57,58). Upon heating H_2O is evolved and the film stress increases (56). At approximately 600°C the stress reaches a maximum (σ_T) and above this temperature stress relaxation occurs. Below the maximum stress temperature no stress relief occurs other than temporary stress relief due to moisture interactions. Thus, substrate materials operating in this temperature region can have sufficient stresses to cause mechanical failure, especially due to stress corrosion over a period of time. Above σ_T, stress is relieved by sintering and structural changes due to evolution of entrapped carbonaceous and hydrogeneous species. It is not clear if stress relaxation occurs by a viscoelastic-relaxation mechanism since the glass transformation temperature (T_g) is 1100°C for bulk SiO_2. Even after annealing the film to 800°C, film stress is not completely relieved as the cooling curve in Figure 8 shows.

Doping of LPCVD SiO_2 films with P_2O_5 and B_2O_3 or co-doping with both oxides is known to reduce intrinsic film stress (59-61). The addition of dopants reduces the glass viscosity as a function of temperature and changes bonding structure. Unfortunately, borosilicate and phosphosilicate glasses are hygroscopic. This may lead to undesirable phase separation, as in the formation of boric acid in a "wet" borosilicate glass. Such interactions are detrimental to an optical film due to decreased durability and increased optical scattering (62-65). Dopants have been used to reduce stress in films made by CVD and PVD (21). Kawakatsu et al. (Toshiba) demonstrated the effect of doping for dipcoated IR films on lamps (6,11).

Little is known about management of stress and the almost inevitable film fracture that occurs in multilayer films. Most of the deposition techniques available do not produce films that are in a stable, stress-free form except by use of a post-deposition anneal. Films rapidly heated to lamp temperatures without such an anneal can fail catastrophically due to film or substrate cracking and spalling. Dipcoating processes essentially incorporate such an anneal into the process by cycling the substrate/film after each layer or pair of layers to decompose the precursors and consolidate the film. This thermal cycle also serves to relieve stress (25). PVD multilayer films also may require an anneal for stabilization (7). Microstructural development and film fracture can be controlled to some degree by careful development of an anneal sequence that is tailored to the filter design and materials used. Parham et al. (29) demonstrated the importance of controlling the rate of Ta_2O_5 crystallization and SiO_2 stress relaxation in a multilayer Ta_2O_5-SiO_2 stack. The crack patterns shown in Figure 6a and 6b are the result of the combination of process (PVD vs. CVD) and material induced multilayer stress and of the methods of post-deposition anneals. The relative sizes of the crack patterns suggest that film stresses were lower in the PVD film.

Total stress and film fracture in a multilayer are affected by a wide range of factors, making it difficult to extrapolate from single layer stress behavior. The complexity of these interactions exceeds the intuition of most technologists. There are no comprehensive models that allow advance prediction of optimal filter designs, material selections or anneal treatment to prevent detrimental film failure. The desire to improve lamp performance will continue to push designers toward thicker filter designs with a higher number of layers. This will increase the importance of reducing single layer film stress and of better understanding stress interactions in multilayer films.

CONCLUSIONS AND FUTURE DIRECTIONS

The prolonged technical efforts of lighting scientists and engineers to develop a more efficient incandescent lamp are now being realized through the application of infrared reflecting coatings. While there are still significant hurdles to overcome, in lamp making as well as in film development, to make this type of product widely available, we believe the technological success of this approach has been ensured by significant progress in all major film deposition technologies. The success of the IR reflecting lamps in the marketplace is encouraging rapidly evolving developments within the lighting industry.

We can expect to see continued improvements in lamp efficacy through better control of lamp/filament geometries and higher performing interference filter designs. A major issue preventing higher performing filters is learning to handle stress induced failures. We also expect to see expanded use of these materials for visible reflecting (cold mirror) filters to allow specific directed light applications using small sources. The use of multilayer dielectric coatings is beginning to be used to obtain colored light. We expect applications in that area to multiply. Finally the use of multilayer coatings to provide protection against UV is a potential development depending on many factors including cost and environmental concerns.

REFERENCES

1) S.O. Hoffman, U.S. Pat. 1,043,009 (1912).
2) L.E. Hoegler, T.K. McGowan, JIES, 13, 165-174 (1984).
3) I.S. Goldstein, R.P. Fontana, L. Thorington and R.P. Howson. "The design, construction and performance of an incandescent light

source with a transparent mirror," Light. Tech. 18, 93 (1986).
4) H. Köstlin, G. Frank, Philips Tech. Rev. 41(7-8), 225-238 (1983/84).
5) R.S. Bergman "Halogen-IR lamp development: A system approach," JIES, No.2, 10-16 (1991).
6) Y. Yuge, 5th Int. Sym. Sci. and Tech. of Light Sources, York, England, September 1989.
7) R.L Martin Jr, Optics News, August 1986.
8) J.D. Rancourt, R.L. Martin, Jr, SPIE 678 Optical Thin Films II: New Developments, 185-191 (1986).
9) A. Kawakatsu et al. U.S. Pat 4,524,410.
10) Y. Yuge, A. Ishizaki, T. Saito and T. Ishii, Pat 4,634,919 (1987).
11) A. Kawakatsu, Y. Yuge, N. Hayama, T. Saito, U. Maeda, U.S. Pat 4,701,663 (1987).
12) G. Kuus, U.S. Pat. 4,734,614 (1988).
13) J. Verheijen, P. Bongaerts, G. Verspui, Proc.10th Int. CVD Conf., 977-981 (1987).
14) H.K. Pulker, Coatings on Glass, Elsevier, Amsterdam, 1984.
15) J.L. Vossen, W. Kern, Thin Film Processes, Academic Press, Orlando, 1978.
16) Deposition Technologies for Films and Coatings, RF Bunshah, Ed., Noyes Pub., 1982.
17) M. Auwärter, BALZERS, Austrian Pat Nr. 192,650 (1952); US Pat. No. 2,920,002 (1960).
18) W. R. Grove, Phil. Trans. Roy. Soc. London, 142, 87 (1852).
19) R.E. Klinger, Photonics Spectra, 123-126, Jan 1989.
20) Hartsough, U.S. Pat 4,430,385 (1983).
21) P.F. Gu, "Deposition of Optical Coatings", Materials and Manufacturing Processes, 5(1), 79-91 (1990).
22) H. Dislich, P. Hinz, J. Noncrystalline Solids, 48, 11-16 (1982).
23) B.E. Yoldas, U.S. Pat 4,346,324 (1981).
24) J.L. Keddie, E.P. Giannelis, Ceramic Bulletin 69(9), 1450-1451 (1990).
25) J.L. Keddie, E.P. Giannelis, to be published Mat. Res. Soc. vol 180 (1990).
26) M.R. Jacobson, SPIE 325, 123 (1982).
27) D.Z. Rogers, SPIE Proc. 1168, 19-24 San Diego (1989).
28) D.Z. Rogers, R.P. Shimshock, SPIE Vol. 1307, 524-532 (1990).
29) T.G. Parham, S.B. Desu, C.D. Tschetter, U.S. Pat 4,949,005 (1990).
30) S.B. Desu, J. Am. Ceram. Soc., 72(9) 1615-21 (1989).
31) K.L. Siefering, et al., Extended Abstracts AIChE Annual Meeting, November 1988, Washington, DC.
32) K.L. Siefering, G.L. Griffin, J. Electrochem. Soc. 137(3), 814-818 (1990).
33) K.L. Siefering, G.L. Griffin, J. Electrochem. Soc. 137(4), 1206-1208 (1990).
34) S.B. Desu, Mat. Res. Soc. Symp. Proc. 168, 349-356 (1990).
35) T.M. Reith, P.J. Ficalora, J. Vac. Sci. Technol. A 1(3), 1362-1369 (1983).

36) H. Demiryont, J.R. Sites, K. Geib, Applied Optics 24(4) 490-495 (1985).
37) S. Schiller, G. Beister, W Sieber, G. Schirmer, E. Hacker, Thin Solid Films, 83, 239-245, 1981.
38) A.E. McHale, H.L. Tuller, J. Am. Cer. Soc., 68, 646-650 (1985).
39) J. Marucco, J. Gautron, P. Lemasson, J. Phy. Chem. Solids, 42, 363-367 (1981).
40) D.C. Cronemeyer, Phys. Rev. 87(5), 876-886 (1952).
41) Mellor et al, Pat 4,891,542, (1990).
42) T. Watanabe, T. Suwa, U.S. Patent 5,093,601 (1992).
43) L.S. Hsu, R. Rujkorakarn, J.R. Sites, C.Y. She, J. Appl. Phys. 59(10) 3475-3480 (1986).
44) Brock et al. U.S. Pat 4,940,636 (1990).
45) C.C. Wang, K.H. Zaininger, M.T. Duffy, RCA Review, 728-741, December 1970.
46) R. Hoffman, "The Mechanical Properties of Thin Condensed Films", Physics of Thin Films, 3, E. Haas, R. Thun, ed., Academic Press, New York (1966).
47) D. Campbell, "Mechanical Properties of Thin Films", Handbook of Thin Film Technology, p. 12-1 to 12-50, R.E. Lang, L. Maissel, ed., McGraw-Hill, New York (1970).
48) R.M. Cannon, R.M. Fisher, A.G. Evans, Mat. Res. Soc. Symp. Proc., 54, 799-804 (1986).
49) M.S. Hsu, M.D. Thouless, A.G. Evans, Acta. Metall. 36(5) 1301-1307 (1988).
50) R.M. Fisher, R.M. Cannon, 17th Int. Conf on Met. Coatings, San Diego, April (1990).
51) M.D. Drory, A.G. Evans, J. Am. Ceram. Soc 73(3) 634-38 (1990).
52) C.D. Tschetter, General Electric, internal report.
53) M.D. Thouless, Thin Solid Films, 181, 397-406 (1989).
54) K.A. Cerqua, J.E. Hayden, W.C. LaCourse, J. Non-Crystalline Solids, 100, 471-478 (1988).
55) M. Nakamura, R. Kanzawa, K. Saka, J. Electrochem. Soc., 133, 1167-1171 (1986).
56) A. Shintani, S. Sugaki, H. Nakashima, J. Appl. Physics, 51, 4197-4205 (1980).
57) B. Bhushan, S.P. Murarka, J. Gerlach, J. Vac. Sci. Tech., B8,1068-1074 (1990).
58) I. Blech, U. Cohen, J. Appl. Physics, 53, 4202-4207 (1982).
59) W. Kern, G.L. Schnable, A.W. Fisher, RCA Review, 37, 3-55 (1976).
60) G. Smolinsky, T.P.H.F. Wendling, J. Electrochem. Soc. 132(4), 950-954 (1985).
61) S.B. Desu, Mat. Res. Soc. Symp. Proc., 168, 221-226 (1990).
62) N. Tohge, A. Matsuda, T. Mihami, J. Am. Cer. Soc., 70, c13-c15 (1987).
63) R.M. Levin, J. Electrochem Soc., 129, 1765-1770 (1982).
64) E. Arai, Y. Terunuma, J. Electrochem. Soc., 121, 676-681.
65) W. Kern, RCA Review, 32, 429-446 (1971).

FTIR DIAGNOSTICS of TUNGSTEN-HALOGEN LAMPS: ROLE of HALOGEN CONCENTRATION, PHOSPHORUS, WALL MATERIAL, and BURNING ENVIRONMENT

Laurence Bigio, Brian Kubicki, and Peter J. Codella
General Electric Corporate R&D Center
P.O. Box 8, Building K1-4C21
Schenectady, N.Y. 12301

The use of Fourier Transform Infrared (FTIR) spectroscopy as a simple and extremely useful tool for making gas-phase measurements of several key species in tungsten-halogen light bulbs has been described previously [1]. This report extends the earlier FTIR investigative work by establishing a 2^4 factorial design matrix experiment to study the effects of four parameters on the HBr and CO levels in glass and quartz halogen lamps as a function of burn time. The four parameters examined are: (1) CH_3Br concentration in the fill gas- nominally 0.05% or 0.15%; (2) presence or lack of ~0.5 μmole P (elemental phosphorus) getter; (3) wall material- hard glass (GE 180) or quartz; and (4) burning environment- open or in a parabolic reflector (PAR). In all cases the main gas fill is 95% Kr and ~5% N_2 at a total fill pressure of 4000 Torr. Four factors, each with two possibilities, constitute a total of 16 matrix cells. Three sets of lamps, each consisting of the 16 matrix combinations, were used in this study to check for reproducibility of the results. In addition, the effects of the four parameters on lamp efficacy and life were also determined.

The reasons for each of the parameters studied vary. Certainly, the halogen concentration is an important variable whose effects should be examined. The presence or lack of phosphorus is an attempt to examine its role in controlling available oxygen. As for the choice of wall material, there are several subtle differences between quartz and the hard glass which may affect the long term HBr concentrations and life. The OH content of the glass is greater than that in the quartz (roughly 300 ppm for the glass and 11 ppm for the quartz). The glass also absorbs slighty more infrared energy than the quartz, making it hotter when the bulb is operating. In addition, the glass has a greater alkali metal content than quartz (~200 ppm of Na and 100 ppm of K for glass; 2.5 ppm of Na and 4 ppm of K for quartz). Released alkali metal atoms may abstract the halogen. The two different burning environments (ie., open and in a PAR) essentially affect the bulb wall temperature and, thus, the diffusion rates of such species as H, OH, and Na in the glass or quartz wall.

An example of the results, Figure 1, shows HBr absorbance signals as a function of burn time for quartz bulbs with 0.05% CH_3Br fill, with no phosphorus, burning open and in a PAR. Clearly, small differences in wall temperature can affect the long term HBr behavior. The major conclusions are summarized as follows:

(A) HBr

1. Going from burning open to burning in a PAR increases HBr by ~100%.
2. Going from glass to quartz increases HBr by ~35%.

(B) CO

1. Phosphorus lowers the CO level by ~30%.
2. Going from glass to quartz lowers the CO level by ~30%.
3. The CO level is oxygen limited, even for bulbs with as low as 0.05% CH_3Br fill.

(C) Efficacy (or lumen maintenance)

1. A minimal amount, ~0.01%, of both HBr and oxygen seems to be necessary to keep the bulb wall clean.
2. Bulbs burned open generally developed slightly darker walls than those burned in PAR's.
3. Glass-walled bulbs stayed slightly cleaner than quartz-walled bulbs.

(D) Life

1. More HBr leads to shorter life.
2. Phosphorus increases life.
3. Burning in open, as opposed to burning in a PAR, increases life.

REFERENCE

[1] L. Bigio and P.J. Codella, "FTIR Diagnostics of Tungsten Halogen Lamps", Nicolet Spect. Lines Vol. 11, 10 (1990).

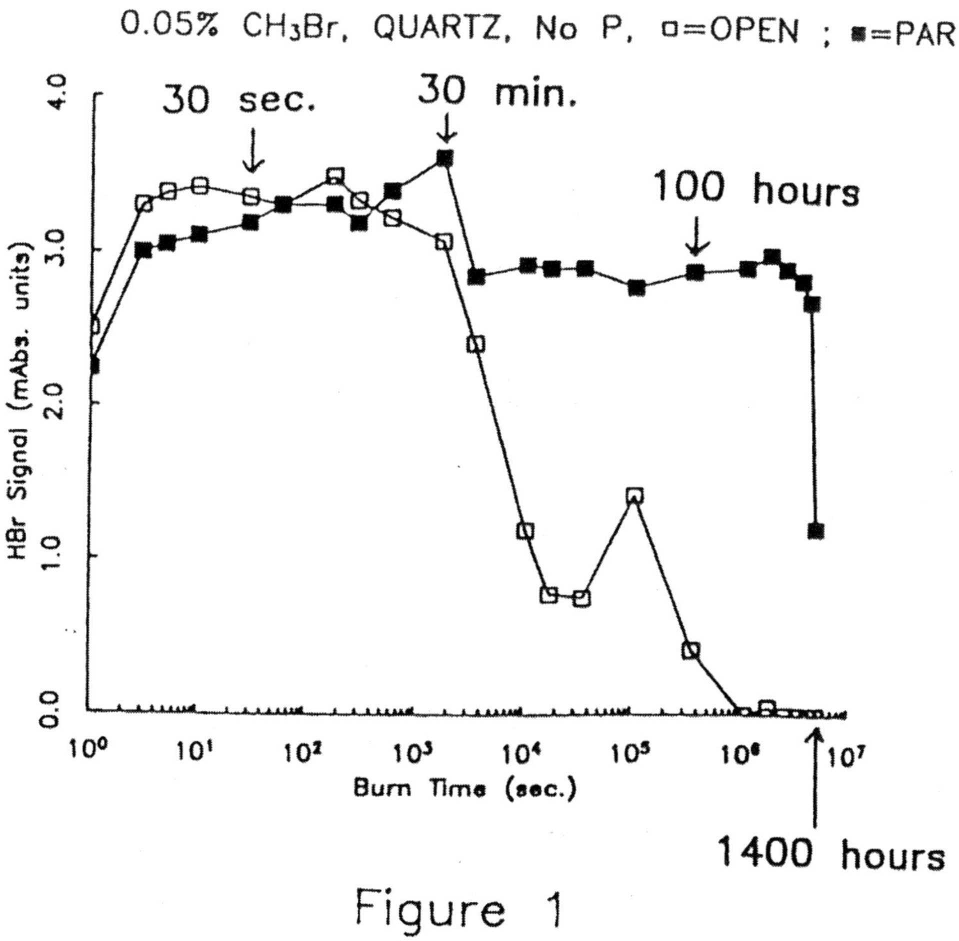

Figure 1

COMPARISON OF PHYSICAL AND LIFE PARAMETERS OF GLS LAMPS WITH AND WITHOUT PHOSPHORUS GETTER

45:P

Éva Dancs
Bródy Imre Research Center, Tungsram Co. Ltd.
H-1340 Budapest, Váci út 77.

The exhausting, pumping and filling technology of lamps which is among the oldest techniques of lighting industry, has now been developed to a highly sophisticated level to insure appropriate performance and life. Nevertheless the application of getters can not be eliminated to avoid occasional life failures.

The investigations discussed in this paper had the objective to study some of the effects of phosphorus getter in the lamp on other parameters. Groups of lamps with and without getters were subjected to physical and life-time measurements.

The results for the mechanical load, the nitrogen afterglow [1], the electrical resistance and its changes during flashing together with the life-time data are summarized in Table 1.
The lamps used for the measurements were taken from the regular production with a slight modification for the group without getter.
The process was first interrupted before flashing. Resistance and nitrogen afterglow at the wavelength of 580 nm were measured for 200-200 lamps. The maximal mechanical load was measured with the breaking of the filament. There was no difference in the maximal mechanical load for the groups before flashing; there is a common average for them with a very large spread calculated from 20 measurements.
The same lamps (except those used in the breaking procedure) were flashed at the production line. Repeated measurement of the electrical resistance revealed the changes due to flashing. No identical changes could be obtained for the two different groups according to about 200-200 measurements again. Possible reasons for the difference may either be the interaction of the getter with the filament or the interaction of the gases influenced by the getter with the filament. Any of them could have a significant effect on the process of recrystallization governing also the resistance itself. Similar results arose from laboratory investigations with thermal treatment at controlled temperatures of about 20-20 lamps.
Resistance after flashing is determined by the original data and the changes. Maximal mechanical load and nitrogen afterglow values are characteristic for the filaments and the gas atmosphere after flashing, respectively. Both parameters differ considerably for the lamps produced with and without getter.
In order to complete the series of the measurements the life-times of some of the lamps (about 40-40) were also determined. Life-tests gave the expected result yielding higher life-times for lamps with getter.

Table 1.
GLS lamp filament, atmosphere and life parameters for groups of lamps with and without getter

Lamp		With getter	Without getter
Step	Parameter		
Before flashing	Max. mechanical load (g)	21.4 ± 9.5	
	Afterglow (AU)	1.1 ± 0.4	2.3 ± 0.5
	Resistance (Ω)	110.9 ± 1.9	110.9 ± 1.2
Flashing	Change of resistance (%)	4.6 ± 0.4	5.4 ± 0.4
After flashing	Max. mechanical load (g)	1.62 ± 0.32	1.36 ± 0.13
	Afterglow (AU)	0.7 ± 0.4	2.4 ± 0.5
	Life-time (h)	910 ± 360	750 ± 220

Conclusion: the role of simple phosphorus getter in GLS lamps is more complicated than mere gettering of oxygen and water in the bulb. Even its participation in the modification of the gas atmosphere is more complex and it may also directly or indirectly affect the recrystallization of the filament thus having further effect on other mechanical and electrical parameters.

Ref.: [1] H.J.Eppig: Appraisal of incandescent lamp filling gas purity J. of IES, October 1982, 34

MORE EFFICIENCY FOR THERMOFORMING PROCESSES

W. Dieudonné, K. Schmitz, R. Stehling
Heraeus Noblelight GmbH, Heraeusstraße 12-14, D-W-6450 Hanau /Germany

All users of thermoforming processes with ceramic- or quartz-glass-emitters require a high homogeneity of temperature on the target surface, a short response time for changing temperature-profils, a high bring out ot the infrared system and a high efficiency.

A high homogeneity is reached with a high resolution of the emitter-field (smaller emitters) and a special heater-geometry.

The short responsetime for a temperatur-profil-change, was reached among others with the construction of the IR-emitter and a high heating-element-temperature (> 900 °C).
The increasing of the efficieny is realised with the rising of the radiation efficiency (see fig. 1) respectively with the reduction of the dissipations, e. g., convection.

$$\eta_{rad.} = \frac{\Sigma\dot{Q}_{rad.}}{\dot{Q}_{el.}}$$

$\eta_{rad.}$ = radiation efficieny
$\Sigma\dot{Q}_{rad.}$ $\doteq$ output: radiation-energy
$\dot{Q}_{el.}$ = input: electrical energy

Another possibility to save energy is a smaller distance between the emitter-field and the target-material (see fig. 3).

The use of all features

1. special IR-emitter geometry and construction

2. high resolution of the emitter-field

3. reduction of the distance: heating field to target-material

4. high radiation efficiency

realises the requirements of the top with following benifits

1. smaller heating-field-size

2. high homogeneity on the target surface

3. high efficiency save energy

Tests with a thermografic-camera-systems show:
the homogeneity, e. g., of a painted iron-sheet (fig. 4, 5 and 6)
and those with a x-y-recorder:
increasing of the efficiency (Fig. 3), e. g., of a GMT-material-test

figures titles in poster:

fig. 1: energy-flow-diagram:

The efficiency ($\eta_{rad.}$) depend mainly from the direct radiation ($\lambda < 4$ µm, e. g., with a emitter-systemborder of quartz-glass)

fig. 2: diagram: distance IR-emitter-field to target-material →

ratio in %: IR-emitter field-area to target-area ↑

To increase a high homogenity at the adges of the target, it is necessary that the emitter-field is larger than the target.

Practically the distance between target and heating-field should be the value which increase the emitter-field-sizes.

e. g. target-size	:	500 x 1000 mm ≙ 0.5 qm
distance	:	100 mm
→		
heating-field-size:		700 x 1200 mm ≙ 0.84 qm

fig. 3: heating-up-curves of GMT-material with two heating-field-distances:

The energy ($\dot{Q}_{el.}$) was 100 % in both cases. The material reached the temperature in a distance of 100 mm faster than in a distance of 200 mm.

fig. 4: iron-sheet in 50 mm distance to the heating-field:

- thermografic-picture
- one color ▲ 5 °C
- size of the painted iron sheet: 500 x 1000 mm

fig. 5: iron-sheet in 100 mm distance to the heating-field:

- thermografic-picture
- one color ▲ 5 °C
- size of the painted iron sheet: 500 x 1000 mm

fig. 6: iron-sheet in 70 mm distance to the heating-field with a power-controll-unit:

- thermografic-picture
- one color ▲ 5 °C
- size of the painted iron sheet: 500 x 1000 mm

RADIATION PROPERTIES OF COILS

I.Gaal
Research Institute for Technical Physics of the Hungarian Academy of Sciences, Budapest,(H-1325 Ujpest 1 POB 76)

The measured radiation coefficients of long coils of practical importance ([1],[2]) can be fitted to very simple formule, since they scale with the pitch ratio (K_P) as $1-1/K_P$ and depend very slightly on the mandrel ratio (K_M). However, the rationalization of this very simple result is difficult because of the radiative interaction between the concave surface elements of the coils. In spite of this, a direct theoretical attack on the radiation properties of the coils is of interest for various reasons: (i) the radiation coefficients of short coils depend markedly on their length, (ii) the wave length dependence of the absorption of the coils is unexplored, (iii) the empirical expressions suggested for the radiation coefficient predict the luminous flux and the temperature with quite different accuracy even for long coils. (The absorption coefficients are needed both at the evaluation of the radiation coefficient of the coiled coils and at the prediction of the efficiency of energy saving at infrared reflection.)

The amount of radiation leaving the inside of the coil without reflection has been deduced in a heuristic way by Aladyshev [3] and even the correct treatment of a coiled thin ribbon by Vukcevich [4] relies on the assumption that the flux of the radiant energy (i.e. the sum of the emitted radiation plus the reflected radiation) is homogeneous on the inside part of the ribbon. The aim of the present work was to evaluate the radiation coefficient of coiled thin ribbons and wires in a mathematically rigorous way in the frame work of the geometrical optics. To this end the balance equation of the radiation has been solved numerically. This equation states that the radiation emitted at a given wave length into a given direction by the surface element dudv is equal to the radiation emitted by the surface element duduv in the said region plus the radiation coming from all the surface elements dxdy and reflected by surface element dudv into the said region:

$$L(u,v) = L_B(u,v)*e+(1-e)\int L(x,y)*K(u,v,x,y)dxdy \qquad (1)$$

and the total radiation leaving the coil is

$$\Phi = \int K(u,v,p,q)L(u,v,p,q)dudvdpdq \qquad (2)$$

where e is the emissivity of tungsten, L_B is the spectral radiance of a black surface element, x,y are the coordinates of that points of the coil surface from which the point u,v on the coil surface can be seen, K(u,v,x,y) is the angle factor of these points and p,q are the coordinates of a closed surface containing the coil.

Vukcevich solved (1) and (2) with the approximation

$$L = L_B*e*[1-(1-e)*\langle\langle K\rangle\rangle]^{-1} \qquad (3)$$

where $\langle\langle K\rangle\rangle$ denotes the surface average of the integrated angle factor

$$\langle K\rangle=\int K(u,v,x,y)dxdy. \qquad (4)$$

In the present work L(x,y) was approximated by a set of unknown numbers by means of which the integral in (1) was expressed and the resulting linear algebraic equation was solved. At the evaluation of (1) we made two approximations: (i) it was assumed that tungsten reflects the radiation in a diffuse way and (ii) the emissivity does not depend on the angle of the radiation. The solution for infinite coils is very much facilitated by their helical symmetry according to which L(u,v) depends only on the axial coordinate of the coiled ribbons and on the angular coordinate perpendicular to the wire axis in wire coils.

The main result was that the integrated angle factor of ribbon coils depends on the axial coordinate very slightly, if $K_P/K_M < 0.2$ and the results of Vukcevich were recovered. Since the amplitude of ⟨K⟩ scales with $(K_P/K_M)^2$ and the average value of ⟨K⟩ scales with $(1-a^*K_P/K_M)$, we have got also a slight mandrel ratio dependence for ribbon coils. Since the number of turns in effective radiative interaction increases with decreasing K_P/K_M, the segment of the coil with inhomogeneous radiance increases as K_M/K_P. At $K_P/K_M=0.2$ the inhomogeneous radiance involves 30 turns.

For coiled wires the strongest radiative interaction takes place between the closest parts of the turns and this interaction gives rise to large variations in ⟨K⟩. Hence the rigorous treatment requires the solution of (1).

[1] Holst,G.,Lax,B.,Oesterhuis,E.,Pirani:
Z.Techn.Phys. 9 (1928) 186

[2] Bergman R.S.,Cox,C.W.,DePriest,D.J.,Faltin,E.W.:
J.Illum.Eng. 19 (1990) 132

[3] Aldyshev,A.N.:
Svetotechnika 22 (1975) 14

[4] Vukcevich M.R.:
IES Meeting, St. Luis (1984)

HIGH QUALITY INFRARED REFLECTING FILTERS ON LAMPS PREPARED BY DIP COATING

Dr. G. Hebbinghaus and G. Frank
Philips Forschungslaboratorium Aachen, Weißhausstraße,
D-5100 Aachen, Germany

The light produced by the heated filament of an incandescent lamp covers a broad spectral range from ultraviolet to infrared. Only a small fraction of this band overlaps with the sensitivity curve of the human eye. By applying an IR reflecting filter to the bulb of the lamp the large infrared part of the emission spectrum can be reflected back to the filament, thus contributing to its heating. The visible light output per electrical power input can be increased and the heat radiation of the lamp can be reduced.

Stable optical filters on halogen lamp bulbs, reflecting selectively the IR radiation, can be realized by interference filters built up of different high temperature stable transparent metal oxide layers having at least two different refractive indexes.

The optical quality of the filter is characterized by high transmission for visible light and high reflectivity for infrared radiation. To achive both, the filter materials have to be free of absorption. The IR reflectivity is determined by the maximal available difference in refractive indexes between the used filter materials and the maximal obtainable thickness of a mechanically stable filter.

An appropriate process for the simple investigation of various filter materials, which also allows for mixing two or more substances, is the dip coating process: Homogeneous thin films are deposited on a substrate by pulling it out of an alcoholic solution of alkoxides, having the desired metal stoichiometry. The layer thickness is controlled by the concentration, viscosity and the pulling rate. Dense oxide films are obtained by annealing the dried gel film at high temperature for some minutes. After the substrate has cooled down, the next layer is added by dipping.

The standard high temperature stable material of high refractive index is TiO_2. It is combined with SiO_2 as material of low refractive index. However, with these materials the total thickness of the filter, prepared by dipping, is limited to ≈2.5 µm. Thicker filters have too many cracks and tend to peel off. Alternative materials of high refractive index are ZrO_2, HfO_2, Nb_2O_5, SnO_2 and Ta_2O_5. The last one shows the lowest thermal expansion coefficient and allows, together with SiO_2, the construction of rather thick filters. The refractive index of Ta_2O_5, however, is considerably lower than that of TiO_2.

Using the sol gel dip coating technique, TiO_2 has been combined with ZrO_2, HfO_2, Nb_2O_5, SnO_2 and Ta_2O_5 creating new materials of high refractive index. By mixing TiO_2 with Ta_2O_5 a new system has been realized which has a refractive index of n= 2.25 at 1.1 µm and which permits to build up filters of up to 8 µm total thickness. By applying this filters to a long linear halogen lamps, the luminous efficacy has been increased up to 70%.

49:P

SPECTRAL IRRADIANCE STANDARD LAMP

LU LUAN YUN

Shenyang Huaguang Lamp Factory

GU YUO ZHONG

National Institute of Metrology

This article breifs you on the structure, the characteristic and the scalar method of the FDZ 500-1 standard lamp, which as a kind precision instrument, can be used to reproduce and preserve the standard value; as a new light source, to transmit the value.

The operating power is 500 wattages at 100 voltage, the colour temperature is 3150 K. The lamp features compacted structure and fixing frame, with the superior transparant quartz glass, filled in with the hydrogen bromide gas, the lamp's rich of ultraviolet ray and shows wide spectral power distribution.

The life test is taken under scalar colour temperature, the variation of irradiance below 1%. The formula Bessel is quated to calculate the relative K.

For the life characteristic, the decline per hour is below 0.08% within 70 hours. For a brief space of continuous 4 hours, the relative standard error does not exceed 0.5%. For the horizontal characteristic in plus and minus 2 degree, the variation is below 0.3% . This article also introduces a special lamp-making method--the rotating aging method and the directed aging method. The scalar method for standard lamps, the lamp structure and the testing principle drawings are presented as well. 6 drawings included.

EFFECT OF TUNGSTEN AND MOLYBDENUM MANDREL WIRE SIZE VARIATION ON INCANDESCENT LAMP PERFORMANCE

Attila NAGY, Csaba László TÓTH, Imre KÁROLYI, Anna KÁLOVICS
GE-TUNGSRAM, Budapest, 1340, Váci út 77., Hungary

Recently Bergman et al. [1] measured some lamp parameters versus coil weight, secondary pitch and length of the non exposed coil leg. The aim of their work was to investigate the effect of process variables on the variance of lamp parameters. The first parameter (coil weight) was varied via the secondary mandrel diameter, so it is a parameter of wire production process. On the other hand, the last two terms reflect the variance of lamp production process. They concluded that it is the coil weight which has the greatest influence on both power, lumen and life time. The relative sensitivities were -0.55, -2.0 and 13.5, resp. (Per cent changes in lamp parameter versus per cent changes in coil weight.)

As coil weight is influenced not only by secondary mandrel but also by tungsten wire and primary mandrel size, we measured the lamp parameter sensitivities on these sizes separately. 250 V 60 W gas-filled coiled coil lamps were made using the all possible eight combinations of "LOW" and "HIGH" sizes of tungsten, primary and secondary MOLYBDENUM mandrel. The wire sizes were measured with a laser diffraction sensor, made by CERSA. The whole length of the wires was checked: there was no overlap between the "HIGH" and "LOW" sizes which were as follows:

TUNGSTEN	"LOW"	2.487	mg/200 mm
	"HIGH"	2.571	-"-
MOLYBDENUM (1)	"LOW"	4.568	-"-
	"HIGH"	4.678	-"-
MOLYBDENUM (2)	"LOW"	78.98	-"-
	"HIGH"	82.58	-"-

(mg/200 mm is a usual measure of fine wire sizes, because the mass of the wire of a given length can be accurately measured.)

100-100 coils were made using each combination of the wires. The coil weight and electrical resistance were measured individually. These values were calculated, as well. Because of the well known deviations between calculated and measured weight and resistance, only the relative values, the "sensitivities" are given here, as follows. The coil weight/resistance sensitivity on tungsten, primary and secondary mandrel sizes are 1.25/-0.75, 0.39/0.39, 0.31/0.31, resp. (theoretical) and 0.96/-0.54, 0.32/0.49, 0.72/0.67, resp. (experimental).

The coil weight and resistance are important test parameters in GE-TUNGSRAM lamp production process. A computer program is used to estimate power, lumen, light efficiency and life time from the mentioned coil characteristics. Following the same philosophy as above, sensitivities can be defined which were determined from experiments, measuring all of the lamps, as well. (The life time was measured only on a fraction of lamps, i.e. on 10-10 pieces.) The power, lumen, light efficiency and life time sensitivities versus coil weight/resistance obtained are 0.14/-0.72, -0.38/-1.91, -0.52/-1.20, 4.99/9.88, resp. (calculated) and 0.03/-0.70, -0.44/-1.66, -0.46/-0.98, 0.95/2.92, resp. (measured).

Combining these sensitivities with those regarding coil characteristics versus wire sizes, one obtains the sensitivities of lamp parameters versus wire sizes. Averaging all of the data but the measured life time we get:

power sensitivity	versus	tungsten wire size	0.6
		primary mandrel size	-0.3
		secondary mandrel size	-0.3
lumen sensitivity	versus	tungsten wire size	0.7
		primary mandrel size	-0.9
		secondary mandrel size	-1.1
light efficiency sensitivity	versus	tungsten wire size	0.2
		primary mandrel size	-0.7
		secondary mandrel size	-0.8
life time sensitivity	versus	tungsten wire size	-0.9
		primary mandrel size	6.1
		secondary mandrel size	7

Our results are in fully agreement with that of Bergman et al. Namely, 1.0 per cent change in secondary mandrel size (mg/200 mm, which is of course 0.5 per cent change in diameter, µm) results in about 0.5 per cent change in coil weight. Thus our sensitivities regarding lamp parameters versus secondary mandrel size must be multiplied by a factor of two for obtaining the sensitivities of the lamp parameters versus coil weight measured by Bergman.

Conclusions

1/ Power is determined mostly by the tungsten wire size.

2/ Light efficiency, and life time are much more dependent on the mandrel sizes than on tungsten wire size.

3/ From the test parameters, coil weight and resistance, the later is a more significant one.

Reference

[1] R.S. Bergman, C.W.Cox, D.J. DePriest, F.W. Faltin: Effect of process variantions on incandescent lamp performance; Journal of Illuminating Engineering Society, Summer, 1990, pp. 132-141.

GAS COMPOSITION INVESTIGATION IN TUNGSTEN HALOGEN LAMPS

Livia Nagy — GE-Tungsram Imre Brody Research Center
H-1340 Budapest Vaci u.77.

Laurence Bigio — GE Corporate Research and Development Center
PO.Box 8, Bldg. K1-4C21, Schenectady,
New York 12301

While the chemical cycle involved in tungsten halogen lamps has been known and exploited for several years, it is believed that some specific side chemical processes not well understood may have a major influence on lamp quality and life. These side processes include the interaction of carbon with adsorbed water, hydrogen loss by wall diffusion, residual metal-oxide/halogen interactions, and desorbed residual gases.

Our recent work has concentrated on the quantitative relationships of theses side reactions. To do so, the gas atmospheres of specially prepared experimental lamps were studied using two different analytical approaches. One approach uses Fourier transform infrared (FTIR) detection inside the intact sample lamps, while the other is a very sensitive and versatile gas chromatographic (GC) technique used for destructive sampling of lamps. Both techniques have been successfully applied separatelyto tungsten-halogenlight bulbs, and the experimental details have been previously described [1-4]. The combined application here of these two techniques yielded an added advantage over each one separately: they partly supplemented, partly confirmed each other.

We studied Tungsram H4 auto-headlight lamp bulbs. In preparing them, special degassing methods were used to minimize the water intake. As a result, very low levels (under 50 ppm) of CO were detected after 20 seconds of burn. The initial filling [redacted] Kr with 660 ppm CH_3Br, at a total room temperature pressure of 4.5 atm. There are two filament coils in the bulb, 60 and 55 W. The 60 W coil was burned at 13.2 V line ac for varying lengths of time, and the gases were measured with the bulb off after each burn period. Thus, with the FTIR method, entire burn time dependencies could be measured for each bulb individually since it is non-destructive. With the GC method, a different bulb was required for each burn time measurement. The GC method however, is much more sensitive.

The main gases measured were CH_3Br, HBr, CO, H_2, CH_4 C_2H_4 and C_2H_2. Interesting observations could be made about the concentration changes of the different species during the life of the lamps. These will be presented in the text of the paper. As an example, the hydrogen and hydrogen bromide results are shown in the figure below.

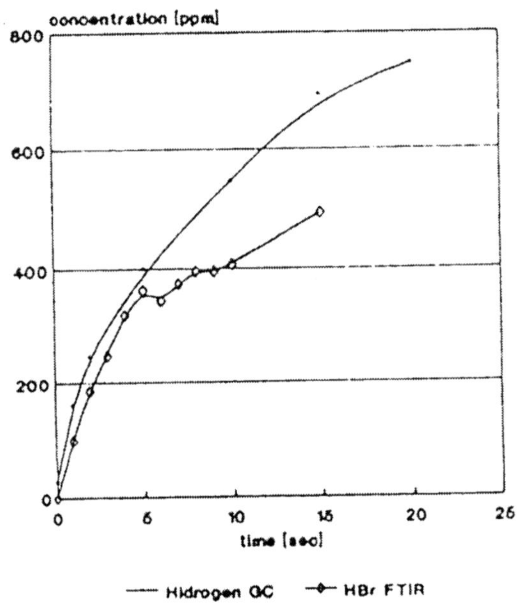

REFERENCES

[1] L.Nagy, E.Biro and J.Balla, Chromatographia, 20, 04 (1988);

[2] L.Nagy, E.Biro and J.Balla, J.Chromatography, 520, 213 (1990).

[3] L.Bigio and P.J.Codella, Nicolet Spect. Lines, 11, 10 (1990);

[4] L.Bigio, B.Kubicki and P.J.Codella, Appl.Spect., 45, 819 (1991);

A SIMPLE LEAD CORROSION MODEL FOR QUARTZ HALOGEN LAMPS

Joseph M. Ranish
Advanced Technology Department

Leonard E. Hoegler
Product Engineering Department

GE Lighting
Cleveland, OH 44112

INTRODUCTION

Typical quartz halogen photolamps are dosed with HBr. During operation, HBr is partly dissociated at the wall and the H_2 formed diffuses through the wall. The elemental H/Br ratio thus decreases as the lamp is operated. Corrosion of the cooler leads and supports occurs in halogen lamps in which the elemental H/Br ratio is less than one[1]. In some cases, the corrosion can cause lamp failure by severing the lead. A simple model describing this phenomena has been developed. Lead lifetimes calculated from this model are compared to lead lifetimes of experimental lamps.

MODEL

The lead corrosion rate is assumed to be controlled by the rate of the heterogeneous W-Br reaction with no mass transfer limitation[2]. Element segregation and oxygen effects are not included and limit the application to "well stirred" lamps having negligible (relative to bromine) gas phase oxygen contents. A numerical time stepping approach is used to calculate the lead life. Starting with the given initial gas dose and lead temperature, thermodynamic equilibrium calculations yield the Br atom concentration. This is used to calculate the lead recession rate using the known kinetics[2]. Next, the known H_2 permeation rate[4] is calculated for the given wall temperature. This permeation rate and a predetermined fractional loss of H_2 are used to set a time interval. The incremental recession of the lead during this time interval is computed and the net recession is computed. Finally, thermodynamic equilibrium calculations are performed at the wall temperature to yield the local H_2 concentration. This combined with the predetermined fractional loss of H_2 is used to adjust the gas dose composition which is used to calculate the Br atom concentration for the next time step. Time stepping is continued until the net recession equals the lead radius.

EXPERIMENTAL

Single ended quartz lamps having approximately cylindrical symmetry with the coil axis coincidental with the cylinder axis, a wall thickness of 0.1 cm, a volume of 7.4 cm^3, and an effective area for H_2 diffusion of 16 cm^2 were filled with a HBr, N_2, inert gas mixture. In some of the lamps, phosphorus was added at a P/Br atom ratio of ~0.3. The lamps were burnt pinch end up, 45° from vertical until failure. During burning, the wall temperature was ~1173 K and the pressure was ~20 bar. All failures were due to lead corrosion. Average lifetimes (~4 lamps/cell) are plotted in Figure 1. If the lamps are burnt pinch end down, the lifetimes are slightly longer but the leads remain intact and the lamps fail by arc out.

RESULTS AND DISCUSSION

Calculated lead lifetimes are more than twenty times longer than the experimental lifetimes. Differences in surface roughening between the literature studies and the experimental lamps are unlikely to explain the discrepancy. Since gettered lamps had lifetimes similar to ungettered lamps, oxygen does not appear to affect the corrosion rate. The dependence of burning position on the lead corrosion rate indicates that perhaps the gas stream reaching the lead has not achieved the composition corresponding to equilibrium at the lead temperature when the lead is above the coil body (pinch end up operation). Such an effect was postulated in the related problem of Mo blackening[5]. Very good agreement between model and experiment can be obtained, Figure 1, if one assumes that the plume from the coil body reaches a Br atom concentration corresponding to the equilibrium concentration at 1673 K, 500 K higher than the lead temperature. The accuracy of this hypothesis will be examined in future studies.

ACKNOWLEDGEMENTS

The authors thank Mr. R. Berki for the HBr analyses. Also, the support of Mr. R. Surmick, Dr. R. Bergman, Dr. F. Ahlgren, and Dr. E. Zubler is gratefully acknowledged.

References

[1] G. R. T'jampens and M. H. A. van de Weijer, Philips Technical Review, 27(7), 173 (1966).

[2] G. Rouweler and B. J. de Maagt, Philips J. Research, 33,(1/2) 1 (1978).

[3] B. J. de Maagt and G. Rouweler, Philips J. Research, 35(2), 95 (1980).

[4] R. W. Lee, R. C. Frank, and D. E. Swets, J. Chem. Phys., 36(4), 1062 (1962).

[5] T. H. Yu and J. M. Ranish, Proc. of the Fifth Intl. Symposium on the Science and Technology of Light Sources, p. 215, Univ. of Sheffield, UK (1989)

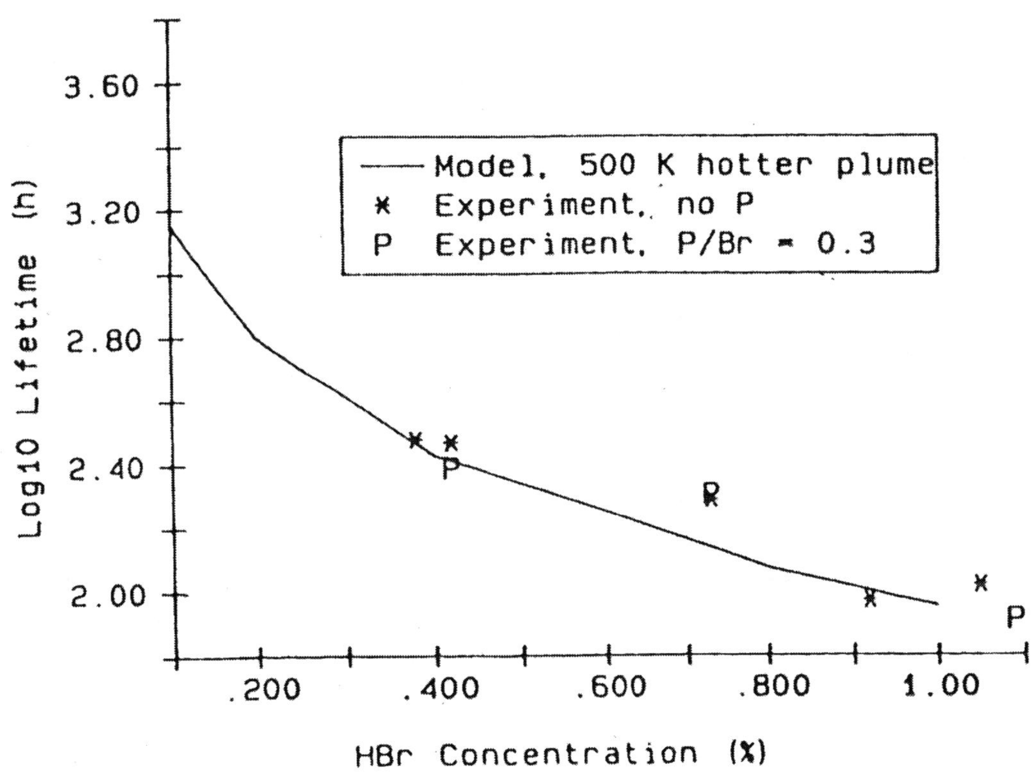

Figure 1: COMPARISON OF LEAD LIFETIMES

COOKING WITH LIGHT

Reinhard Kersten
Philips GmbH Forschungslaboratorium
Postfach 1980
D-5100 Aachen

Electric cooking systems which are popular in many countries have a noticable disadvantage: They have a high thermal inertia and hence a poor controllability. This holds for both common systems, the cast iron plate and the glass ceramic cooking fields. Reasons for the slow reaction are:

- a high heat resistance of the air gap between the hot plate and the pot's bottom combined with a typically large power density (5 W/cm^2 and more)
- hence a high temperature level of the heating system and a high stored energy

This blocking thermal resistance between the hot plate and the pot's bottom holds even with halogen cooking units as long as the glass ceramics absorb well over 50 % of the lamp's thermal radiation: The air gap between plate and pot is the most critical heat resistance factor. Essential points for improving the halogen cooking system are:

- the use of glass ceramics transparent for the radiation of halogen lamps (AE88). Radiation crosses any air gap.

Additionally:

- reduce visible light in order to avoid dazzling
- equalize power distribution with two halogen lamps
- fulfill the existing rules for the disturbances on the mains (flickering, harmonics and high frequency noise)

Optical properties of glass ceramics

There are three classes of different spectral material behaviours with respect to the thermal radiation of halogen lamps. Most common glass ceramics on the market (Schott 8557 and 8573) have a total transmittance well below 50 % and hence halogen cooking systems react like common systems having no reduction of their thermal inertia. Important contributions are coming from new materials (Schott 8575 and NEG Neoceram-Black in Japan). They have transmittances well above 50 % and hence more than 50 % of the power is immediatly available above the plate when switched on. Though they transmit quite differently at the maximum of the halogen lamp's spectrum all materials mentioned above behave similar in the visible range, dampen the light by a factor of 20 - 30 and look red-(brown). The highest transmitting materials on the market (clear white, Schott Robax, NEG Neoceram-O) are dazzling in this application and can't be accepted.

Aluminium reflector

The lower absorption of the new glass ceramics facilitate the use of aluminium reflectors in this application which is not feasible up to now for reasons of temperature limits. This is advantageous for several reasons:

- The power distribution is equalized with **two** linear lamps, where three or four lamps (AE88) were needed up to now.

- The aluminium reflector acts as a thermal insulating shield.
- The reflector is simultaneously the lamp's holder itself.

With the help of a two dimensional ray tracing program reflector shapes could be found in different planes perpendicular to the lamp's mounting direction. The total reflector shape was calculated by an interpolation process. The resulting measured spatial power distribution is better than in any other halogen plate known to us so far.

Electronic power control

A mechanical on-off control with long cycling times is unacceptable in a hob with a quick thermal reaction; e.g. simmering would result in sequentially sharp boiling and cooling phases. Because of the high lamps starting currents, there would also be difficulties in meeting the rules for low disturbances on the mains (e.g. IE82). A special half wave control (KE89) for a two-lamp-system produces a lot of quasi-stationary power steps having a mixture of the half wave patterns "2 lamps in series", "1 lamp only" and "2 lamps in parallel". Intermediate power levels are attained by a short cycling (1 - 3 seconds) between adjacent stationary steps. The advantages are:

- All existing rules are met concerning disturbances on the mains.
- Any desired power step between 5 % and 100 % is possible.
- There is a "soft start" of the lamps.

Results

The most noticable improvement of this new design is the quicker thermal reaction in comparison with common electric plates. The consumer benifits are:

- Improved control: Milk doesn't overflow after switching off. The pan cools down faster after the steak is brown.
- Approx. 20 % saving in energy because of less overshoot and finer power control when simmering.
- The power distribution is excellent even at low wattages.
- All kinds of pots may be used, even glass pots which are more critical with common electric cooking units. Uneven pot bottoms are less critical. Shiny metal pots may also be used - in the worst case they still act better than with common electric cooking plates.
- There is a direct reaction for the user: One sees the light and immediately feels the heat radiation if the device is switched on.

AE88 Cooking by Light- A Revolutionary Approach to Electric Hobs, F.J. Adams & O.T. Evans. IEEE Transact. on Industry Appl., Vol. 24, No. 1, p. 33 -37

IE82 IEC555-3 (1982) Disturbances in supply systems caused by household appliances and similar electric equipment, Part 3: Voltage Fluctuations, Genève

KE89 Small Voltage Fluctuations by the Management of Short Multicycle Control, R. Kersten, Proc. of the 3rd Europ. Conf. on Power Electronics and Applications (Aachen 1989), Vol. II, p. 771 - 776

NON-EQUILIBRIUM EFFECTS IN THE RECIRCULATING GAS FLOWS IN INCANDESCENT LAMPS.

M. Makhlouf, R. Devonshire and H.F. Boysan*.
High Temperature Science Laboratories,
Chemistry Department, Sheffield University,
Sheffield S3 7HF, U.K.
*Fluent Europe Ltd., Hutton's Building, 146 West Street,
Sheffield S1 4ES, U.K.

In a prior investigation [1], coherent anti-Stokes Raman scattering (CARS) thermometry was used to reveal the presence of non-equilibrium effects between the vibrational (vib) mode and the translational/rotational modes (tr) in recirculating N_2 gas flows present in an optical cell containing an incandescent filament (N_2 is a constituent of the filling gas in GLS incandescent lamps). The CARS experiment revealed large, spatially dependent, differences in the mode temperatures T_{vib} and T_{tr} of the nitrogen gas.

The stationary non-equilibrium condition established in such optical cell appears to provide a unique opportunity to explore quantitative aspects of non-equilibrium effects which are found in a variety of situations such as electrical discharges and supersonic nozzle flows.

We have thus carried out a modelling investigation of the recirculating fluid flows in these extreme conditions of temperature and temperature gradient to see if the magnitudes and the spatial distribution of the energy mode non-equilibrium effects could be understood and reproduced.

The governing hydrodynamic equations for such systems were formulated on the basis of Andersen's [2] formal treatment for a system where the vibrational temperature is different from the translational temperature. The continuity and momentum equations are identical to those for conditions of local thermodynamic equilibrium (LTE), and depend on the translational/rotational temperature (T_{tr}). The energy equation, however, is split into two mode-specific equations coupled by a source term (S) that accounts for intermode energy transfer. The gas-filament interface is a potentially an additional source of mode specific effects. The gas flow near the filament is slightly rarefied, therefore, velocity slip and temperature jump effects were included in the model as an option for treating the boundary conditions at this interface.

The hydrodynamic equations were solved by patching an appropriate subroutine into the FLUENT [3] fluid flow and heat transfer modelling code. This approach has the advantages of using FLUENT's general purpose solver, user friendly menu-driven interface, and powerful built-in graphics.

The modelling problem involves several interesting areas of energy transfer phenomena both in the bulk gas phase (mode-specific transport properties, intermode relaxation times, etc) and at the gas-filament interface (mode-specific Knudsen effects). Amongst the range of input data necessary to solve the hydrodynamic equations, the mode-specific thermal conductivities pose a particular problem because their theoretical basis is inadequately developed.

Initial predictions of the model of the ΔT ($\Delta T = T_{vib} - T_{tr}$) distribution which used the expressions of Thomson [4] for calculating λ_{vib} and λ_{tr} gave poor agreement with experimental results. We have investigated the reasons for this failure by looking more closely at the specific roles of the following in influencing the ΔT distribution: (i) intermode relaxation times; (ii) mode-specific thermal conductivities; (iii) mode-specific Knudsen effects.

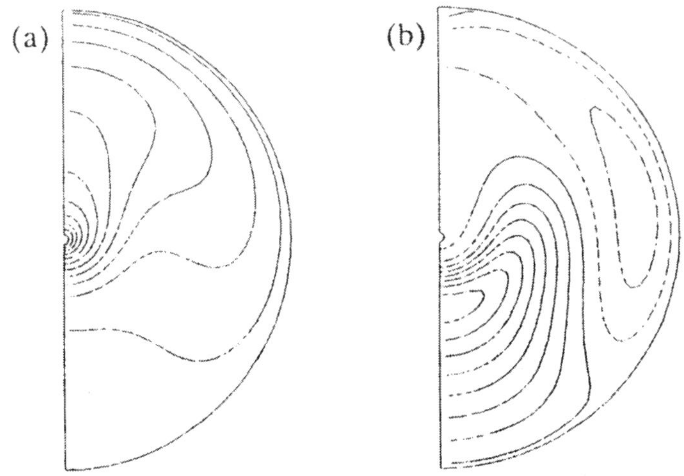

Figure 1: (a) Contours of T_{tr}; max=2670K, min=382K, spacing=254K. (b) Contours of ΔT; max=380K, min=-112K, spacing=55K (dashed contours represent negative values of ΔT).

Our investigations show clearly that the mode-specific thermal conductivities have a dominant effect on both the spatial distribution and magnitude of the non-equilibrium effects. **Figure 1** shows a set of results for the T_{tr} and ΔT spatial distributions where the total equilibrium thermal conductivity (λ_{eq}) over the entire temperature range of interest was divided between the modes by an arbitrary, uniform, percentage. The results in **Figure 1** are for the case where $\lambda_{vib} = 40\%\lambda_{eq}$ ($\lambda_{tr} = 60\%\lambda_{eq}$), the case which on the basis of this simple arbitrary division of λ_{eq} gives the best agreement with the experimental results.

The present model promises to be of considerable value as it provides a sound basis for the design of special devices for experimentally accessing, for the first time and in a quantitative manner, mode-specific gas transport properties and mode-specific accommodation coefficients. The model is also a basis for interpreting studies of chemical non-equilibrium effects in light sources and related systems.

References

[1] I.S. Dring, R. Devonshire et al, Chem.Phys.Let., 132 (3), (1986), 283-90.

[2] H.L. Andersen in 'Kinetic Processes in Gases and Plasmas', ed. A.R. Hochstin, Academic Press, 1969.

[3] FLUENT is available from Fluent Europe Ltd., Sheffield, U.K.

[4] R.M. Thomson, J.Phys.D: Appl.Phys., 11, (1978), 2509-2516.

Session 5

Chairman: D. O. Wharmby

Modelling/Diagnostics

55:I **B. Nyíri**
Notes on local equilibrium in high pressure discharges

56:P **K. E. Brown, L. Cifuentes, D. A. J. Mottram, S. A. Mucklejohn, B. Preston**
The role of modelling in new product development

57:P **P. Y. Chang, W. Shyy**
Three-dimensional natural convection in horizontal high pressure discharge arc

58:P **K. Charrada, G. Zissis, M. Stambouli, A. Asselman**
On the possibility to simulate the mercury discharge lamp during the start-up phase

59:P **S. Drawert**
Barium absorption spectroscopy in the electrode region of fluorescent lamps

60:P **S. Holló, B. Nyíri**
On the cathodic modes in discharge lamps

61:P **H. Kawahara, K. Wani**
Time-dependent computer model of a high-pressure mercury lamp

62:P **I. Pérès, L. C. Pitchford, J. P. Boeuf, H. Gielen, P. Postma**
Current – voltage characteristics in low pressure cold cathode argon discharges

63:P **M. Stambouli, A. Asselman, K. Charrada, N. Sewraj**
Experimental parameters determination in the start-up phase of a high pressure mercury discharge

64:P **J. F. Waymouth**
Integration of the Elenbaas-Heller equation

65:L **A. J. Page**
A study of electrical characteristics of deuterium lamp cathodes

66:L **D. Karabourniotis, E. Drakakis, A. Palladas**
Determination of excited-state densities in non-LTE high pressure discharges

Notes on Local Equilibrium in High Pressure Discharges

B. Nyiri
Tungsram, LPD., Váci út 77. H-1340 Budapest, Hungary

1. Introduction

In the physics and chemistry of discharges a vast amount of non-equilibrium phenomena are known. These include chemical reactions in different phases, multi-temperature systems (electron temperature), excited atomic and molecular (radical) species, radiation (optically thick or thin), electric field induced processes, various non-equilibrium (electric, inertial, transient) sheaths and sub-sheaths, ...

Statistical/kinetical methods proved to be very flexible in handling all details of this great variety of phenomena. It is noteworthy, however, that a considerable part of the efforts is exerted in the sophisticated determination/calculation of coefficients of 'good old' continuum equations. So, in most of this work we would like to concentrate on some simple thermodynamical concepts, observe the texture of equations rather than refine each and every term and coefficient in them. To this end we will even overlook important non-linearities, having a crucial role e.g. in the process of discharge mode switching, and use the so called quasilinear approximation of thermodynamics[1].

1. Non-Equilibrium and Temperature

One of the simplest formulations of a body being in equilibrium is that its state does not change if it is left alone. Problems can arise only when we look kinetically at what is meant by 'left alone' and 'state'; not mentioning how it can actually be done. Here what is 'left alone', are energy and matter content, besides unchanged geometrical boundary.

The next step is the equilibrium of two bodies.

A relatively simple interaction (relative e.g. to 'chemical') is thermal interaction. It is obvious that if any two bodies are brought into thermal, and only thermal, interaction then either there is a heat flow from one to the other, or there is not. Now, if we consider only such bodies that 'do not change their state if left alone' - i.e. that are in equilibrium - then thermal interaction gives an ordering on these bodies: if A is hotter than B and B is hotter than C than A is hotter than C; if A is as hot as B, and B as C then A is as hot as C ...; so, hotness can be represented by numbers and their relations 'greater than' and 'equal'.

We cannot, of course, expect such a decent behavior from bodies that are not in equilibrium themselves. We have to specify further details of the interaction: it does matter which end of the fire iron I grab!

There are more complicated non-equilibrium cases, however, when regarding a sufficiently small part of the body does not resolve the ambiguity. To forget about lamps, let us just put a piece of porous platinum into an $H_2 - O_2$ mixture: the platinum gets much hotter than a piece of, say, iron would in its place (as it catalyses the oxidation of H_2). Note that neither the platinum nor the iron is altered chemically. It is the environment that is out of equilibrium. Just like when we stick our thermometer out the window on a bright summer morning: depending on the reflectance of its surface (black or white) the value shown, the temperature of the thermometer, can be anywhere in a, say, 15 C^o range.

The bodies used as thermometers here were in equilibrium bodies. The experimental temperature was interaction dependent, so non-unique. One may say that these examples are too artificial, and that it is intuitively clear that there is a single temperature in each case. Behind such argumentation, however, there is usually a special thermometer, relying on some carefully chosen interaction: a piece of iron in the first, shielded mercury in the second case. What is to be stressed is that we *never* have such problems in the case of equilibrium bodies. For non-equilibrium systems there is no temperature, in the sense that the direction of energy flow between interacting body pairs does not give an ordering on this set of bodies. Which, of course, does not mean that nothing can be done: we can assign a temperature to each different type of thermal interaction. It is just harder to keep the fact hidden that all definitions are made on purpose.

Electron temperature is perhaps a little bit different. It is a statistically defined quantity, a given functional of the distribution function, so it is not *automatically* related to a thermal interaction of some kind.

***

We say that a body is in equilibrium if keeping its geometrical boundaries, energy and material contents unchanged it does (or would) not change its state. A body is said to be in local equilibrium at a point if a sufficiently small neighborhood of the point is in equilibrium.

The state of an equilibrium (fluid) body is entirely determined by its energy contents and volume, or temperature and pressure.

3. Thermodynamics of Non-Equilibrium Gases

In order to be able to discuss thermal and chemical equilibrium - non-equilibrium properties of high pressure discharges let us assume the entropy function

$$S = S(U_g, U_e, V, N_i) \qquad i = 0, 1, ..., n; \tag{1}$$

where the internal energy U is split into a part U_g belonging to the gas, and a part U_e attributed to the electrons ($U = U_g + U_e$); V is the volume, and N_i are the different components of the gaseous mixture: including not only the chemically different species, electrons and ions, but allowing distinction between ground level and excited states as well. Now, it seems plausible to define the gas and electron temperature, T_g and T_e respectively, as

$$\left(\frac{\partial S}{\partial U_g}\right)_{U_e} = \frac{1}{T_g} \qquad \left(\frac{\partial S}{\partial U_e}\right)_{U_g} = \frac{1}{T_e} \tag{2}$$

Admitting that in the continuum manifold of possibilities there must be certainly many other choices, which e.g. put more emphasis on applications, or conform some principles better, here we shall regard T_g as the temperature and denote it by T. This choice can be expressed by writing the entropy function in the form

$$S = S(U - U_e, U_e, V, N_i) \tag{3}$$

whence

$$\left(\frac{\partial S}{\partial U}\right)_{U_e} = \frac{1}{T} \qquad \left(\frac{\partial S}{\partial U_e}\right)_{U} = \frac{1}{T_e} - \frac{1}{T} \tag{4}$$

So, introducing the quantity ξ as

$$\xi = \frac{T_e - T}{T_e} \implies \left(\frac{\partial S}{\partial U_e}\right)_{U} = -\frac{\xi}{T} \tag{5}$$

the differential of entropy becomes

$$dS = \frac{1}{T}dU - \frac{\xi}{T}dU_e + \frac{p}{T}dV - \frac{\mu_i}{T}dN_i \tag{6}$$

(Summation to subscripts i, j, k, α, β in factors of products is implicit; i, j, k always refer to components, while α and β to reactions.) Hence the change of internal energy is

$$dU = TdS + \xi dU_e - pdV + \mu_i dN_i \tag{7}$$

whereas that of free enthalpy reads

$$dG = \mu_i dN_i - SdT + Vdp + \xi dU_e \tag{8}$$

Since, perhaps, U_e is not as convenient a variable as the electron temperature or ξ, it may be useful to introduce further thermodynamical potentials by the usual Legendre transformation technique:

$$G^* = G - \xi U_e, \quad and \quad H^* = G^* + ST \tag{9}$$

Thus, for example,

$$dG^* = \mu_i dN_i - SdT + Vdp - U_e d\xi \tag{10}$$

The chemical potentials μ_i, which normally are the partial molar free enthalpies, now become partial molar G^*, at constant gas and electron temperature and pressure.

***

The introduction of the entropy density s, in order to apply the second law to a continuum, is made rather easy by the fact that the entropy function (1) is a homogeneous linear one:

$$s = s(u - u_e, u_e, n_i) = S\left(\frac{U - U_e}{V}, \frac{U_e}{V}, 1, \frac{N_i}{V}\right) \tag{11}$$

To build a balance equation for the entropy we need to define the entropy current density [2,3], which we take as

$$J_s = \frac{1}{T} J_q - \frac{\xi}{T} J_{u_e} - \frac{\mu_i}{T} J_i \tag{12}$$

The currents J in this expression are defined by the next energy and component balances

$$\frac{\partial u}{\partial t} + div(J_q) = \sigma_u, \qquad \sigma_u = Ec_i J_i \tag{13}$$

$$\frac{\partial u_e}{\partial t} + div(J_{u_e}) = \sigma_{u_e} \tag{14}$$

$$\frac{\partial n_i}{\partial t} + div(J_i) = \sigma_i, \qquad \sigma_i = \nu_{i\alpha} r_\alpha \tag{15}$$

Here E is the electric field, c_i are the charges of the components, $\nu_{i\alpha}$ are the stoichiometric coefficients, and r_α are the rates of chemical reactions. In these equations, as everywhere in this work from here on we ignore convection and consider all but one, the independent, concentrations only.

The general formulation of the Second law of Thermodynamics for a continuum can be given by

$$\frac{\partial s}{\partial t} + div(J_s) = \sigma \geq 0 \tag{16}$$

which in our case, after some calculation, leads to

$$\begin{aligned}\sigma &= J_i\left(-\nabla\frac{\mu_i}{T}+\frac{e_iE}{T}\right)+J_q\nabla\left(\frac{1}{T}\right)-J_{u_e}\nabla\left(\frac{\xi}{T}\right)- \\ &-\sigma_{u_e}\left(\frac{\xi}{T}\right)-r_\alpha\left(\frac{A_\alpha}{T}\right)\geq 0 \end{aligned} \tag{17}$$

where the A_α chemical affinities are defined as

$$A_\alpha = \nu_{i\alpha}\left(\frac{\mu_i}{T}\right) \tag{18}$$

Since the entropy production must be non-negative, the process rates J_i , J_q , J_{u_e} , σ_{u_e} , r_α in the above expression cannot be independent from their coefficients, called thermodynamical forces. Still due to the non-negativity of the entropy production the forces must vanish in equilibrium, which, on the other hand, is defined by the vanishing of the process rates. So the behavior of a system for which our assumptions hold is described, in linear approximation, by constitutive equations of the form

$$J_q = L^v_{qq}\nabla\left(\frac{1}{T}\right)-L^v_{q\xi}\nabla\left(\frac{\xi}{T}\right)-L^v_{qk}\left(\nabla\frac{\mu_k}{T}-\frac{e_iE}{T}\right) \tag{19}$$

$$J_{u_e} = L^v_{\xi q}\nabla\left(\frac{1}{T}\right)-L^v_{\xi\xi}\nabla\left(\frac{\xi}{T}\right)-L^v_{\xi k}\left(\nabla\frac{\mu_k}{T}-\frac{e_iE}{T}\right) \tag{20}$$

$$J_i = L^v_{iq}\nabla\left(\frac{1}{T}\right)-L^v_{i\xi}\nabla\left(\frac{\xi}{T}\right)-L^v_{ik}\left(\nabla\frac{\mu_k}{T}-\frac{e_iE}{T}\right) \tag{21}$$

and

$$r_\alpha = -L^s_{\alpha\beta}\left(\frac{A_\beta}{T}\right)-L^s_{\alpha\xi}\left(\frac{\xi}{T}\right) \tag{22}$$

$$\sigma_{u_e} = -L^s_{\xi\beta}\left(\frac{A_\beta}{T}\right)-L^s_{\xi\xi}\left(\frac{\xi}{T}\right) \tag{23}$$

The Onsager Reciprocity Relations require that the coefficient matrices of L^v and L^s be symmetric, while the second law demands that these matrices be positive definite.

Local Thermodynamical Equilibrium means in our case that ξ and all A_α vanish, i.e. the electron and gas temperatures are equal and the affinities are zero. The first condition can be called local thermal equilibrium, while the second local chemical equilibrium.

These equations are, from the thermodynamical point, quasilinear equations: the coefficients L are allowed to depend on the state variables only, and not on the thermodynamical forces [1]. This is a restriction with respect to everyday practice in the case of chemical kinetics only. Yet, even there the power of the linearised equations is underestimated sometimes. To get the usual chemical kinetic equations for ideal mixtures in local thermal equilibrium the affinity A_β in the first term of (22) should be substituted with the expression $(\exp(A_\beta) - 1)$ [4]. This is, however, beyond our present scope.

***

It is perhaps easier to start with the discussion of equations (22) and (23). The $L^s_{\alpha\beta}$ terms describe chemical reactions. The diagonal terms represent the 'normal' flow of reactions, while the off diagonal terms stand for coupled processes. The latter can be rather important in gas discharges, where a lot of very active, 'energy rich', species can be present (ions, excited atoms and molecules). The collision with such an excited or recombining particles can result, beside the main process of deactivation or recombination, in the production of an active particles of different kind, or can facilitate some sort of dissociative process. The kinetical background of coupling of chemical processes to the thermal equilibration of the electron gas ($L^s_{\xi\beta}$) is qualitatively also rather trivial.

Equations (19), (20), and (21), stand for coupled heat conduction and diffusion. Perhaps the separate conduction equation for electron energy (equ. (20) for J_{u_e}) is worth a few comments. Since J_q in equation (19) represents all of the transported energy, the role of the second equation (i.e. of 20) is to account for the transport of (thermal) non-equilibrium. In thermal equilibrium, however, when ξ and its gradient both vanish, equation (20) is not an independent one, J_{u_e} is a linear combination of J_q and the J_i-s.

***

In order to get equations resembling more the conventional equations of diffusion and heat conduction we need to introduce a few quantities and go through some mathematical manipulations. These derivations can be regarded as an extension of the classical one (see e.g. [5]) adding effects of the presence of the different electron temperature characteristic to plasmas.

Let us define the quantities Q_i and C_i by the equations

$$L^v_{i\xi} = L^v_{ik} Q_k \qquad L^v_{i\xi} = L^v_{ik} C_k \tag{24}$$

With their help equations (21) can be written as

$$J_i = L^v_{ik} \left[Q_k \nabla\left(\frac{1}{T}\right) - C_k \nabla\left(\frac{\xi}{T}\right) - \nabla\left(\frac{\mu_k}{T}\right) + \frac{e_i E}{T} \right] \tag{25}$$

If we multiply these equations with Q_i and subtract them from eq.(19), repeat the same procedure using C_i and eq.(20), then, after executing some of the differentiations, we get

$$J_q - Q_i J_i = \frac{L^*_{qq} - \xi L^*_{q\xi}}{T^2} \nabla T - \frac{L^*_{q\xi}}{T} \nabla \xi \tag{26}$$

$$J_{u_e} - C_i J_i = \frac{L^*_{q\xi} - \xi L^*_{\xi\xi}}{T^2} \nabla T - \frac{L^*_{\xi\xi}}{T} \nabla \xi \tag{27}$$

The quantities Q_i can still be interpreted as heats of transfer [5,6], while C_i have similar meaning for that part of internal energy which is attributed to the electrons, u_e. (The coefficients L^* are simple, but lengthy, expressions of L^v , Q and C.)

In our case the chemical potentials μ_i are functions of the temperature T, pressure p, the concentrations n_i, and of the electron temperature, so

$$\nabla \mu_i = -s_i \nabla T + v_i \nabla p - u_{e,i} \nabla \xi + \frac{\partial \mu_i}{\partial n_j} \nabla n_j \tag{28}$$

where the partial molar entropy s_i, the partial molar volume v_i, and the partial molar electron energy $u_{e,i}$ are defined by the corresponding second derivatives of G^* (see equ. 10). So, with the use of the partial molar enthalpy $h_i = \mu_i + T s_i$ the diffusion equations (21) become

$$J_i = L^v_{ik} \left(\frac{h_k - Q_k + \xi C_k}{T^2} \nabla T + \frac{u_{e,k} - C_k}{T} \nabla \xi - \frac{v_k}{T} \nabla p - \frac{1}{T} \frac{\partial \mu_k}{\partial n_j} \nabla n_j + \frac{e_i E}{T} \right) \tag{29}$$

Hence it is seen that the deviance of the heats of transport Q_k and C_k from the partial molar heats h_k and $u_{e,k}$, which belong to local processes (chemical reactions), is a measure of thermal diffusivity.

4. The Effect of Temperature Gradient on Local Equilibrium

Since the affinities and ξ can be regarded the measure of local non-equilibrium, we express their gradients by the currents and the temperature gradient. With equation (27) there is not much to do:

$$\nabla \left(\frac{\xi}{T} \right) = \frac{L^*_{q\xi}}{L^*_{\xi\xi}} \nabla \left(\frac{1}{T} \right) - \frac{J_{u_e} - C_i J_i}{L^*_{\xi\xi}} \tag{30}$$

To express the gradients of affinities from equation (25) we first multiply it with the inverse matrix R^v_{ik} of L^v_{ik}, then with the stoichiometrical coefficients $\nu_{i\alpha}$. Thus, making use of the notation

$$Q_\alpha = \nu_{k\alpha} Q_k \qquad C_\alpha = \nu_{k\alpha} C_k \tag{31}$$

and of the fact that

$$0 = \nu_{k\alpha} e_k \tag{32}$$

viz. charge is conserved in chemical reactions, too, we can write

$$\nabla\left(\frac{A_\alpha}{T}\right) = Q_\alpha \nabla\left(\frac{1}{T}\right) - C_\alpha \nabla\left(\frac{\xi}{T}\right) - \nu_{i\alpha} R^{v}_{ik} J_k \tag{33}$$

Let us consider now the near wall region of a discharge lamp. If the wall is impenetrable to material fluxes and the possibility of chemical reactions on the wall surface is ignored then the material fluxes J_i vanish there. In local equilibrium the gradients of affinity and ξ are also zero. According to equations (30) or (33) this is possible only if the temperature gradient is zero as well. From this we can immediately conclude that in the presence of temperature gradient there can be no local equilibrium in the neighborhood of a heat conducting, but otherwise insulating wall.

This argumentation is not seriously weakened by allowing reactions to take place on the wall surface, since the rate of these depends only on the value of the temperature, and not on its gradient.

It is interesting that, due to the conservation of electric charge, the electric field cannot directly cause affinity gradient, chemical non-equilibrium.

***

It is not too easy to see directly the quantitative consequences of the fundamental set of equations (13)...(15), (19)...(23),on these near wall phenomena. Let us [illegible]

In a one dimensional closed steady state system the material currents must obey the same constraints as the source densities do, so, analogously to the second group of equations (15) we can write

$$J_i = \nu_{i\alpha} J_\alpha \tag{34}$$

So, denoting the inverse of the matrix $\nu_{i\alpha} R^{v}_{ik} \nu_{i\beta}$ by $\Lambda_{\alpha\beta}$, and making use of equations (15) and (33) we get for the quantities measuring local non-equilibrium (ξ, A_α) the system of equations

$$\nabla \cdot \left[\Lambda_{\alpha\beta}\left(Q_\beta \nabla \frac{1}{T} - C_\beta \nabla \frac{\xi}{T} \nabla - \frac{A_\beta}{T}\right)\right] = -L^{s}_{\alpha\beta}\left(\frac{A_\beta}{T}\right) - L^{s}_{\alpha\xi}\left(\frac{\xi}{T}\right) \tag{35}$$

$$\nabla \cdot \left[C_\alpha \Lambda_{\alpha\beta}\left(Q_\beta \nabla \frac{1}{T} - C_\beta \nabla \frac{\xi}{T} \nabla - \frac{A_\beta}{T}\right) + L^{s}_{q\xi} \nabla \frac{1}{T} - \frac{L^{s}_{\xi\xi}}{T} \nabla \frac{\xi}{T}\right]$$

$$= -L^{s}_{\xi\beta}\left(\frac{A_\beta}{T}\right) - L^{s}_{\xi\xi}\left(\frac{\xi}{T}\right) \tag{36}$$

This last one can be omitted if the electron temperature is equal to the gas temperature. What is important from these, is that if the gradients are not too great the coefficients can be regarded as constant. Then these equations for A/T and ξ/T beside a given $1/T$ distribution are a matrix equivalent of the equation

$$l^2 y'' = y + f(x) \tag{37}$$

The diffusivities and rate constants together determine a set of characteristic distances similar to l of (37). The inhomogeneity of the equation (the perturbation $f(x)$) is proportional to the product of the temperature gradient and the reaction heat.

Trying to make an analogy, we can say that as the mean free path is the result of the process pair 'free flight' and 'collision', so are these characteristic distances the result of process pairs 'diffusion' and 'reaction'.

5. Examples

The best developed example of these temperature gradient 'induced' chemical sheath phenomena is one presented by Fischer in his model of high-pressure discharge lamp model. The reaction process is the ionization - recombination pair in the near cathode region. The equivalent of equation (35) is 'hidden' in his differential equation for the electrical conductivity. The inclusion of non-equilibrium ionization was necessary to get a realistic description beside the 10^9 K/m temperature gradient [7].

* * *

The characteristic length belonging to the heat conduction in the electron gas and the energy exchange of the electrons with the heavy particles is [8]

$$l = l_0 \cdot \sqrt{\frac{5\pi}{32} \frac{M}{m_e}}, \tag{38}$$

as expressed with the mass m_e and mean free path of electrons l_0, and the mass of heavy particles M. This uncoupled equation predicts a few hundred mean free path for l. However, a much thicker region of non-equilibrium electrons can be found in high-pressure sodium lamps, which is caused by the coupling, the energy exchange with (radiation) excited sodium atoms [9].

***

We can try to apply equation (35) to the decomposition of sodium iodide (n_3) to atomic sodium (n_2) and iodine (n_1). Taking $L^*_{\alpha\alpha}$, which is the absolute reaction rate in equilibrium ($k^+n_3 = k^-n_1n_2$), and the diffusion of the reaction partners into account, one gets for the characteristic length

$$l^{-2} = k^+n_3 \left(\frac{1}{n_1D_1} + \frac{1}{n_2D_2} + \frac{1}{n_3D_3}\right) \tag{39}$$

Still in the linear approximation the deviance of sodium concentration from its equilibrium value in case of insulating and non reacting wall is

$$D_in_i^\bullet = l\left(\frac{1}{n_1D_1} + \frac{1}{n_2D_2} + \frac{1}{n_3D_3}\right) \Delta H \mathrm{grad}\left(\frac{1}{T}\right) \tag{40}$$

Here ΔH is the reaction heat, the partial molar enthalpy change h_α.
For the calculation of k^+ the hard sphere model was used, each collision with sufficient energy leading to dissociation. This is an overestimation of the reaction rate, likely, by a factor of few tens [10]. The consequence is the under estimation of both the characteristic length and the distance from equilibrium. The consideration of wall reactions, on the other hand, moves results towards equilibrium, while it cannot alter the characteristic length, of course.

On fig.1. the characteristic length is plotted as a function of temperature for mercury pressures of 80, 320, 1280, and 5120 torr, the shorter length belonging to the higher pressure. It is noteworthy that at 1000K (even at 5120 torr) the 'action radius' of chemical perturbations reaches and surpasses the centimeter range. It is well above 2500-3000K, depending on the pressure, where this reaction can be handled locally.

On fig.2. the five curves versus temperature represent the wall concentrations of sodium, as predicted by the previous model. The lower running curves belong to the higher pressures, and the lowest is the equilibrium one. Although it is obvious that at the assumed $6 \cdot 10^5$ K/m temperature gradient the linearised model cannot be quantitatively valid, it is sure from the results that the deviance from equilibrium cannot be small.

Modifying the previous results by assuming a reaction taking place on the wall with an apparent activation energy 35% of the bulk process, the curves on fig.3. are obtained, all lying within a single order of magnitude range of equilibrium. Note that even if the rate constant of the surface reaction is small there can still be a significant absolute reaction rate if the amount of reactants on the surface is great. So, e.g. over the sodium iodide pool one would expect this reaction to be in equilibrium, although this view has been challenged elsewhere [11].

***

As a rather direct experiment, one can take two similar sodium iodide containing metal halide discharge lamps. Operate one of them under the usual circumstances, that results in usual enormous temperature gradient at the wall. Bake the other one in an oven, isothermally, at the wall temperature of the first, operating one; and see the difference. The iodine liberated by the wall reaction in the oven baked sample was one third of the amount found in the operated one.[12]

***

Varying the rare gas pressure in sodium iodide containing metal halide lamps one gets a higher wall temperature for the higher pressure. The local equilibrium approach predicts a higher iodine liberation rate for the higher temperature, and it is insensitive to the rare gas pressure. The actual result, despite the approximately 100 K temperature increase, is just the contrary, as it can be seen in fig.4. [13]. The tendency fits perfectly into the non-equilibrium theory, since the increase in the inert gas pressure decreases the diffusion coefficients and increases the reaction rate. Both moves the sodium density at the wall down towards the equilibrium value. The effect is apparently so strong that it surpasses the upward motion of the equilibrium value.

***

It is experimentally clear from the ionic mechanism of sodium loss that a considerable amount of sodium ions must reach the wall of a metal halide lamp [14]. Considering the temperature of the wall and the ionic dissociation energy of sodium iodide, this mechanism is rather unlikely in an isothermal system. But beside the quite huge temperature gradient the ionic to bonded sodium ratio can be orders of magnitude higher than the equilibrium value.

References

[1] Gyarmati I.:On the Wave Approach of Thermodynamics, and some Problems of Non-Linear Theories.*J. Non-Equilib. Thermodyn.,* **2** *(1977) 233.*

[2] Verhas J.: On the Entropy Current.*J. Non-Equilib. Thermodyn.* **8** *(1983), 201.*

[3] Nyiri B.: On the Entropy Current.*J. Non-Equilib. Thermodyn.* **16** *(1991) 179.*

[4] Lengyel S., Gyarmati I.:Constitutive Equations and Reciprocal Relations of Non-Ideal Homogeneous Closed Chemical Kinetic Systems. *Acta Chim. Hung. 122 (1986)*

[5] de Groot S.R.: Thermodynamics of Irreversible Processes.*North-Holland Publishing Company, Amsterdam, 1952.*

[6] de Groot S.R., Mazur P.: Non-Equilibrium Thermodynamics. *North-Holland Publishing Company, Amsterdam, 1962.*
[7] Fischer E.:Modelling of Low-Power High-Pressure Discharge Lamps *Philips J. Res.,* **42** *(1987) 58-85*
[8] Ingold J. : personal communication
[9] de Groot J.J.,van Vliet J.A.J.M: The High-Pressure Sodium Lamp.*Philips Technical Library, Kluwer Technischen Boeken B.V.- Deventer - Antwerpen 1986. pp. 60-61*
[10] Jordan P.C.: Chemical kinetics and transport. *Plenum Press, New York, 1980.*
[11] Cifuentes L., Forsdyke G.M. and O'Brien N.W.: Corrosion of tungsten in metal halide lamps : a review. *Accepted for publication in Corrosion Science (1992).*
[12] Dakin J.T., Roberts V. et al.: personal communication
[13] Hollo S. : personal communication
[14] Nagel F. : On the Ionic Mechanism of Sodium Loss. *Proceedings of the Am.El.Chem.Soc. High Temperature Lamp Chemistry meeting. pp. 18-21, (1988), Atlanta.*

fig.1.

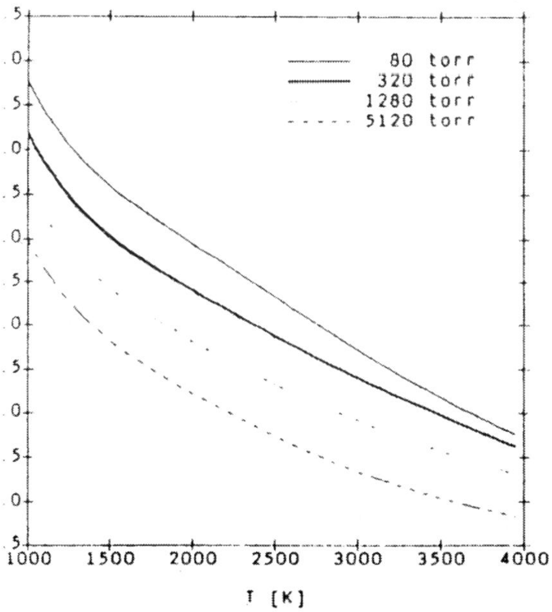

fig.2.

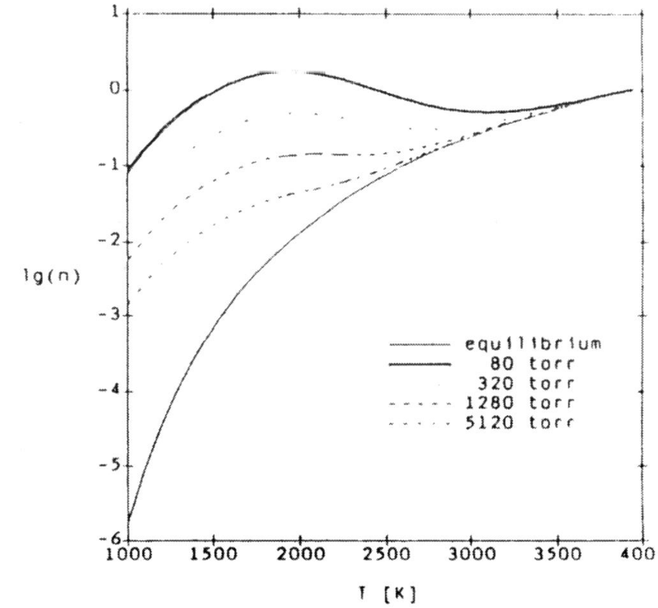

fig.3.

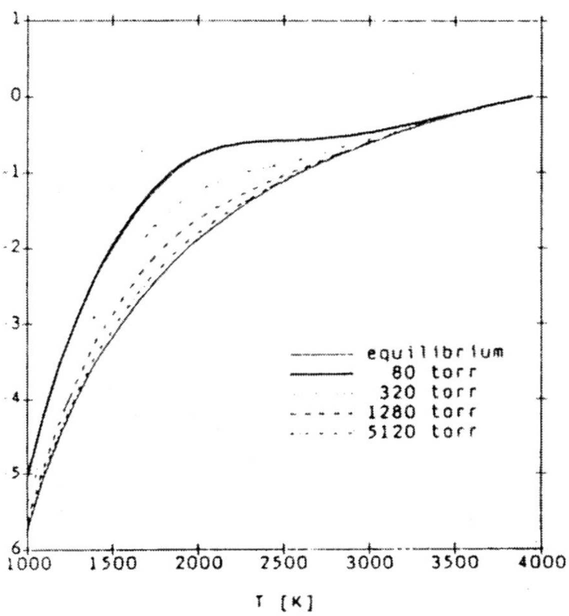

fig.4.

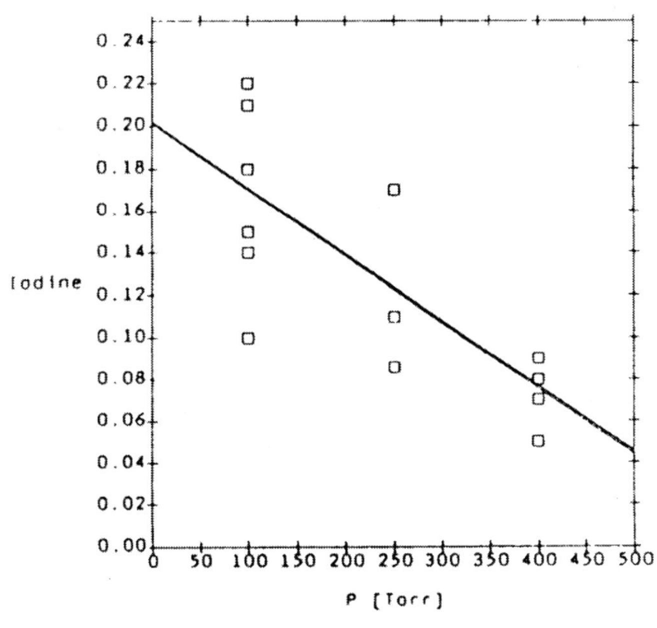

THE ROLE OF MODELLING IN NEW PRODUCT DEVELOPMENT

K.E.Brown, L.Cifuentes, D.A.J.Mottram, S.A.Mucklejohn and B.Preston
GE Thorn Lamps Limited, Technology & Engineering Division,
Melton Road, Leicester LE4 7PD, U.K.

This paper aims to illustrate the way in which a range of modelling techniques have been used to support development of a new high-pressure discharge lamp (1). The required properties of the new source were:

power	250 W
efficacy	>75 lumens per Watt
CCT	4000 K
R_a	>70
burning position	horizontal
dose	Hg, Na, In, Tl, Br, I

The original questions to answer were:

(a) can the dose elements satisfy the required light output characteristics ?
(b) can they give the necessary spectral efficacy ?
(c) can the dose produce the required vapour composition ?
(d) will the lamp performance be sensitive to sodium loss from the dose ?
(e) which lamp parameters most affect the lamp efficiency ?
(f) what is the effect of arc tube geometry ?

Points (a) and (b) were tackled by spectral mixing; (c) to (e) by a combination of thermodynamic modelling of multicomponent metal halide systems and calculated predictions using a 1-D spectral model of a high-pressure discharge and (f) by 3-D modelling of the same system. Theoretical modelling was carried out in support of selected experimental work throughout.

Spectral mixing: Four lamps were made and dosed with {Na-In-X}, {Na-Tl-X}, {In-Tl-X} and {Na-In-Tl-X} (X = Br or I). The spectral outputs of these lamps were measured at a number of power loadings. By careful matching of spectra, it was possible to extract the contributions of the single components Na In and Tl. These unary spectra were then combined by computer in varying proportions and the colour properties of the resulting spectra calculated. The results suggested that the required colour rendering, the target colour temperature and the target efficacy were just achievable.

Thermodynamic modelling: The partial pressures above the dose were derived from models for multicomponent metal halide systems (2). The behaviour of the condensed phases in multicomponent systems can be predicted from the critically assessed thermodynamic properties of the corresponding binary systems. The equilibrium partial pressures above the multicomponent system can then be calculated for a given temperature and composition.

1-D modelling: The calculated elemental abundances and total pressure were used as input to the next step. A radial spectral model (3,4) of a high-pressure discharge lamp was used to carry out predictions of lamp efficiency given several arc tube bores, cool-spot temperatures, power loadings and mercury pressures. The weights of different lamp parameters in determining the efficiency of the lamp were estimated using Taguchi analysis (5). The sensitivity of the discharge to sodium loss was also predicted by the same model. Results showed that cool-spot temperature and power loading were the most influential of the selected parameters. Thermodynamic modelling showed that the partial pressures were relatively insensitive to sodium loss from the dose (up to 30 per cent depletion). Similarly, the corresponding changes in the predicted efficiencies were small.

3-D modelling: A 3-D energy transport model (3,6) is being used to establish which design features have the strongest influence on lamp stability and reproducibility. The effect of variations in arc current, voltage, arc tube bore and arc gap are being investigated.

Predictions were, in general, supported by experimental observations. They gave confidence to embark upon substantial lifetesting and reduced the number of prototype variants tested. This project produced a lamp, the 250 W MBI-T, with the performance outlined below:

lamp efficacy	86 lumens per Watt
spectral efficiency	320 lumens per Watt
lamp efficiency	27 per cent
CCT	4200 K
R_a	70

thus illustrating the benefits to be gained from the close interaction of experimental and theoretical approaches.

REFERENCES

(1) K.E.Brown, *Proceedings of the National Lighting Conference*, Manchester, The Chartered Institution of Building Services Engineers, (1992).

(2) S.A.Mucklejohn and A.T.Dinsdale, *6th Int. Symp. Science & Tech. Light Sources*, Budapest, Hungary (1992).

(3) L.Cifuentes, *Lighting Research & Technology*, 23, 161 (1991).

(4) D.A.J.Mottram, L.Cifuentes and S.A.Mucklejohn, *5th Int. Symp. Science & Tech. Light Sources*, editors: R.Devonshire, J.Meads and D.O.Wharmby, The University of Sheffield, p.299 (1989).

(5) *Taguchi techniques for quality engineering*, J.P.Ross, McGraw Hill (1988).

(6) L.Cifuentes, D.A.J.Mottram and S.A.Vasquez-Malebran, *5th Int. Symp. Science & Tech. Light Sources*, editors: R.Devonshire, J.Meads and D.O.Wharmby, The University of Sheffield, p.281 (1989).

THREE-DIMENSIONAL NATURAL CONVECTION IN HORIZONTAL HIGH PRESSURE DISCHARGE ARC
(arc centering by magnetic field)

P.Y. Chang and W. Shyy*
Advanced Technology Department, G.E. Lighting, Ohio 44112, USA
*Department of Aerospace Engineering,Mechanics and Engineering Science,
University of Florida, FL 32611, USA

It is well known that the natural convection plays a very important role in the performance of high pressure discharge lamps. It influences the stability and location of the discharge, and transports energy away from the discharge. In a classic study of convection in a vertical arctube, Kenty [1] measured the convection velocity. However, experimental measurements of the velocity and temperature fields inside a discharge arctube are, in general, difficult and are time consuming to obtain.

The advent of fast and powerful computers, and the advanced computational technology for numerically solving the nonlinear equations governing fluid motion, have enabled the lighting industries to rely more and more on numerically modeling of the physics and chemistry processes inside a light. Since many of the lamps are burned in a horizontal position a 3D computer model is essential to capture the intrinsic physical mechanisms involved. Several papers appeared at LS:5 which utilized either an in-house or a commercial code based on first principles to analyze the thermal and flow fields inside a high pressure discharge arc [2 – 5] or inside an incandescent lamp [6].

In [2], we reported that with a straight 400W high pressure mercury discharge arctube placed horizontally, the buoyancy effect creates an upward bowing arc and a nonuniform temperature distribution. It was demonstrated that a properly curved arctube with offset electrodes can be used to adjust the temperature field and produce a nearly uniform temperature distribution.

Kenty also reported in [1] the conditions under which rotation or an externally applied magnetic field would cause arc straightening. Since LS:5, we have extended our code to include the effects of an externally applied magnetic field and rotation. The magnetic field enters simply as the $J \times B$ source term in the momentum equations. Rotation is introduced by assigning rotating velocity to the inner quartz wall boundary. Thus the no-slip boundary condition becomes no-slip relative to a moving arctube surface. The extended code described in detail in [7] has been shown to emulate, both qualitatively and quantitatively, the observations related to arc straightening in a horizontal arctube made by Kenty.

In the present paper, the extended code is used to induce arc centering by magnetic fields. We present the following three cases:

1. Arctube with geometry and operating condition similar to [2].
   To facilitate the discussion, Fig.1 includes both the straight arctube design and the

optimized curve arctube design of [2]. Fig.1c is a straight design subject to an externally applied magnetic field. A pair of contra-rotating recirculating eddies is observed in all of them. Figures 1a and 1b basically have similar flow patterns induced by gravity. However, the curved design accommodates the intrinsic buoyancy induced transport through geometrical modification and thus achieves a nearly uniform temperature distribution. In Fig.1c, the most noticeable feature is that the particles in the middle of arctube are convected along the downward direction, i.e., a reversal of convection direction results from the imposed magnetic field and the original pair of eddies induced by the buoyancy effect has been pushed toward a region very close to the side wall and weakened. With an externally applied magnetic field of 3 gauss, the arc is completely centered. Figures 1b and 1c show two effective ways in reaching the design goals of a uniform temperature distribution.

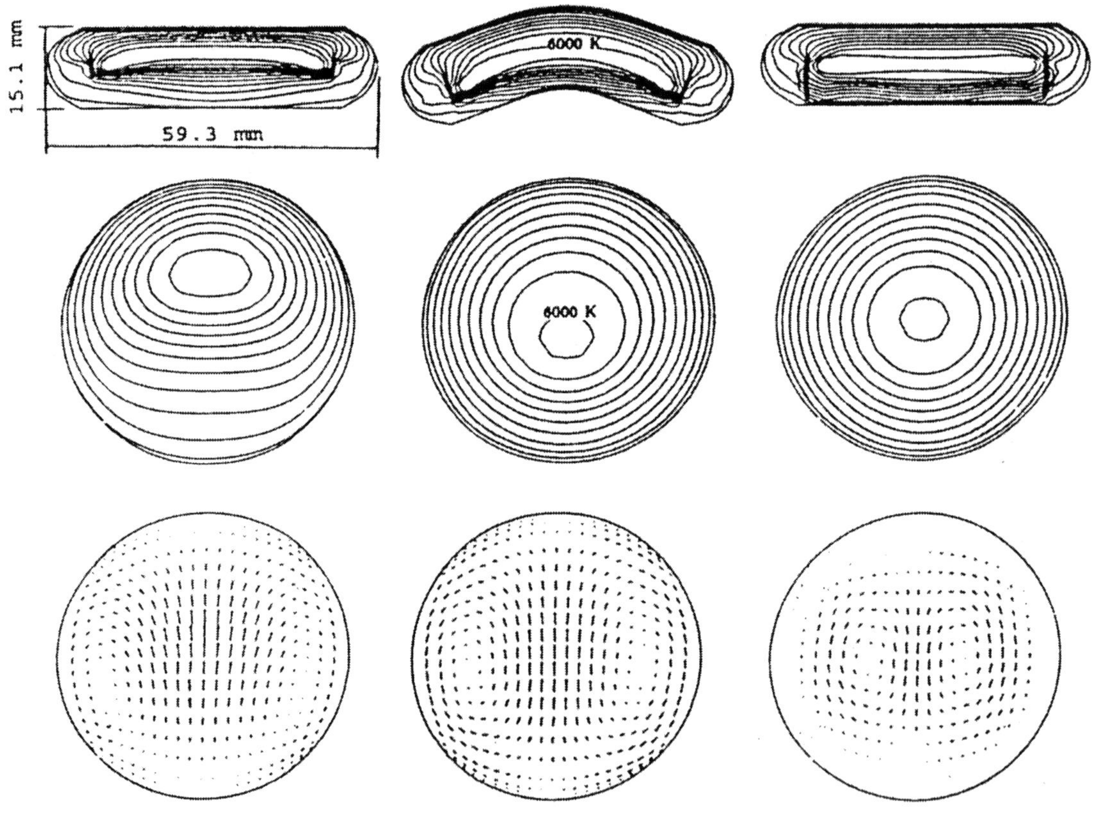

(a) $-4.5 \le V_y \le 10.5$ cm/sec magnetic field = 0.

(b) $-5.7 \le V_y \le 8.3$ cm/sec magnetic field = 0.

(c) $-7.4 \le V_y \le 6.4$ cm/sec magnetic field = 3 gauss

FIG.1 Implications of design options on thermal and convection characteristics. (a) base design at 4 atm (b) design optimization through geometrical modification (c) design optimization through external magnetic field of 3×10^{-4} Tesla.
The most inner temperature contours are 6000K and contour intervals are 500K

2. Arctube with geometry as in Fig.1c and with varying operating pressures - 3 atm, 4atm, 5 atm, and 6atm

By varying the pressure inside the same arctube, the analysis shows that the hot core remains centered as long as the magnetic strength is appropriately adjusted. Furthermore, it appears that there exists a very simple linear relationship between the tube pressure and the required magnetic strength.

3. A small 30W elliptically shaped high pressure xenon discharge

A schematic view of a small 30W high pressure xenon discharge arctube is shown in Fig.2a. The arc gap is 2.5 mm and the operating pressure is 56 atm. The numerical analysis predicts that with a magnetic field strength of 9 gauss, the arc is straightened and centered. Figs.2b and 2c show the temperature contours with and without externally applied magnetic field.

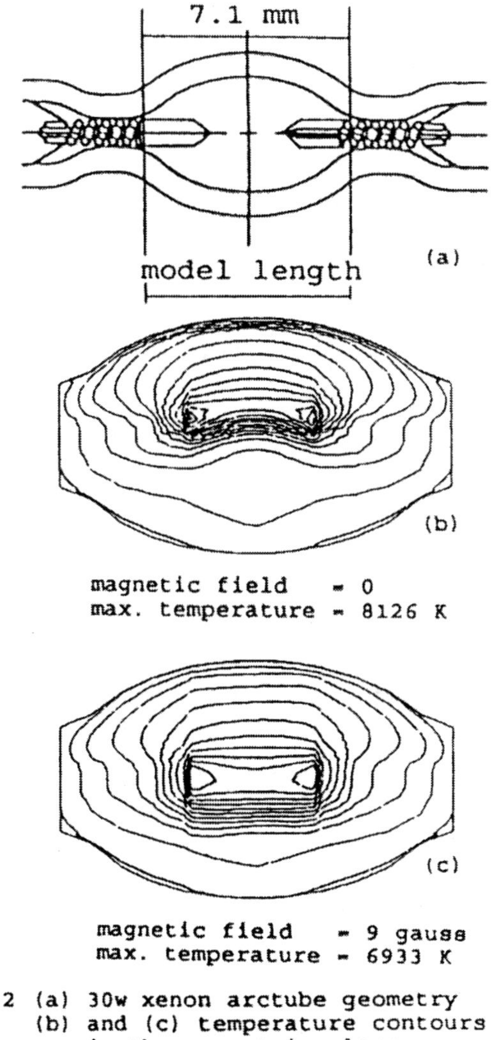

FIG.2 (a) 30w xenon arctube geometry
(b) and (c) temperature contours in the symmetric plane
(contour intervals are 500K)

REFERENCES

[1] C. Kenty, On convection currents in high pressure mercury arcs, *J. Applied Physics* 9, 53-66 (1938)

[2] P.Y. Chang, W. Shyy, J.T. Dakin, and M.E. Duffy, A study of three-dimensional natural convection in horizontal high pressure mercury lamps, *5th Int. Symp. on the Science and Technology of Light Sources*, York-England, Sept. 1989

[3] L. Cifuentes, Computer predictions of flow patterns and temperature contours in high pressure discharge lamps, *5th Int. Symp. on the Science and Technology of Light Sources*, York-England, Sept. 1989

[4] L. Cifuentes and D.A.J. Mottram, Three-dimensional computer prediction of a high pressure discharge lamp, *5th Int. Symp. on the Science and Technology of Light Sources*, York-England, Sept. 1989

[5] P.A. Vicharelli and T. Fohl, Computational fluid dynamics of high intensity arcs, *5th Int. Symp. on the Science and Technology of Light Sources*, York-England, Sept. 1989

[6] D.G. Green, R. Devonshire, S.A. Mucklejohn, and N.W. O'Brien, Fluid flow modelling as an aid in the design of incandescent light sources, *5th Int. Symp. on the Science and Technology of Light Sources*, York-England, Sept. 1989

[7] P.Y. Chang and J.T. Dakin, Effects of rotation and magnetic field on a horizontal high-pressure Hg arc, *J. Applied Physics* 69, 3763-3765 (1991)

ON THE POSSIBILITY TO SIMULATE THE MERCURY DISCHARGE LAMP DURING THE START-UP PHASE

K. Charrada, G. Zissis, M. Stambouli*, A. Asselman

Laboratoire des Décharges dans les Gaz (U.R.A.-277), 118 rte de Narbonne, 31062 Toulouse, FRANCE
* E.N.S.E.T., 5 Av. Taha Hussein, 1008 Tunis, TUNISIE.

Light sources based on high pressure electric discharge in pure mercury need an important start-up time (some minutes). The rapid evolution of the discharge's typical features during this phase may have serious consequences on the wearing out of the components of the network distribution as well as on certain applications. In this study we deal with preliminary numerical simulations of the discharge start-up.

During this phase, the mercury pressure in the discharge changes strongly, e.g. 7 orders of magnitude between the ignition moment (some 10^{-1} Pa) and the nominal discharge operation (10^{6} Pa). Discharge properties change fundamentally. Thus we deal, at the beginning with a low-pressure mercury-rare gas discharge controlled by ambipolar diffusion, and at the end with a high pressure pure Hg discharge near to the Local Thermodynamic Equilibrium. A self-consistent model describing the whole phenomenon should be very complicated because of the numerous and various physical processes that should be accounted for. In fact, in order to simplify the simulation, it is possible,as a first approximation, to divide the start-up phase in three parts:

(1) The first one is situated just after the ignition. In this part the amount of mercury in the positive column is sufficiently low. Mercury pressure is that of the saturated vapour one controlled by the tube's cold spot temperature. Recombination occurs only near the tube wall because the rare gas pressure (Ar) introduced in the lamp in order to facilitate the ignition is also very low (~10 Torr). Hence, Shottky's regime can be assumed.

(2) The second part is characterized by a rapidly increasing electric field. During this period mercury pressure is still the saturated vapour one for the tube's cold spot temperature. However, at the beginning of this region, radiative recombination and later on, three-body processes become important. Inelastic Hg/Hg collisions should be taken into account too.

(3) In the third region, the discharge is closer to the Local Thermodynamic Equilibrium conditions. An important quantity of the mercury introduced in the tube is evaporated.

In the present work we simulate only the phases (1) and (3) of the start-up by using two models describing essentially the main physical phenomena taking place during these periods. Thus, a classical self-consistent collisional-radiative model (S.C.C.R.) can be used for the description of the phase (1) and the phase (3) can be simulated by a model based on the L.T.E. concept. Assuming that the system's evolution is quite slow it is possible to use steady-state models only. Thus the time evolution of the discharge can be simulated by a succession of independent steady state situations. However, the time interval between two successive calculations must be larger than the system's characteristic times.

The S.C.C.R. model is based on the following assumptions:[1] A uniform steady state positive column established in a circular cross section tube is considered. The discharge is supposed to operate vertically with neither stationary nor running striations, nor oscillations. Under these conditions, the positive column is supposed to be axisymmetric. The discharge tube contains a constant amount of argon and some mercury vapour. The total pressure is assumed to be constant along the radius. Both mercury and argon may be (de-)excited by collisions with electrons. In this model we take into account the fundamental $Hg(6^1S_0)$ level, the first five $Hg(6^3P_0, 6^3P_1, 6^3P_2, 6^1P_1, 7^3S_1)$ excited states. We also include the fundamental state of Ar and a virtual excited state Ar* at 11.65 eV which regroups the $1s_2$, $1s_4$, $1s_3$ and $1s_5$ Ar excited states. The maintenance of electrical conductivity in the discharge plasma requires production of electron-ion pairs by e/Hg and Hg*/Hg* inelastic collisions as fast as they are lost by recombination at the tube wall, electron temperature is controlled by this balance. Then, the discharge characteristics can be deduced by solving an integrodifferential equation set which includes population and energy balance equations as well as the Ohm's and perfect gas laws. The validity of this model is verified in several cases of low pressure Hg-Ar discharges for which experimental data could be found in the literature.

Phase (3) is simulated by means of L.T.E. model based on the following hypotheses: similarly to phase (1), an axial uniform positive column is considered and the external electric field is purely axial. As before, the total pressure is constant along radius. Temperature profile is calculated from the local energy balance between the input energy and the energy losses from conduction and radiation. The local net emission depends on the temperature profile. In order to determine the local net emission we use an analytic approximation which gives the former as a function of the temperature profile, the mercury pressure and the arc diameter.[2] We also solve, the continuity equation in order to calculate the radial velocity. The local electric field strength is determined for any input current from Ohm's law. For this model the required material functions are the thermal conductivity, electric conductivity and specific heat of the arc plasma. Assuming that the discharge is in L.T.E. these quantities are known as a function of the local temperature only. The above equation set is solved, by using finite element techniques and the implicit numerical schemes. This model has been validated experimentally by using the data of Asselman et al. during the last phase of the start-up.[3]

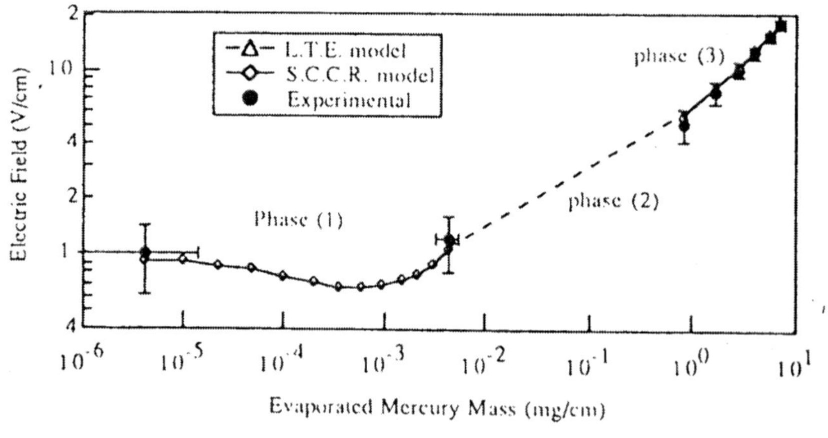

Figure 1: Electric field as a function of the evaporated mercury amount

Figure 1 illustrates our calculation results for the electric field in the positive column relative of the evaporated mercury amount. A good agreement between experimental data and calculations can be observed.

On the one hand, we observe that the S.C.C.R. model is limited in the region for which $m_{Hg}<0.05$ mg/cm. In fact, when the mercury amount in the plasma increases the Shottky's regime assumed by the model is no more valid. Hence, the S.C.C.R. model can not converge, because the e-ion loss rate as calculated under the ambipolar diffusion approximation is not sufficient and the convergence criterion (e-ion production rate = e-ion loss rate) is never satisfied.

On the other hand, the validity of the L.T.E. model is limited to higher mercury pressures ($m_{Hg}>1$ mg/cm). This is mainly due to the fact that material functions and radiation have evaluated by using the one-temperature concept imposed by the L.T.E. Furthermore, the analytical relationship used for the radiation calculations restricts the model application range too.

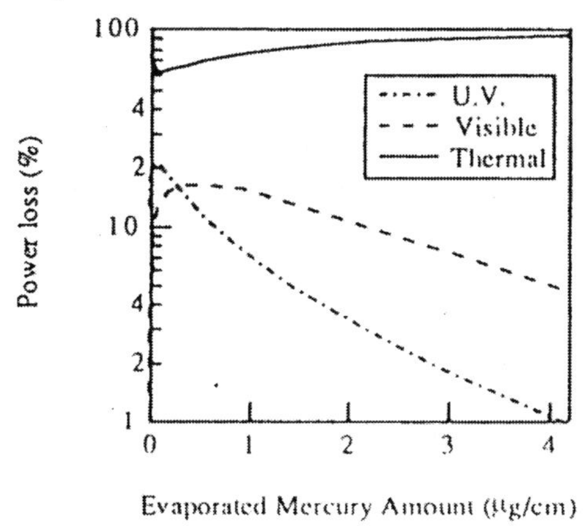

Figure 2: Power Losses during the phase (1) of the start-up

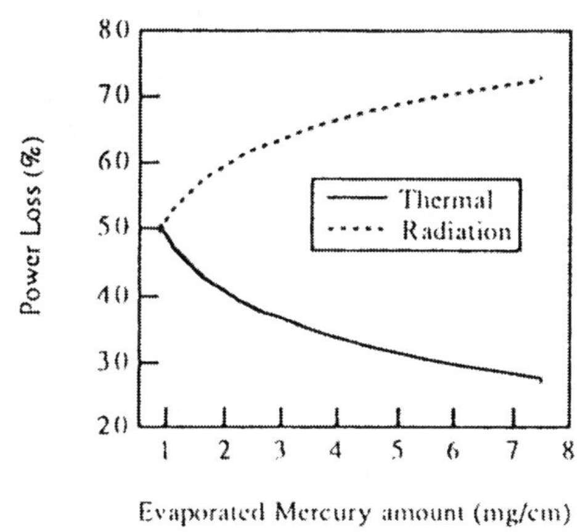

Figure 3: Power Losses during the phase (3) of the start-up

Figures 2 and 3 show the calculated principal power losses in the discharge positive column during the phases (1) and (3). We observe that thermal loss is the main process in the phase (1) but the radiation losses become the most important mechanism in the phase (3). This suggests that discharge thermalisation occurs during the phase (2). This particular phase should be studied in more details. We intend to go deeper into this study in a near future.

The authors are indebted to Professor J.J.Damelincourt for his useful remarks and E.D.F. for its financial support.

<u>References</u>

[1] G. Zissis, P. Bénétruy, I. Bernat, Phys. Rev. **A45**, 1135 (1992)

[2] H.P. Stormberg, R. Schäfer, J. Appl. Phys. **54**, 228 (1983)

[3] A. Asselman, M. Stambouli, M. Aubès, J. Salon, Proc. Xth Int. Conf. Gas Discharges Appl., Swansea (1992)

BARIUM ABSORPTION SPECTROSCOPY IN THE ELECTRODE REGION OF FLUORESCENT LAMPS

Drawert, S.
Ilmenau Institut of Technology, Dep. of Lighting Engineering and Technical Optics, 6300 Ilmenau, Germany

The determination of the life time of fluorescent lamps is increasingly studied by spectroscopical techniques. On the base of the fact that the life time of a fluorescent lamp is mainly determined by the quality of its electrodes several authers have investigated the processes of emitter loss during the preheating and the ac discharge operation /1/ /2/ /3/ /4/.

In this work the barium sublimation enthalpy and the barium spatial distribution is studied in preheating operation. It was used an atomabsorption technique from /5/, which is characterized by a simply and cheep construction. A barium hollow cathode lamp served as a resonance radiation source at λ = 553.6 nm. By this technique it is easy to measure free barium atomes which are evaproated from an electrode or dissociated in vicinity of an electrode.

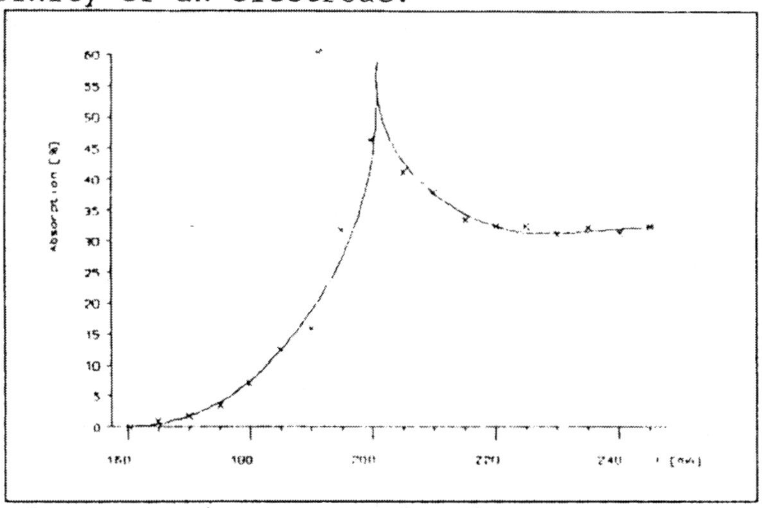

Fig. 1 Barium absorption in preheating of an 8 watt fluorescent lamp

Figure 1 shows the absorption of barium in dependence on the preheating current. It was measured at an 8 watt staff fluorescent lamp with coiled-coil filament. It is clearly to be recognized the exponential shape which is sharp broken due to discharge along the electrode during the preheating at current upper 200 mA.

The sublimation enthalpy ΔH calculated at this measurments amounts ΔH = 234 kJ/mol (56 kcal/mol), that was in good agreement with the measurements from /1/, which received ΔH = 247 kJ/mol (59 kcal/mol).

Futhermore it was measured the barium spatial distribution at a preheating current of 180 mA. This measurments allow to calculate the barium evaporation rate. So calculated values lie in the order of 1 μg/h. Unfortunately this values have an insignificant importance for discharge operation.

References

/1/ Bhattacharya, A. K.: Measurment of barium loss from a fluorescent lamp electrode by laser induced fluorescence. J. Appl. Phys. 65(12) 1989, p. 4595-4602

/2/ Bhattacharya, A. K.: Measurement of barium ion density in the vicinity of fluorescent lamp electrodes. J. Appl. Phys. 65(12) 1989, p. 4603-4607

/3/ Hilbig, R.; Chittka, U.: Barium Diagnostics in Front of Heated Fluorescent Lamp Electrodes. Paper at 5th Int. Symp. on the Science and Technology of Light Sources, York 1989

/4/ Khettal, A.: Speltralanalytische Diagnostik an Bogenentladungslampen zur Charakterisierung der Erosionsprozesse der Elektroden. Diss. A, Humboldt Uni. Berlin 1990

/5/ Drawert, S.; Riemann, M.; Jordanow, W.: Atomabsorption von Barium in Leuchtstofflampen. Contrib. Plasma Phys. 31(1991) 4,457-460

ON THE CATHODIC MODES IN DISCHARGE LAMPS

S. Holló, B. Nyiri
TUNGSRAM, LPD., Váci út 77. Budapest, H-1340 HUNGARY

There are several well distinguishable operational modes of the cathode - plasma interface in discharge lamps. All these modes can be observed in discharge lamps either in steady-state or in transient – e.g. during ignition (Table 1.) [1].

There has made considerable effort recently to reveal the details of the highly nonequilibrium processes in plasmas (e.g. Monte-Carlo simulation [2], various approximation of Boltzmann's equation [3], etc.). Our purpose was to set up a clear macroscopic framework that makes it possible to introduce the result of the novel microscopic investigations. The way of doing this is to use the basic balance equations in integral form for the sheath:

Particles

$$\frac{dN_e}{dt} + J_e^c - J_e^p = J_e^c\mathcal{M}; \qquad \frac{dN_i}{dt} - J_i^c + J_i^p = J_e^c\mathcal{M}$$

where N stands for the particle numbers in the sheath, $N = \int_c^p n\,dx$ with n being the local particle density (lower scripts e, i referring to the electrons or ions), J for the current densities (upper scripts referring the cathode or plasma side), $\mathcal{M}$ is the ion-electron multiplication factor.

Momentum of ions

$$p_i^p - p_i^c + m_i J_i^{p\,2}/n_i^p - m_i J_i^{c\,2}/n_i^c = \int_c^p e n_i E\,dx - \int_c^p k_i J_i\,dx$$

where p is the pressure, E is the electric field, k_i is the friction coefficient. Similar form applies to the electrons.

Poisson's equation

$$\epsilon_0(E_p - E_c) = N_i - N_e$$

Energy

$$\Psi_p - \Psi_c = U_c e(J_i - J_e)$$

where Ψ is the heat current, U_c is the cathode fall. Boundary conditions: at the plasma J_e^p, J_i^p are determined by the positive column mechanisms; at the cathode $J_e^c = J_{em}(T_c, E_c) + \gamma J_i^c$ with J_{em} being the emitted electron current.

This system of equations for the macroscopic quantities is not closed mathematically. However, the integrated form of the microscopic results can be used to evaluate the integrals and source terms.

In [4] the above system has been solved for the glow type sheath, using a linear approximation for the electric field, integrating the well known Townsend's formula for $\mathcal{M}$, using the simplest drift approximation for the ion momentum balance. No thermal effects were included. Radial effects were included by calculating the radial ion drift. The model properly gave back all well known macroscopic glow phenomena, both for steady state and transient.

To describe the arc type modes (i) cathodic thermal effects are to be included, (ii) and the integral of the Townsend ionization formula is to be corrected by inertial effects for thin sheath of the abnormal glow and arc. So we modified the multiplication function $\mathcal{M}$ by using data from [2], and included heat conduction in the cathode.

Fig.1 shows typical steady state voltage and temperature characteristics. The different discharge modes from the dark discharge (breakdown) to the thermionic arc can be clearly distinguished.

The pressure has opposite effect on the very low current and the high current region of the voltage characteristics. At low pressure it is easy to start the lamp but the glow arc transition is difficult. At higher pressure ignition needs higher voltage but once started the discharge turns into arc.

Fig.2 illustrates the effect of the heat conductivity of the cathode. The lower heat conductivity reduces the critical voltage for the glow arc transition, as expected..

As far as the dynamics of the glow - pure thermionic arc transition is concerned there are two basic characteristic times involved: space charge relaxation time (typically $< 1\ \mu s$), and the thermal time constant of the electrode (typically 0.1 - 1s). Obviously the dynamics is determined by the thermal time constants. So we studied the transient by solving the steady space charge equation for a given anomal glow state and put the calculated current density values to the time dependent heat conduction equation. Typical time intervals are obtained from a few tenths of second to a few seconds. This is consistent with measurements [5].

However, there is another type of glow to arc transition with typical characteristic time $\ll 1\text{ms}$ [1]. This process can not be explained by pure thermionic mechanism. Most probably this transition is governed by field emission. This is supported by the very fact that the final arc state of this type of transient is a hot spot arc where the role of field emission has been proved.

	Glow			*Diffuse arc*	*Hot spot*	
	subnormal	*normal*	*anomal*		*steady*	*unsteady*
Dominating emission process	secondary			thermionic	thermo-field	
Active area size on the cathode	small	proportional to current	whole cathode surface	moderate	very small	very small
Typical occurrence in discharge lamps	HID starting			HPS steady, fluorescent steady	HPM, MH steady	HID starting

Table 1.

Basic cathode modes in electrical discharge lamps

REFERENCES

[1] Holló S, Szántó J, Nyiri B *Proceedings of the Symposium on High Temperature Lamp Chemistry II.* Atlanta, Georgia 1988. pp 272

[2] Pérès I, Pitchford L C, Boeuf J P *XX. International Conference on Phenomena in Ionized Gases*, Pisa 1991, Contributed Papers pp 131

[3] Ingold J, *Phys. Rev.* **A40**, 3855 (1989)

[4] Holló S, Nyiri B *XX. International Conference on Phenomena in Ionized Gases*, Pisa 1991, Contributed Papers pp 482

[5] Luijks G M J F, van Vliet J A J M *Lighting Res. Techn* 1988, **20** pp 87

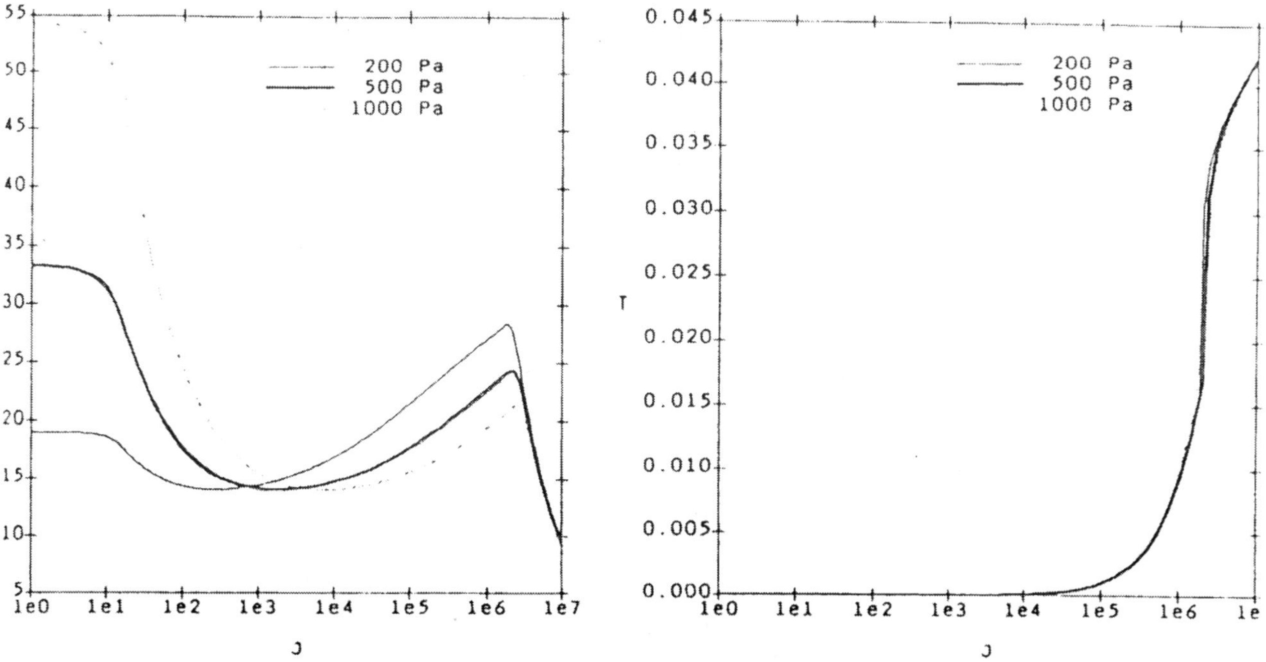

Fig.1. Steady state voltage-current density and temperature-current density characteristics at different pressures. Nondimensional units on the axis. For a typical activated tungsten cathode - argon plasma system the maximum of the voltage scale corresponds to appr. 350 V; temperature to 1900 K; current density to 3 A/cm^2.

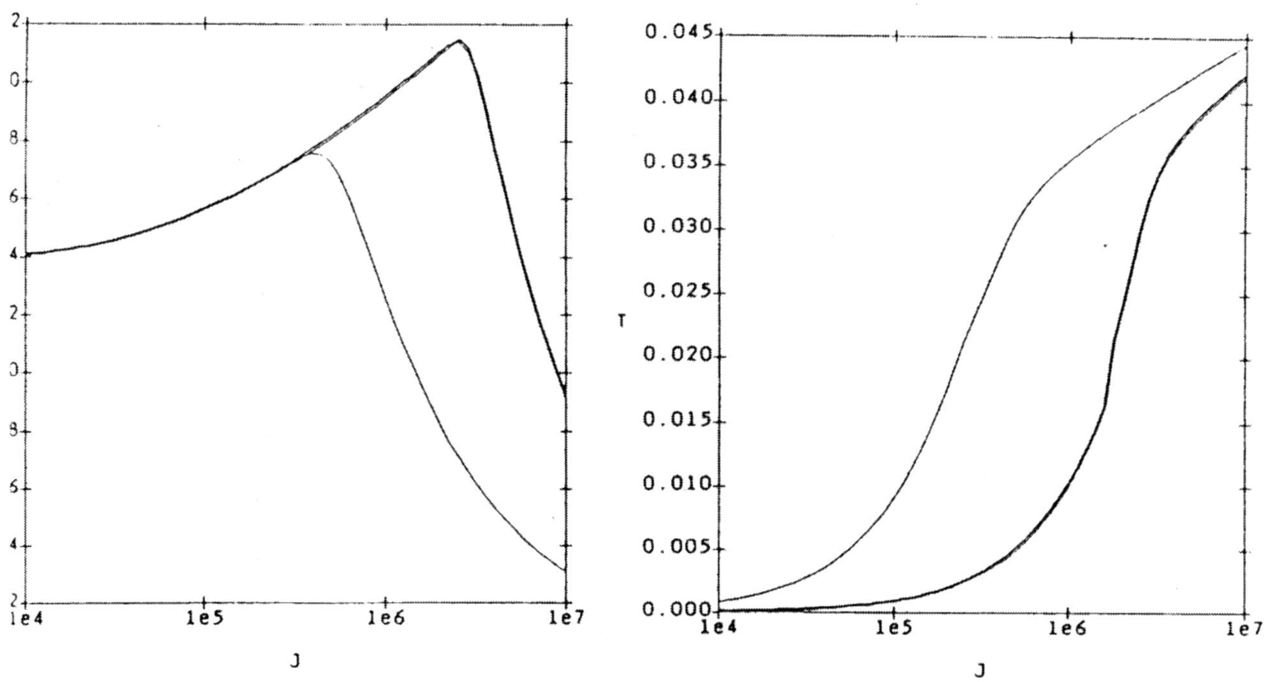

Fig.2. Steady-state voltage-current density and cathode surface temperature-current density characteristics at different cathode heat conductivities. Same units as in Fig.1. Single line: low, double line: high heat conductivity; the ratio of the two conductivities is 10.

TIME-DEPENDENT COMPUTER MODEL OF A HIGH-PRESSURE MERCURY LAMP

H. kawahara and K. Wani
Electronics Research Laboratory, Matsushita Electronics Corporation
1-1 Saiwai-cho, Takatsuki, Osaka 569, Japan

Recently, modeling of HID lamps have been studied for predicting a lamp performance in a stationary state.[1-3] For commercial HID lamps which are operated on an a.c. power source, temporal change of plasma characteristics is to be analyzed to obtain more detailed information of the lamp. In this paper, the authors will present a time-dependent model of lamp characteristics in a vertical operated high-pressure mercury lamp.

A finite differential method was employed to solve the time-dependent model equation. A local thermodynamic equilibrium can be assumed for the whole plasma in the lamp tube. Thus, the electrical and optical characteristics are governed by the gas temperature at each element. To obtain a gas temperature distribution in a lamp tube, energy balance between heat conduction, convection, and radiation was taken into account. These heat transfers are described by three differntial equations which consist of continuity equation, momentum equation, and energy equation.[4] Ohmic heating, radiation loss, and absorption gain are included in the energy equation while electrical field and current direction are left out of consideration. The simultaneous equations for a whole lamp tube are solved by employing a finite differential method software called SUNSET.[5] Starting from a set of trial initial values, we solved the time-dependent equations until the solutions show a periodicity for a sine current wave form.

An example calculation is applied for a 400(W) vertical mercury lamp. (Fig.1) Discharge current is assumed to be 60(Hz) sine wave and effective current value is 3.0(A). The result shows a periodic change of temperature at the symmetrical axis of the tube following the current cycle. (Fig.2) The model is expected to be usuful to predict the lamp characteristics of an a.c. operated HID lamps.

References

1. P.Y.Chang et al. 5th International Symposium on the Science &Technology of Light Sources, York-England, 1989
2. L. Cifuentes et al. 5th International Symposium on the Science &Technology of Light Sources, York-England, 1989
3. J.T.Dakin, 5th International Symposium on the Science &Technology of Light Sources, York-England, 1989
4. S.V.Patankar, "Numerical heat transfer and fluid flow" Japanese edition, Morikita Syuppan, 1988
5. M.Katsuki and A.Nakayama, "Numerical simulation of fluid flow" in Japanese, Morikita Syuppan, 1988

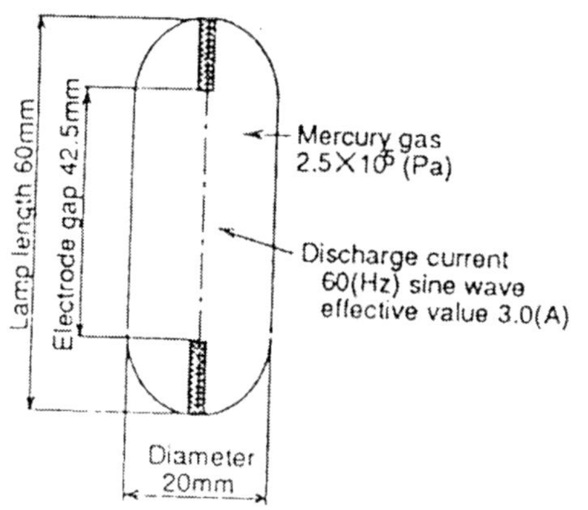

Fig.1 400(W) Mercury lamp for calculation

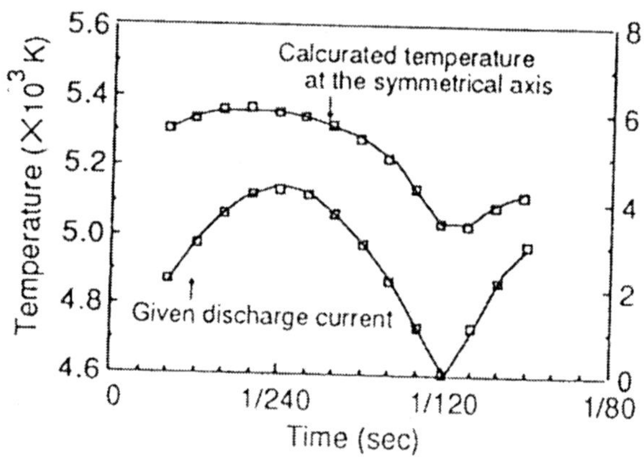

Fig.2 Temporal change of temperature at the symmetrical axis

CURRENT-VOLTAGE CHARACTERISTICS IN LOW PRESSURE COLD CATHODE ARGON DISCHARGES

I. Pérès*, L.C. Pitchford*, J.P. Boeuf*, H. Gielen**, P.Postma***

* Laboratoire des Décharges dans les Gaz, CNRS URA 277, Toulouse, France

** Philips Lighting, Lamp Dept, Eindhoven, The Netherlands

*** Philips GmbH Forschungslabor, Aachen, Germany

Calculations of current-voltage characteristics using a 1D, hybrid fluid particle model are presented and compared to experimental results and data from the literature, for low pressure cold cathode argon discharges. Possible gas heating is also checked by introducing a temperature equation in the model. The range of parameters reported below is appropriate to the start-up phase in discharge lamps.

Hybrid fluid particle model :

The spatio-temporal evolution of the distribution of charged particle densities and electric field in the discharge is obtained with a 1D fluid model, by the simultaneous resolution of continuity and transport equations and Poisson's equation. This fluid model is coupled to a particle model (Monte Carlo simulation), for the accurate determination of the ionization source term. The basis of the fluid model and of the Monte Carlo algorithm can be found in ref 1 and 2. The temperature of the gas was included in the fluid model, assuming that gas heating is due only to the motion of positive ions in the electric field. The temperature profile obtained is introduced in the Monte Carlo calculation of the ionization source term, through the correction of the neutral gas density.

For comparison with different experimental set-ups, two sets of boundary conditions were used in the calculations :

* cooled electrodes : the temperature at each electrode is fixed and equal to 300 K.
* cooled anode : the anode temperature only is fixed, and a null temperature gradient at the cathode is assumed.

Experiment :

The experiment was designed to allow comparison with the 1D model results : the discharge vessel consists of two aluminum circular electrodes with a diameter of 4 cm and a spacing of 1 cm. The current-voltage relation was determined for a filling pressure of 1, 5 and 10 torr. In order to minimize gas heating, voltage pulses of approximately 4 ms were applied. The value of the current could be chosen by adjusting a resistor in combination with the height of the voltage pulse (between 0 and 2kV), generated by the pulse generator. The discharge current was determined by measuring the voltage drop over the resistor, and voltage and current were measured using a digital storage oscilloscope. For low values of the current a glow discharge developed that maintained for the whole period of 4 ms.

Results :

Calculated and experimental current-voltage characteristics are presented on figure 1. The secondary emission coefficient at the cathode, γ, is a free parameter of the model, and the values of 0.08 and 0.05 indicated on fig. 1, chosen for a good fit with the experimental data, are within the range of previously reported results. Because the model is one-dimensional, the part of the characteristics corresponding to the normal glow regime (constant voltage - increasing current) cannot be reproduced. The shape of our experimental characteristics is correctly described with this model, when a Monte Carlo simulation is used to determine the ionization source term.

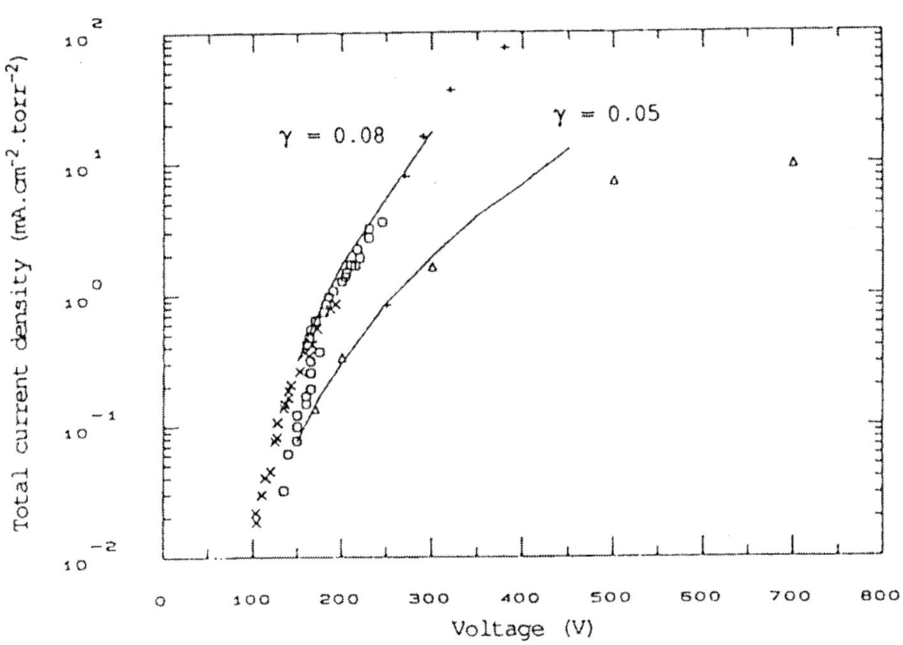

Figure 1: Current voltage characteristics in argon. Calculations : solid lines ; experiment : (+) 1 torr, (o) 5 torr, (*) 10 torr ; (Δ) data from Güntherschulze [3].

Agreement is not so good at high voltage, but the differences are perhaps due to the effect of gas heating. Calculations including the gas temperature equation were performed in conditions comparable to those of Güntherschulze's experiment. The temperature and electric field profiles obtained in the steady-state for the different sets of boundary conditions are shown on fig. 2a and 2b. The maximum increase of gas temperature is 130 K if both electrodes are cooled. This result can be compared to that of Güntherschulze, who estimated a maximum increase of gas temperature of about 62 K for 300 V and 160 K for 500V, in similar conditions. If the cathode is not cooled, the calculated increase of temperature reaches 416 K after 1 ms .

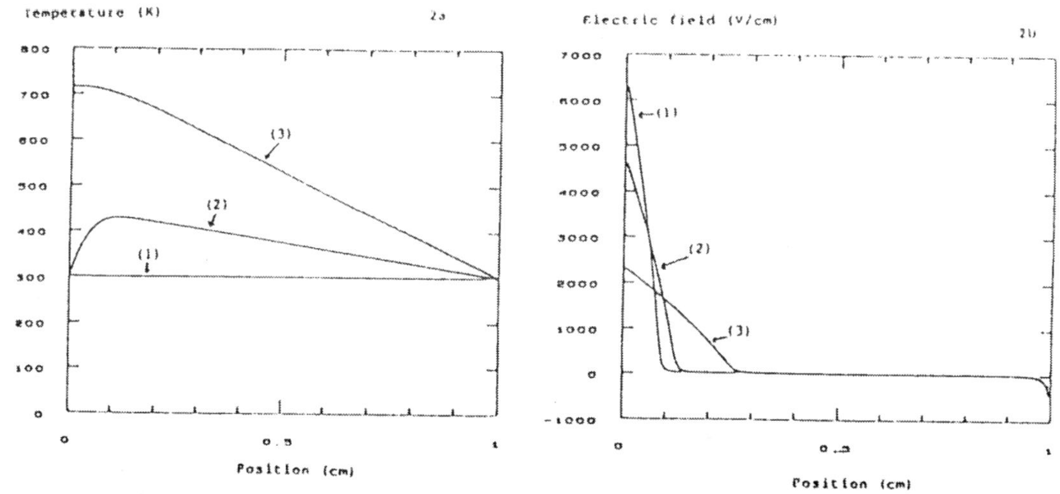

Figure 2: (a) Temperature profile, (b) electric field profile, in the steady-state. Interelectrode distance : 1 cm, pressure of the gas : 1 torr, applied voltage : 350 V, value of the secondary emission coefficient : 0.05. (1) cold gas, (2) cooled electrodes, (3) cooled anode.

The steady-state value of the total current for cold gas (temperature of the gas equal to 300K everywhere in the discharge) was equal to 3 mA/cm2. Calculations including gas heating performed in the same conditions gave 1.45 mA/cm^2 for cooled electrodes, and 0.25 mA/cm^2 if the anode only was supposed to be cooled. Moreover, it is expected that the decrease of the total current increases with increasing voltage, as cold-gas steady-state electric field values are higher for higher voltages.

References :

1. Boeuf J.P., Phys. Rev. A, 36, 2782 (1987)
2. Boeuf J.P., Marode E., J. Phys. D : Appl. Phys., 15, 2169 (1982)
3. Güntherschulze A., Z. Physik, 49, 358 (1928)

63:P

EXPERIMENTAL PARAMETERS DETERMINATION IN THE START-UP PHASE OF A HIGH PRESSURE MERCURY DISCHARGE

M. Stambouli*, A. Asselman, K. Charrada, N. Sewraj

Laboratoire des décharges dans les Gaz, 118 rte de Narbonne, 31062 Toulouse, FRANCE
*E.N.S.E.T., 5 Av. Taha Hussein, 1008 Tunis, TUNISIE

INTRODUCTION

During the start-up of a high pressure discharge the physical processes taking place in the plasma are strongly modified by the increase of the total pressure. In order to identify the most typical phases of the plasma warm-up, it seems necessary to evaluate some specific parameters of the discharge. This paper deals with the determination of the evolution of the electric field strength and excitation temperatures during warm-up of a high pressure mercury lamp.

The electric field strength evolution is obtained from measurements performed on a set of lamps of different lengths filled with the same net mercury amount per unit inter-electrode length. Beside the inter-electrode length, all the geometrical characteristics of these lamps are identical to the 400W commercial one (internal diameter of 1.85 cm). The argon filling pressure is 10 torr. The operating current was maintained at 3.2 A. Table 1 resumes the discharges characteristics. The reference lamp is the D7.2 one.

Table 1: Discharge characteristics

Discharge	D4	D7.2	D10	D13
Hg (mg)	39	70	97	126
Inter-electrode length (cm)	4	7.2	10	13

Excitation temperatures are determined from the total intensities of mercury and argon lines. On the one hand, by plotting in the Boltzmann's diagram the total intensities of the 407.8, 434.7 and 577nm mercury lines, a linear curve leads to the excitation temperature of the 7^1S_0, 7^1D_2, 6^3D_2 mercury levels respectively. On the other hand, a direct evaluation of the excitation temperature of the $2p_2$ argon level is derived from the total intensity of the 696.5nm argon line. The experimental results are compared to those obtained from a self-consistent model based on the L.T.E. hypothesis.[1]

EXPERIMENTAL ARRANGEMENT AND RESULTS

The experimental setup consists of a Jobin-Yvon HRP monochromator equipped with a 1200 groove/mm diffraction grating and a C4351 Hamamatsu photodiode array detector at the lateral exit. Covering around 30nm in the spectrum this device allows the simultaneous measure of the 407.8nm and the 434.7nm mercury lines. The calibration is provided by a standard tungsten ribbon lamp. In order to drive the monochromator diffraction grating and lamp positioning, a PC format card electronic device was elaborated. The whole apparatus is driven by a PC computer through an adequate Software. The so described device allows the follow up of the total intensity evolution of the studied lines during the start-up phase.[2]

As stated in a previous work[3], for discharges having the same geometrical characteristics but filled with different mercury mass, measured intensities for each line are identical provided the arc voltage is the same. Thus, we suppose that for a given intensity, the discharges are in the same state (temperature, density profiles, electric field). Assuming the correctness of this hypothesis, we measured, for the same line intensity, the arc voltage of lamps given in table 1 and deduced the electric field from a linear regression.

The results concerning the time-varying electric field strength relative to the D7.2 lamp are plotted in Figure 1.

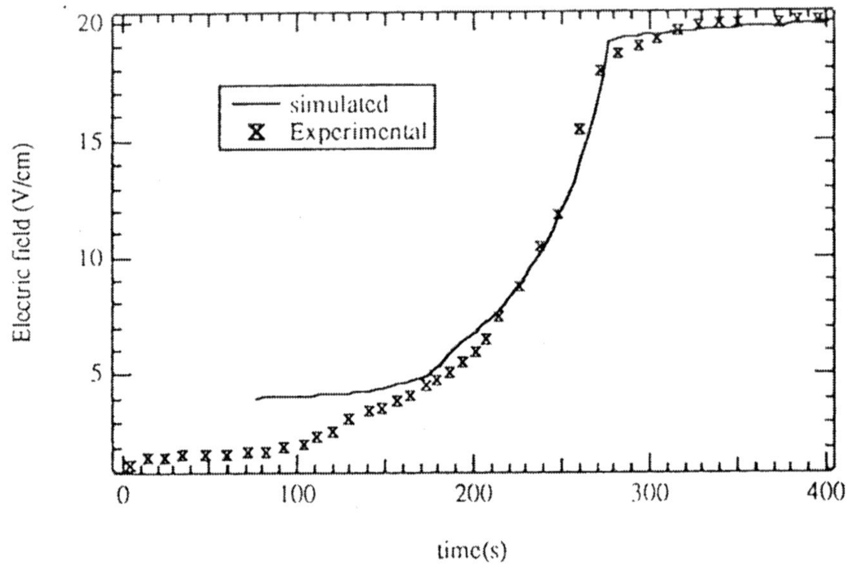

Figure 1 : Time-varying electric field strength in the D7.2 lamp at start-up

The excitation temperature is obtained by assuming that only the central part of the discharge efficiently contributes to the emission. Besides, a corrective factor is used in order to account for the temperature profile. The transition probabilities used are given in Table 2.

Table 2: Spectral line characteristics

λ(nm)	Upper level	E_e (eV)	g_e	A_{ea} (MHz)
Hg 407.8	7^1S_0	7.92	1	2.8
Hg 434.7	7^1D_2	9.55	5	9.0
Hg 577.0	6^3D_2	8.85	5	27.
Ar 696.5	$2p_2$	13.32	3	6.7

Mercury and argon excitation temperatures are plotted in Figure 2. We also give the temperature derived from the L.T.E. model. For electric field strength above 4.5 V/cm, these results show a good agreement between the L.T.E. model and measurements. From Kenty's[4] and Elembaas's[5] formulae, these electric field strength values correspond to a pressure and a mercury mass greater than 0.1 atm. and 0.2 mg/cm respectively. At the beginning of the start-up, mercury and argon excitation temperatures are different since the plasma is far from thermodynamic equilibrium. For electric field strength values greater than about 2, the excitation temperatures seem to rapidly converge to an equilibrium.

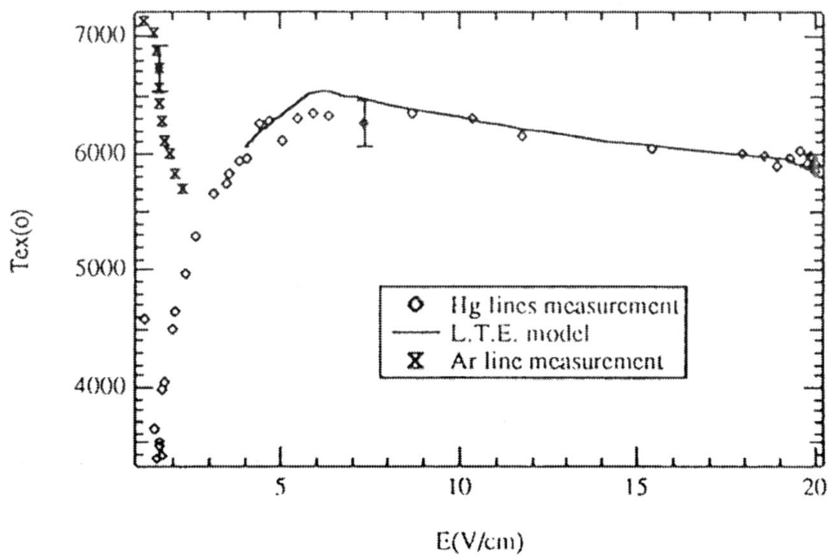

Figure 2 : Axial excitation temperature as a function of electric field strength

Acknowledgments

The authors are indebted to Professor J.J.Damelincourt for his usefull remarks and to Mr J. L. Bonneval for technical support. Financial support is assumed by E.D.F " Etudes et Recherche.

REFERENCES

[1] K. Charrada, G. Zissis, M. Stambouli, A. Asselman, 1992, 6TH Int. Symposium on the Sci. & Tech. of Light Sources, Budapest

[2] Salon J., Courteix M., Asselman A,. 1992, Xth Int. Conf. Gas Disch. Appl.,Swansea.

[3] Asselman A, Stambouli M., Aubès M. and Salon J, 1992, Xth Int. Conf. Gas Disch. Appl., Swansea.

[4] C. Kenty, 1942, Physical Revue **61**, 545

[5] W. Elenbaas, 1951, "The High Pressure Mercury Vapor Discharge", North-Holland Publishing Company, Amsterdam.

64:P

INTEGRATION OF THE ELENBAAS-HELLER EQUATION

John F. Waymouth, Consultant
16 Bennett Road
Marblehead, MA 01945 USA

This report describes a simple stepwise integration routine I have developed for solving the Elenbaas-Heller Equation for an LTE plasma to determine the radial temperature profile. I express the differential equation as a difference equation, and stepwise integrate the temperature from the wall inward towards the axis. The independent variables are the wall temperature and the temperature gradient at the wall. Alternatively, the wall temperature and heat conduction loss to the wall may be chosen. for ease of matching the solution to a heat transfer equation governing the system for cooling the wall. The Elenbaas-Heller equation is a characteristic-value equation. Where energy input to the plasma is by electric-field-driven flow of current, in the methodology above there is only one value of electric field for which the temperature profile reaches the axis with zero slope. Electric field is assumed and the integration to the axis carried out. The assumed electric field is too small if temperature gradient at the axis is negative, too large if it is positive. On successive integrations the characteristic-value solution for the electric field can be bracketed more and more closely until the correct value is chosen. I have implemented this technique using a spread-sheet program on the Macintosh LC computer (4 Meg RAM). I have applied it to an optically-thin mercury arc, using empirical material functions matching those of Zollweg et al[1]. Figure 1 shows

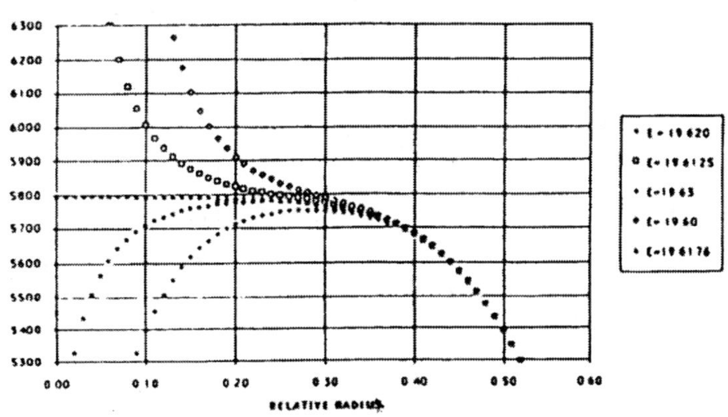

FIG 1

the approach to solution by adjusting the values of electric field. Figure 2 shows temperature profiles calculated for a range of power inputs, and Fig 3 plots electric field vs. power input.

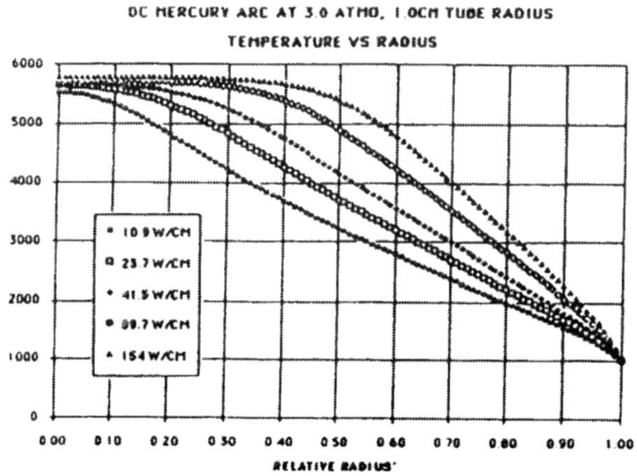

FIG 2

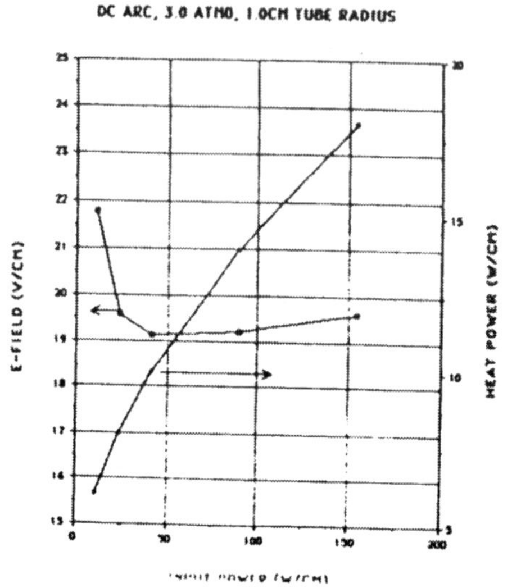

FIG 3

I have also extended the methodology to an optically-thick case in which radiation absorption figures strongly in the energy balance. I assume two hemispherical fluxes, one ingoing, the other outgoing, with a radiation transport equation for each, expressed as difference equations. Absorption of radiation from both fluxes appears in the net emission coefficient of the Elenbaas-Heller equation. The Elenbaas-Heller (energy balance) equation and the two radiation-transport equations are integrated stepwise simultaneously from the wall inward. The outward flux at the wall is an additional characteristic value to be determined, such that inbound and outbound fluxes are equal at the axis, but the poorly-converging iterative solutions of the temperature profile hitherto employed for this type of problem are unnecessary. For this problem, I have assumed a black-body type radiation source function varying as T^4, together with the same electrical conductivity and heat conductivity as before. With this combination of material functions, the arc is radiatively-constricted in the absence of absorption because radiation source strength varies more slowly with temperature than electrical power input. Calculated temperature profiles for various radiation source strengths and a small value of optical depth are given in Fig 4, while the variation of temperature profile with absorption coefficient (optical depth τ) is shown in Fig 5.

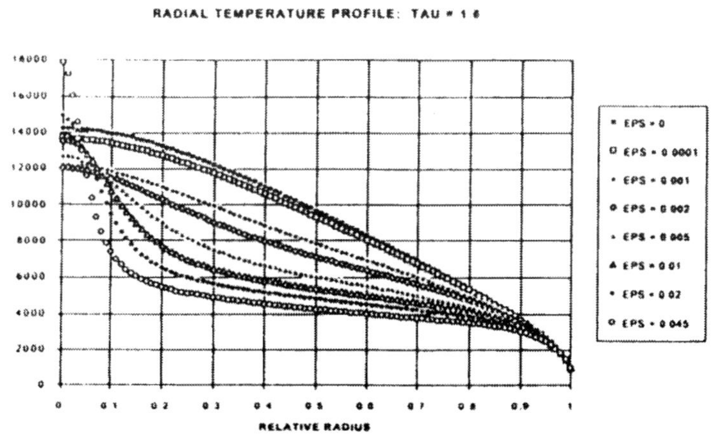

FIG 4

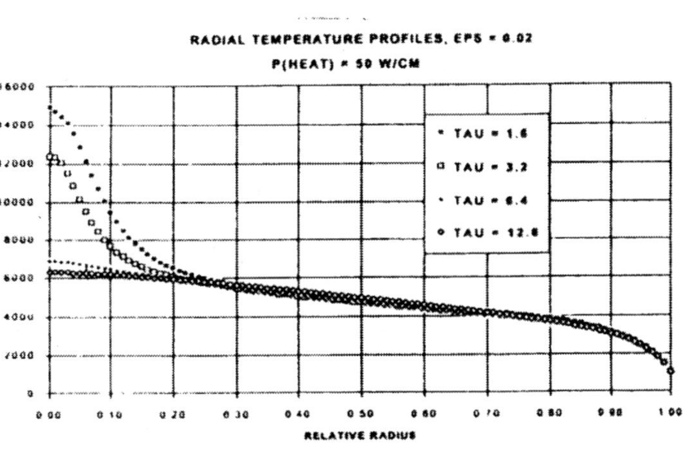

FIG 5

) R. J. Zollweg, J. J. Lowke, R. W. Lieberman, J. Appl Phys **46**, 3828 (1975)

A STUDY OF ELECTRICAL CHARACTERISTICS OF DEUTERIUM LAMP CATHODES

A J Page Cathodeon Ltd, Nuffield Road, Cambridge CB4 1TW, UK

Abstract

The stability of a deuterium lamp in its applications is becoming of increasing importance. This in turn is placing emphasis on the stability of its cathode.

There are many parameters in the production of the lamp which have an effect on the final lamp / cathode performance. The only way of determining the scale of these effects has been to construct batches of lamps and test them over many hours.

This paper reports some of the attempts to construct cathode diagnostic techniques to reduce the amount of time involved in assessing cathode performance and to help provide a better understanding of the oxide cathode in this application.

If the deuterium lamp is operated with an AC voltage across the heater / cathode and a DC constant current discharge between the anode and cathode then the following can be applied.

The technique presented, simply consists of monitoring the AC heater voltage in conjunction with the AC component of the anode voltage. This has been done using an XY digital storage oscilloscope. It has been shown that in the deuterium lamp this provides several features which are individual lamp / cathode dependent. From the gradient of the AC components of heater and anode voltages a relationship has been derived for the position of the cathode emission "spot". This has been confirmed by video monitoring and inspection of many operating lamps.

The AC heater voltage frequency has also been varied to provide a range of characteristics. This can be used to separate the thermal and electrical properties of the cathode. The techniques have also involved both end and centre tapped "Earthing".

The effects of varying the set lamp current have also been described.

By increasing the heater voltage to a level approaching the maximum recommended operating level a characteristic distortion curve can be obtained. See fig.1. This has been determined to be due to the movement, during the cycle, of the cathode emission "spot". Simple models have been constructed in order to show that the main features of the characteristic can be explained by "spot" movement. The latter model involves a more realistic representation of the cathode geometry and movement, again backed by optical video monitoring of operating cathodes.

Fig.1 - Characteristic distortion curve

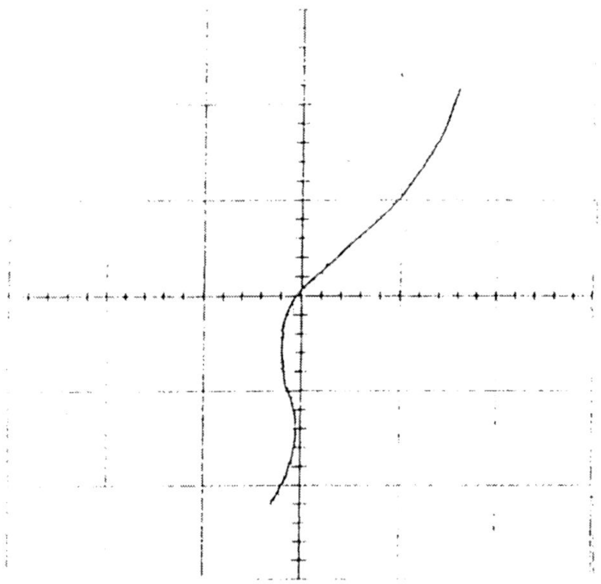

DETERMINATION OF EXCITED-STATE DENSITIES IN NON-LTE HIGH PRESSURE DISCHARGES

D. KARABOURNIOTIS, E. DRAKAKIS AND A. PALLADAS

Low Energy Plasma Laboratory, Physics Department, University of Crete, 71409 Heraklion Crete, Greece

We have developped a modified absorption technique which allows the measurement of the line optical thickness profile of a plasma within transparent walls avoiding any data on the mirror reflection coefficient and the wall transparency. The effect of the presence of a uniformly absorbing continuum background in the line optical thickness measurement is also eliminated by this technique. It is possible to determine the mean radial density population of the lower level of an optically thick line by integrating the optical thickness profile within the line limits. This is independent of any model assumptions (LTE or others).

The described technique was applied to the measurement of the mean radial density of the Hg 6^3P_1, the Tl $6^2P_{3/2}$ and the Na $3^2P_{3/2}$ levels using the Hg 4078-Å, the Tl 5350-Å and the Na 8183-Å lines emitted by a Hg, Hg-TlI and Hg-NaI discharges, respectively.

High-precision spectral intensity measurements were required for a reliable determination of the optical thickness within optically thick lines with $\tau > 2$. A computer controlled data acquisition system was used for the time-resolved intensity measurements during the ac cycle. It allowed us to achieve time dependent measurements of the optical thickness values up to $\tau \approx 6$.

Determination of the mean radial density has also been made using the CDK technique. The results agree fairly well with those obtained by applying the absorption technique and prove the validity of the CDK technique in determining the mean radial density, at least within our experimental conditions.

Comparison with the results obtained by optically thin lines and assuming LTE conditions proved the existence of deviations from excitation equilibrium. The magnitude of these deviations is different for each species and depends on the experimental conditions.

The divergence between the results may be explained by the associative ionization reaction which affects mainly the low-lying excited states.

The experimental lamps have been prepared by OSRAM-Munich (Hg-TlI lamp) and Philips Lighting-Eindhoven (Hg-NaI lamp). This work was supported by the Greek Ministry of Research and Technology.

Session 6

Chairman: A. G. Jack

HID

67:I **T. Hiramoto**
Compact HID lamps for liquid crystal projectors

68:P **M. Goodman, R. Snellgrove**
New emissive materials for unsaturated vapor sodium lamps

69:P **A. Itoh, K. Okamura**
Evaluations of sodium reduction in HPS arc tube

70:P **G. M. J. F. Luijks**
Sodium-PCA interaction in unsaturated HPS lamps

71:P **W. P. Moskowitz, W. M. Keeffe**
Microgravity simulation studies of metal halide lamps for space station applications

72:P **J. M. Strok**
A practical method for HPS arc tube efficacy

73:P **K. Tomokiyo, Y. Kitahara, T. Watanabe, T. Uemura**
Effects of cold spot temperature and operation frequency on the spectral properties of metal halide lamps for LCD projectors

74:L **J. T. Dolan, M. G. Ury, C. H. Wood**
A novel high efficacy microwave powered light source

75:L **K. S. Haagsma**
Non-passive failure of high pressure gas discharge lamps

COMPACT HID LAMPS FOR LIQUID CRYSTAL PROJECTORS

Tatsumi Hiramoto
Ushio Inc.
2-6-1 Ohtemachi, Chiyodaku, Tokyo, Japan

1. Introduction

In these ten years liquid crystal displays(LCDs) were developed for various uses with making best use of their outstanding merits of small size in thickness and lightness. Especially, liquid crystal light valves with thin film transistors(TFT) have recently emerged on the scens. LCDs with TFT have made the image quality of LCDs equal to that of the conventional cathode-ray tubes(CRTs). The size of the LCDs is distributed from 0.7 inch for a view-finder of vide cameras to 14 inches for office automation machines(OAs). These are used as the direct-view type displays. Small size-fluorescent lamps are used in the above applications as back-light sources.

The LCDs are used otherwise as light valves of image projection system, in which electronic signal control ON-OFF of the light valves, and the image is optically enlarged and projected on a screen. Although the present LC panel size is 1.3 to 5.5 inches, the screen size covers 25 to 100 inches. Compact HID lamps are adopted as the projection light sources. The electronic signals are those of the video, television or others. As the optical principle is same for all signal modes, the term "LCD Projectors (LCDPs)" will be representatively used in the present paper. The LCDPs are now in an infant stage. Further development is essential in order to reach the growth stage.

Some market forecasts of LCDs and LCDPs are given first of all in this paper. Then, requirements for lamps from LCDPs are described. Problems on lamps to be solved in the near future are finally explained.

2. Market forecasts of LCDs

Most of LCDs are produced in Japan and supplied to world-wide countries at present. Figure 1 shows the forecsts of production amount of LCDs in comparison with the historical trends of semiconductor devices and ICs(1). It is expected that LCDs run after semiconductor devices with the time lag of about 15 years. Prospective trends in major application fields are shown in Fig. 2(1). In 1995 the use for TV will share 20% of the total market.

Two types of the LCDPs have been published: front type and rear type. In the latter, image is shown with transmitted light through the screen. On the contrary, in the front type image is given with reflected light on the screen. Another forecast gives the trend of projection displays in Fig. 3(2). The front type LCDPs expand their market. The rear type LCDPs will gradually invade the share of the CRT rear type projection TVs. Although the share of the rear type LCDPs is still low in 1995 as seen in Fig. 3, this will be dominant at the beginning of the next century, since this will be able to become a home-use TV with a large screen, due to compactness and lightness.

As shown in Fig. 1 to 3 some optimistic forecasts have been done on the growth of LCDs and LCDPs. These forecasts are based upon the prospects of successful future development of constituent technologies. Lamp technology is one of them.

3. Requirements for lamps and problems on the present lamps

(a) Present status of the lamps

A metalhalide burner is combined with a dichroic parabolic mirror to form a lamp. Metalhalide lamps with 150W input power are mainly used at present, in the projectors with a medium number of pixels. However, 250W metalhalide lamps are adopted for HDTVs and HD-projectors with LC panels of a larger number of pixels such as 1.2×10^6, because the transmittance of the LC light valves decreases with the increase in the pixel density.

The rear-type TVs need higher screen brightness and use 250W metalhalide lamps, since they are set in brighter rooms. The electode distances(arc length) of such metalhalide lamps are designed to be about half centimeter to increase both of an arc luminance and an optical collection rate of a parallel light beam incident on LCD panels. The lamps are operated at the one of the horizontal or vertical position with an electrode-stabilized arc.

Filling elements into the lamps are Dy, Nd and Cs or Li with Hg for a group of the lamps(3). In another group, the filling elements are Sn, Dy, In, Tl and Hg(4).

In both groups the efficacy is about 80lm/W and Ra is 80 to 90. The color temperature of the lamp is adjusted to a spectral property of the dichroic mirrors for the color separation in LCDs. The typical spectra of the first and the second groups of the lamps are shown in Fig. 4(3) and 5(4) respectively. The color reproductivity and efficacy of the spectra are good for the present metalhalide lamps. Lamp life is nearly 2000 hours for both types, which is very short compared with one required by LCDP engineers. Behaviors of electrode-stabilized metalhalide lamps are not sufficiently known. A method for estimating the spectrum of the electrode-stabilized arc with use of 2-step temperature distribution model has been proposed for plasmas composed of multi- filling elements(5).

(b) Requirements and related problems

The characteristics of a target lamp required by engineers of LCDs are given in Table 1 in comparison with those of a metalhalide lamp used in LCDPs and a Xe lamp. The most important problem to be solved on the way to the target lamp is how to lengthen the lamp life. It is absolutely necessary for the lamp life to compete with that of CRTs, about 10000 hours or longer, in order for LCDPs to gorow home-use TVs. However, as this goal is too far, it is better to settle upon a tentative goal before reaching the final goal. The life of 5000 hours may be reasonable. The phenomenon at the life end is the reduction in screen-brightness, 50% of the initial value. This comes mainly from the decrease in the optical collection rate of the parallel light beam due to a deterioration of the inner surface of the vitreous quartz envelope. In order to increase the color reproductivity and efficacy of lamps, the envelope temperature is kept high, e.g. near 1000℃, in the metalhalide lamps used now in the LCDPs. It is well known that the devitrification due to some rare-earth elements proceeds acceleratedly with the rise of the envelope temperature. Developmental efforts should be made in order to solve this problem. The followings may give the key points:

(1) Enhancement of a spectrum in special wavelength regions and increase of efficacy, with adding some emissive species(4,6).

(2) Ceramic envelope(10) Saphire envelope may be better for this application, although poly-crystalline alumina is experimentally used in the reference(7).

(3) Coating the inner surface of the vitreous quartz envelope with chemically resistive materials.

(4) Investigation on new rare-earth halide complexes possessing higher vapor pressure.

The second problem is how to increase the rate of the optical collection of the parallel beam incident on LC panels. As seen in Table 2, the rate is relatively low. This may be increased by shortening the electode distance or designing a non-parabolic mirror. When the distance becomes small, the region of the cathode spot dominates the whole volume of the arc. As the spot temperature becomes higher, the spectral lines of mercury become strong, but the lines of rare-earth elements, which have relatively low excitation levels do not become so strong as mercury lines. Anyway, shortening of the electode distance will cause unfavorable spectrum similar to that of mercury, and a power supply of high current. Since the parallel light beam is essential for a present twist-nematic LC and a polymer dispersion LCDs under development, one should search for favorable spectra of short arcs. On should utilize the merit of a high temperature of the spot. A Xe short-arc lamp is well known to emit a continuous spectrum which is favorable for this application. Moreover, as shown in Fig. 6, the luminance of the cathode spot is very high compared with that of a metahalide lamps adopted in LCDPs (8). Xe-based short-arc metalhalide lamps many have one possiblility.

The third problem is color difference on the screen. This is solved by raising the vapor presure of the emitting species(9). Start-up time and re-start time are required to be short especially for TVs, because ones are used to operate the conventional TVs without any trouble on these times.

Another big problem is a drastic improvement on the low transmittance of LCDs shown in Table 2. New LCDs should be developed. One example is a polymer dispersion LCDs, which have no polarizer(10). These displays are promising as LCDPs, since the light transmittance becomes three times higher effectively in comparison with the present one. An optical system converting one of polarized component of the light into another component has been demonstrated. This may double the utilization rate of the light beam incident on the polarizer.

Conclusion

At present, LCD projectors are in the infant stage. Lamp techbnology has a key point for them to grow further.

The most important item of developing lamps for this application is lengthening the lamp life from 2000 at present to about 10000 hours without any deterioration in color reproductivity.

Next item is increasing the rate of the optical collection of the parallel light beam. At last, it is essential for lamp development that lamp scientists and engineers should do research and development in cooperation with engineers developing LCDPs.

References

1. A. Asada; Flat panel display '92, P 117(1991), Nikkei Microdevice.
2. Flat panel display '91, P 60(1990), Nikkei Microdevice
3. M. Sugiura; Kogaku 19(9) 581(1990)
4. K. Honda; Personal communication
5 T. Hiramoto; 6th Symp. Charact. and Appl. Light Sor., Aug.(1991)
6. M. Narita et al.; 5th Symp. Sci. Techn. Light Sour., Sept.(1989)
7. K. Takatsu et al.; Symp. Backlight and Lamps for Projectors(LS-90-4) Feb.(1990)
8. I. Iwai et al.; ibid.(LS-90-3)
9. Y. Kitahara et al.; Illum. Eng. Inst. Jap. Ann. Conf., April(1991)
10. J. L. West et al.; J. Appl. Phys. 70(7) 3785(1991)

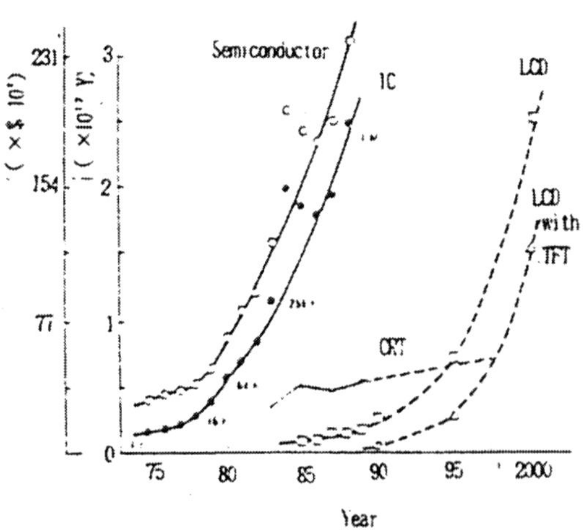

Fig.1 Past result of semiconductor devices and forecasts of LCD and CRT

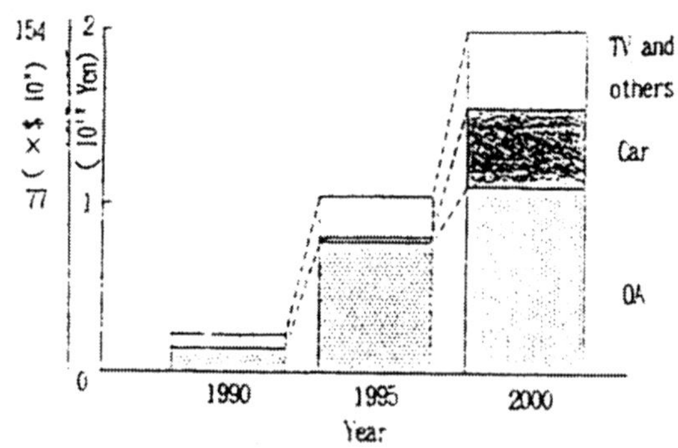

Fig.2 Forecast of LCD in application fields

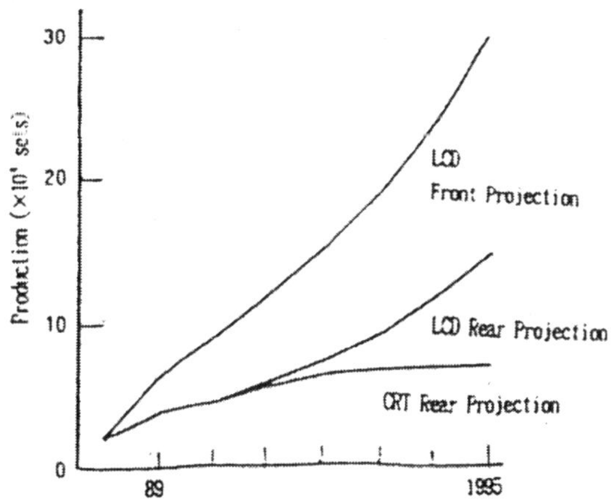

Fig. 3 Forecasts of projection displays

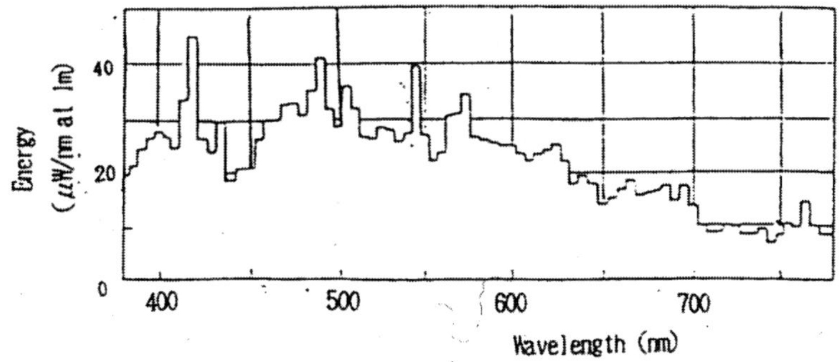

Fig. 4 Spectrum of a metalhalide lamp containing Dy-Nd-Cs-Hg elements

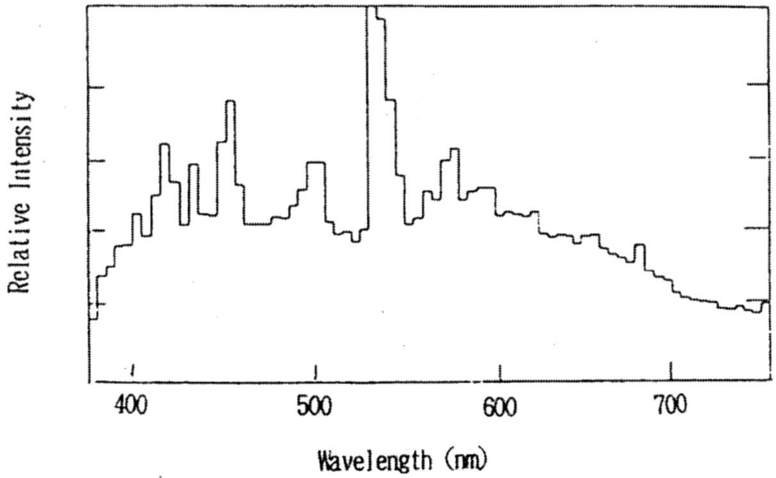

Fig. 5 Spectrum of a metalhalide lamp (250W) containing Dy-Sn-In-Tl-Hg elements

(1) Metalhalide lamp (max. luminance=4 ×10^8 nt)

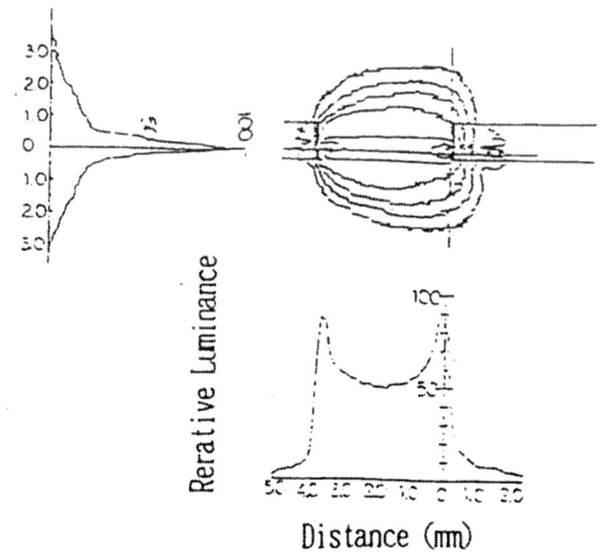

(2) Xe lamp (max. luminance =30×10^8 nt)

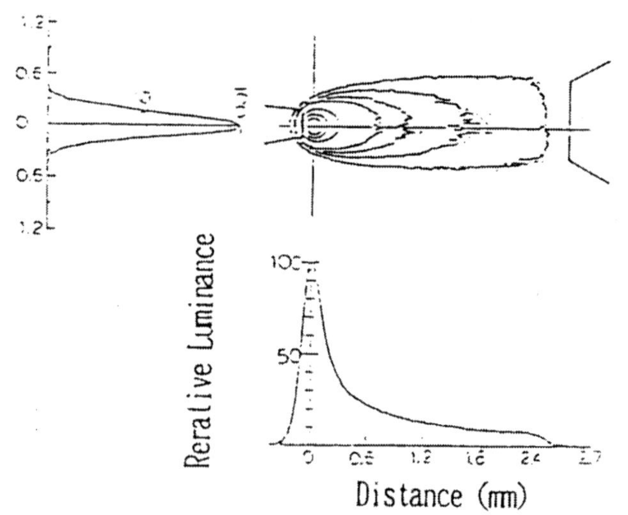

Fig. 6 Luminance distributions of a metalhalide lamp(150W) and a Xe lamp(150W)

Table 1 Characteristics of the target, present metalhalide, and Xe lamps

Lamp / Characteristics	Target lamp	LCDPs	Metalhaide lamp	Xe lamp
Life	10000hrs (Tentatively 5000hrs)	No replacement of lamps	~ 2000hrs	~ 2000hrs
Efficacy	$\gtrsim$ 80 lm/W	Low heat production	~ 80 lm\W	~ 30 lm/W
Luminance	high	High collection rate	~ 10^8 nt	~ 10^9 nt
Color rendering index	$\gtrsim$ 90	High color reproductivity	~ 90	90
Electrode distance	small	High collection rate	~ 4mm	~ 1mm
Start-up time	$\lesssim$ 5sec	Especially short time is required for TVs	~ 0.5min	0
Re-start time	$\lesssim$ 5sec		~ 0.5min	0
Ignition voltage	low	For safety	~ 15KV	~ 20KV

Higher luminance and shorter electrode distance bring about a higher optical collection rate of a parallel beam.

Table 2 Light transmittance through each component of an LCDPTV

light source	100%
optical collection and parallel beam	0.30
color separation (dichroic mirror)	0.75
inlet polarizer	0.40
LC light valve	0.45
outlet polarizer	0.80
color synthesis	0.85
projection lens	0.85
Total	0.0234 (2.34%)

NEW EMISSIVE MATERIALS FOR UNSATURATED VAPOR SODIUM LAMPS

David Goodman and Richard Snellgrove

GTE Electrical Products
100 Endicott Street,
Danvers, Massachusetts 01923 USA

Some emissive materials commonly used in standard high pressure sodium lamps react sufficiently with sodium that they are unusable for unsaturated vapor (USV) operation. The USV lamp contains only 20-100 μg of sodium, depending on wattage, and very little reaction with the emitter can be tolerated. In addition to low reactivity, USV emitters should also provide a low work function for the electron emitting surfaces of the electrodes throughout the life of the lamp, dispense emitter to these surfaces without being excessively volatile, and be easy to process and store.

Yttria or baria plus yttria combinations can be used for USV sodium lamps, but we have found that formulations in the larger system Ba-Sr-Y-O give superior results. In particular, the compound $(Ba,Sr)Y_2O_4$ provides good lumen output with little loss of sodium. All components of this solid solution compound are active emitters with barium being the most active. By varying the ratio of barium to strontium in $Ba_xSr_{1-x}Y_2O_4$ it is possible to adjust and balance the desired properties. Substitution of other ions for the yttrium provides even further control over properties like vaporization rate to help optimize maintenance of both the emitter itself and the sodium.

$Ba_xSr_{1-x}Y_2O_4$ compositions with $0 \leq x \leq 0.8$ are stable with respect to phase transformations and melting for temperatures encountered in processing, sealing and lamp operation. Over this range of x values, the barium and strontium form a continuous solid solution with linear variation of the orthorhombic unit cell dimensions. The vaporization rate of the alkaline earth components increases with increasing barium content. After a short induction period, weight loss is linear in the square root of the time consistent with a diffusion controlled process. Figure 1 shows the barium dependence of evaporation of $(Ba,Sr)Y_2O_4$ in comparison to corresponding hafnates which appear to have too low a rate of dispensation. The highest barium content yttrate, $Ba_{.75}Sr_{.25}Y_2O_4$, while possibly too volatile at 1600 °C, is much in line with the other compositions at 1400 °C. For a given composition, barium is lost more rapidly than strontium and the emitter is converted partly to yttria and partly to $Ba_xSr_{1-x}Y_2O_4$ compositions of increasing strontium content (i.e. decreasing x). Substitution of smaller ions for yttrium decreases the rate of vaporization.

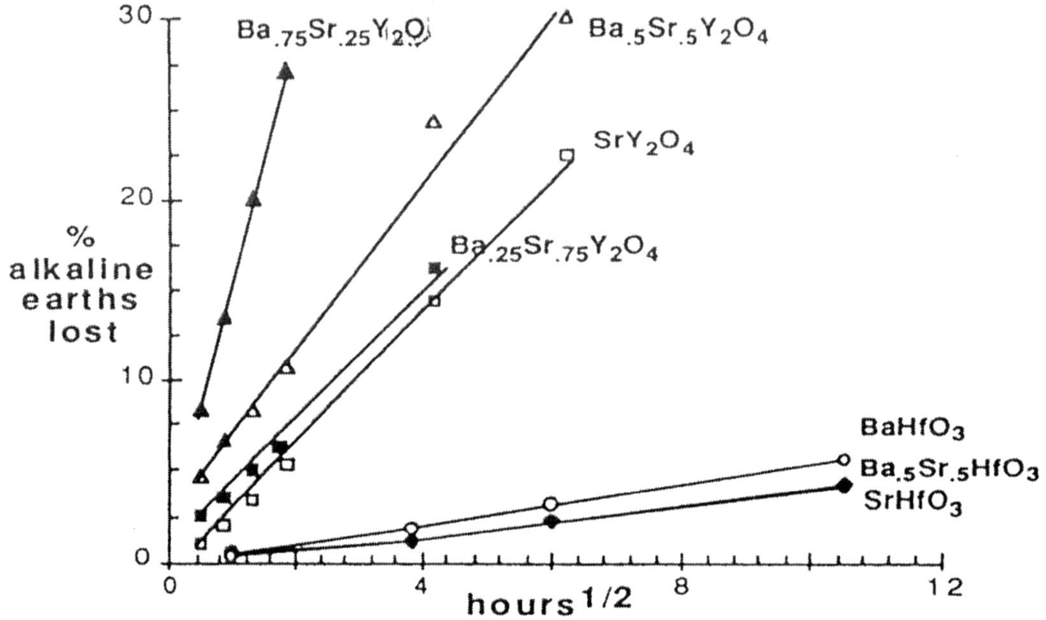

Figure 1. Comparison of vaporization rates of yttrate and hafnate solid solutions at 1600 °C in vacuum.

Moisture pickup from the atmosphere is also, like vaporization, a function of the barium content of $Ba_xSr_{1-x}Y_2O_4$. The rate of pickup increases modestly with barium for low barium concentrations but jumps significantly as the barium fraction x exceeds a value of about 0.6. High barium content solid solutions may therefore prove more awkward to use in USV lamps because even a small amount of water contamination can be disastrous.

Measured electrode temperatures for $Ba_xSr_{1-x}Y_2O_4$ compositions are not a strong function of x for unaged lamps. Temperatures of the first coil of a 400W electrode fell in the range of 1350 to 1400 °C for all ratios of barium to strontium. The corresponding temperature was 300 °C higher for yttria and over 100 °C higher for barium hafnate. While early in life there does not seem to be any great emissivity advantage of barium over strontium, Auger surface analyses of electrodes aged more than 1000 hours show a much higher concentration of barium over strontium on the emitting surface. And in these aged lamps electrode temperatures are definitely lower for higher barium content so that as time of lamp operation increases, barium appears more effective than strontium. This is born out by measurements of lumen maintenance and electrode fall in aged lamps. Examination of electrode cross sections confirm that residual barium remains greatest at the hot end of the electrode for the high barium end members even though the loss of barium is greatest for these compositions.

Lumen maintenance for three USV emitters is shown in figure 2. The comparison is made using the same size electrode, arc length, and construction.

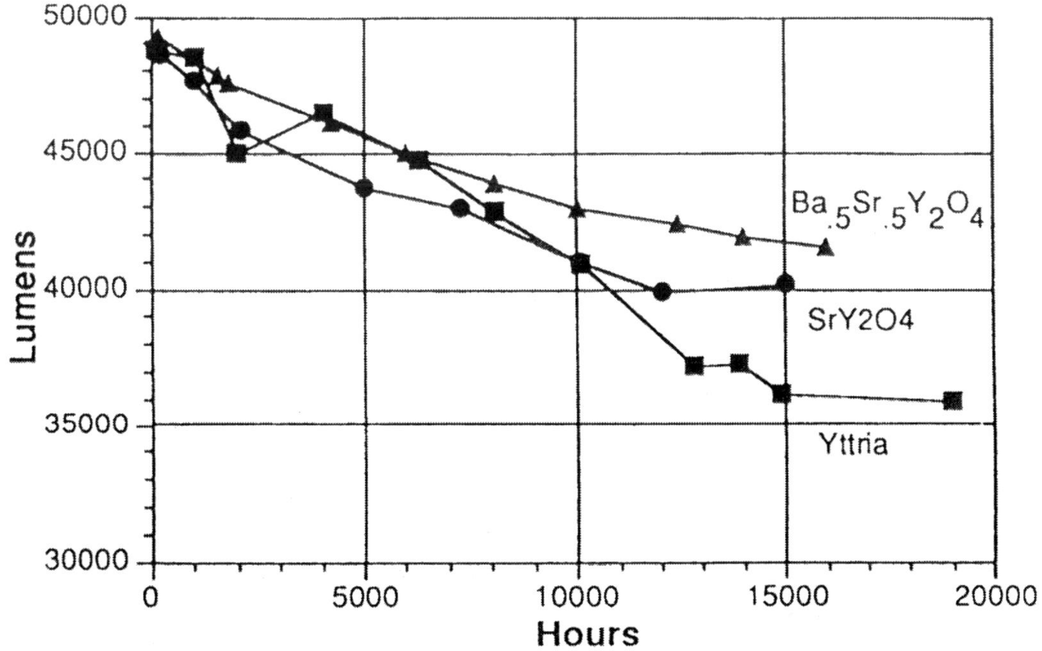

Figure 2. Lumen maintenance in 400W unsaturated vapor sodium lamps.

The barium containing emitter provides higher lumens. This is due partly to better sodium maintenance and partly to reduced wall darkening. The depicted performance can be improved for all of the emitters by optimizing lamp design parameters, but $(Ba,Sr)Y_2O_4$ still seems to provide higher lumens over life.

The authors wish to acknowledge the assistance of John Plumb in obtaining the lamp data.

EVALUATIONS OF SODIUM REDUCTION IN HPS ARC TUBE **69:P**

Akira Itoh, Kazuyoshi Okamura
Toshiba Lighting & Technology Corporation
1-201-1 Funakoshi, Yokosuka 237, Japan

1. Introduction

Sodium reduction in HPS arc tubes has been studied by many researchers all the time. Most of those reports focused on reactions between sodium and alumina. Very few of them studied the reaction in operating HPS arc tubes [1]. In this report, the reduction rate of sodium in operating arc tubes is tried to evaluate quantitatively about PCA tubes, sealing frits and emission oxides.

2. Experiment

Lamp voltages of HPS lamps with unsaturated vapour are sensitive to sodium dose. Influence of the sodium on the electric field strength in HPS arc tubes is much greater than that of the mercury (Fig.1). During operation, the sodium reacts with alumina, seaing frits, emission oxides and other impurities in the arc tubes, but the mercury does not. Therefore, the lamp voltages gradually decrease during their lifetime and the sodium reduction is deduced from the relationship between the sodium-mercury dose and the electric field strength in figure 1.

The 360W HPS lamps are used in our evaluations. Their ratings are as follows; lamp voltage: 130V, lamp current: 3.3A, lamp wattage: 360W and lumen output: 50000lm. The arc tube is 7.25mm of bore dia., 0.75mm of wall thickness, 85mm of electrode spacing and 4.56cc of volume. The sodium and mercury dose is 0.072mg and 2.31mg respectively. The xenon gas pressure in the arc tubes is 120 or 200 torr at room temperature.

3. Reaction with PCA tubes

To evaluate the influence of PCA tubes on the sodium reduction, several kinds of the PCA tubes are prepared, which physical properties are shown in table 1. The tube B has a larger area exposed to sodium vapour than the tube A, because of the higher roughness value of its surface. The tube C is expected to have less $MgO/CaO-Al_2O_3$ compounds on its grain boudaries to react with sodium [2]. The tube D-F have different MgO contents and grain sizes. Hence, the tube A-F are expected to show different reduction rates. Yttrium oxide is used as emission materials, because its influence on the sodium reduction is estimated to be small. A frit of $Al_2O_3-CaO-BaO$ compounds is used as a sealing material. In advance, it is confirmed that the suitably sealed frit material hardly has any influence on the sodium reduction, altering sealed area exposed to sodium vapour. The arc tube temperatures are monitored with an IR photometer.

Figure 2 shows the sodium reduction rate per 1000 hours on unit inner surface area of those PCA tubes. It appears that the sodium reduction rate is supposed to be sensitive to MgO contents, but insensitive to area exposed to sodium vapour, grain sizes and amount of $MgO/CaO-Al_2O_3$ compounds on the PCA tube grain boundaries.

Reaction of sodium with PCA tubes is supposed to have three different processes, which are the reaction with Al_2O_3 in the absence of oxygen, the one with alumina in the presence of oxygen and the one with MgO/CaO-Al_2O_3 compounds in the grain boudaries. According to the result in the figure 2, the second reaction is supposed to have more influence on the sodium reduction than the others. If the first one had major influence, the tube B should show the higher reduction rate, because of its larger reaction site area. The oxygen is supposed to be released mainly from Al_2O_3 and MgO of solid solution in Al_2O_3 under high temperature condition. Since MgO has higher oxygen equilibium pressure, the tube E and F are supposed to show the higher reduction rate.

4. Reaction with sealing materials

To evaluate the influence of sealing materials to the sodium reduction, several kinds of Nb heatshields are provided to the sealed portions of the PCA tube A. Those Nb heatshields give different operating temperatures to the sealed portions, but do not vary their lamp voltages, because of the unsaturated vapour.

Figure 3 shows required times for sodium penetration through the 1mm thick frit. The sodium penetration is detected as an arc tube leakage. The frit temperatures are monitored at an upper end of arc tubes measured with Pt-Rh thermocouples. Figure 4 shows an example of lamp voltage decreases of the lamps, which caused arc tube leakages at 4200 hour and 5000 hour respectively. The arrows indicate the lamp voltages after their arc tube leakages. Since any differences of the lamp voltage decreases are found between leakage lamps and unleakage lamps, it is supposed that the frits have little influence on the sodium reduction. Neither are any sodium signals detected from the frit of the leakage lamp in the EPMA inspection.

Apart from these experiments, strong sodium signals are detected from an unsuitably sealed frit of high color renditon HPS lamps, which sodium pressure is several times higher than conventional HPS lamps. It is found that the sodium penetrates through the specific crystal phase, probably C3A, exchanging with calcium. A calcium spectrum from 422.6 nm is also detected.

5. Reaction with emission oxides

Finally, the influence of the Ba_2CaWO_6 emission oxide is evaluated. Taking into account that emission oxide temperatures depend on electrode shank diameters, electrodes of 0.9mm and 1.2mm shank dia. are provided with the PCA tube A. Table 2 shows that the Ba_2CaWO_6 emission oxide on the electrode of 0.9mm shank dia. reduces 50 micrograms of the sodium in 1000 hours. On the other hand, the Y_2O_3 emission oxide reduces little, because there is no differnce of the sodium reduction between 0.9mm and 1.2mm shank dia.

6. Summary

During a 1000 hours operation of the conventional 360W HPS lamp, the sodium reduction is estimated to be roughly 4 micrograms with the PCA tube, 50 micrograms with the Ba_2CaWO_6 emission oxide and very small with the frits.

[1] W. Reints Bok, J. Appl. Phys. 59 (12), 15 (1986)
[2] G. de With, P. J. Vrugt, and A. J. C. Van de Ven, J. Mater. Sci. 20 (1985)

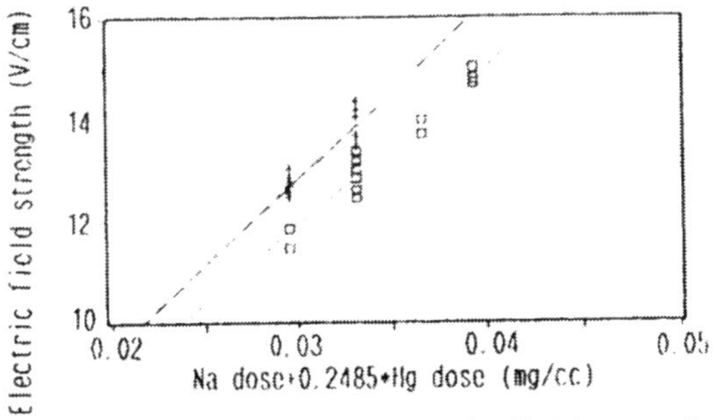

Fig.1 Variations of the electric field strength with sodium-mercury dose
□: Xe 120torr; +: Xe 200torr

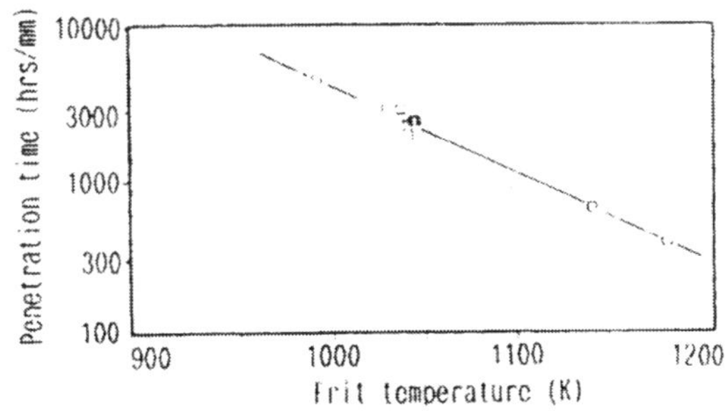

Fig.3 Sodium penetraton rate in the 1mm thick frit

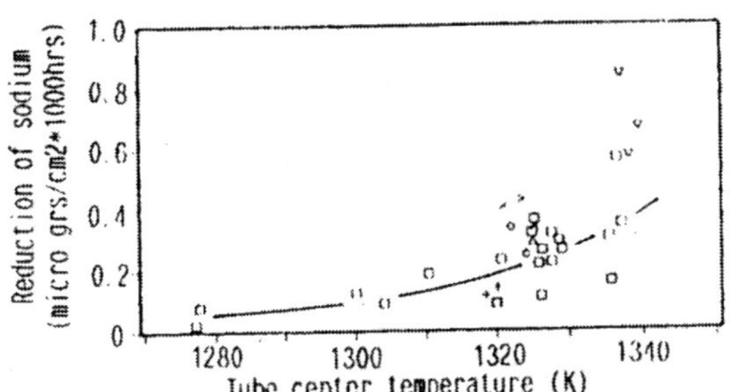

Fig.2 Sodium reduction rate per 1000 hours on unit surface area of the PCA tubes

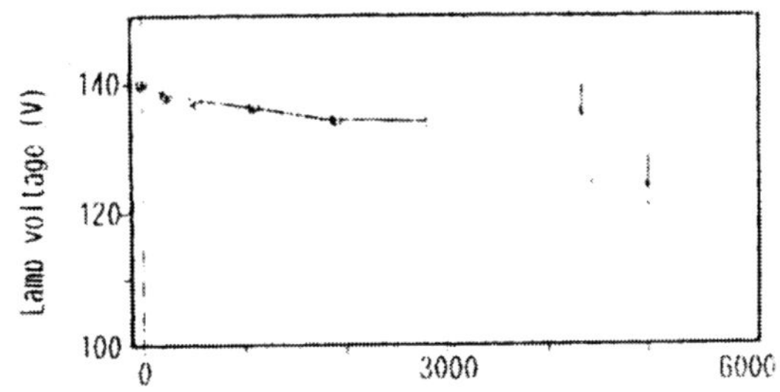

Fig.4 Voltage variation of the lamps having the leakage trouble

Table 1 Test lamp list

	A	B	C	D	E	F
tube in-line(%)	45	32.8	32.8	16.7	56.2	11.8
grain size(μm)	34.8	29.7	29.7	21.3	52.5	14.5
roughness (μm)	0.26	0.54	0.54	0.25	0.25	0.25
MgO (ppm)	360	380	380	480	740	650
CaO (ppm)	<10	<10	<10	<10	<10	<10
remarks	-	-	*1	-	-	-

*1 The PCA tubes were 1600°C pre-heated for 8 hours in high vacuum.

Table 2 Reduction of dosed sodium (micro grs/1000 hrs)

emission oxide	Ba2CaWO6	Y2O3	
shank dia. (mm)	φ0.9	φ0.9	φ1.2
reduction rate	56	3	3

SODIUM - PCA INTERACTION IN UNSATURATED HPS LAMPS

G.M.J.F. Luijks
Philips Lighting BV, p.o. box 80020, 5600 JM Eindhoven, The Netherlands

Ever since the introduction of HPS lamps in the sixties, it has been suggested to operate this type of lamp in the unsaturated vapour mode just like high-pressure mercury lamps [1].
The main advantages of unsaturated over saturated lamps are easy electrical stabilisation and insensitivity to external influences.
However, the very low sodium dose required for unsaturated HPS lamps (in the order of 0.1 mg) makes the lamp characteristics very sensitive to sodium loss. Reactions of sodium with impurities and lamp materials such as the PCA arc tube, sealing frit or emitter, make it very difficult to obtain sufficient life performance.

Japan Storage Battery Company (Nippon Denchi) was the first to announce a practical unsaturated HPS lamp, which was brought onto the Japanese market [2].
Recently, Iwasaki and Philips have introduced unsaturated HPS on the US market for retrofit application on mercury ballasts [3,4].
In addition to the known advantages of HPS lamps in this application (viz. a nearly doubled luminous efficacy and 10% energy savings compared to mercury lamps), unsaturated HPS lamps offer the extra feature of operating on all types of mercury ballasts. In contrast, saturated HPS lamps do not stabilize properly when operated on constant wattage (CWA) mercury ballasts in the US.

In the present contribution, experimental results are presented obtained in a study on sodium loss effects under unsaturated HPS conditions at increased sodium pressure (P_{Na} = 40 kPa).
Two types of experiments have been carried out:

a. model experiments, where a PCA vessel with an unsaturated sodium fill is exposed to different temperatures in an isothermal oven;
b. lamp experiments, where practical unsaturated arc tubes with their non-isothermal temperature profile are operated at different wattages.

Real time monitoring of sodium loss is possible via a transmission/absorption technique in the model experiments and via the self-reversal width of the sodium D-line ($\Delta\lambda_{total}$) in the lamp experiments.

For a proper interpretation of the sodium loss data in the lamp tests, the Sodium Total Analysis has been developed, based on Atomic Absorption Spectroscopy and Neutron Activation Analysis techniques.
It distinguishes three sodium loss channels:

a. PCA "mono"-corrosion, i.e. formation of sodium mono-aluminate
b. PCA "ß"-corrosion, i.e. formation of sodium ß-aluminates
c. sealing frit corrosion, i.e. sodium-sealing frit reactions.

The arc tubes are cut into 7 pieces, representing different temperature sections of the PCA. Each PCA segment is analysed separately.
In all cases the total amount of recovered sodium is found to be well in agreement with the initial sodium dose, within the experimental accuracies.

As far as sodium-PCA interactions are concerned, the results of the analyses show that:

PCA "mono"-corrosion occurs mainly at the cold arc tube ends; increasing the lamp wattage decreases the corrosion rate;

- PCA "ß"-corrosion occurs mainly in the hot arc tube center; increasing the lamp wattage increases the corrosion rate.

Fig. 1 shows a typical example of results obtained from test lamps operated at different wattages and thus at different temperatures.

The results of the experiments are compared with the thermo-dynamical model on sodium-PCA interactions, developed by Vrugt [5]. It is concluded that:

- PCA "mono"-corrosion is dominated by thermodynamics;
- PCA "ß"-corrosion is dominated by kinetics.

Apparently, unsaturated HPS lamps under the present test conditions require a reduced PCA center-temperature and an increased PCA end-temperature, i.e. a nearly isothermal lamp design. These insights have been realised due to the team work of various people within Philips Lighting.

[1] K. Schmidt, High pressure sodium vapour lamp, US patent 3,248,590 (1966).

[2] Y. Hida, H. Furukubo, Y. Takeji and K. Takatsuka, High pressure sodium lamp of unsaturated vapour pressure type, GS News 38,(2), 92 (1979).

[3] T. Okada, H. Yokome, A. Kosasa, T. Habu and Y. Yamamoto, High pressure sodium vapour lamp having unsaturated vapour pressure type characteristics, US patent 4,755,721 (1984).

[4] J. Ravi and G.M.J.F. Luijks, Characteristics of unsaturated HPS lamps on mercury ballasts, paper submitted for the IES conference in San Diego, Aug. 1992

[5] P.J. Vrugt, Compatibility of alumina (sapphire) and sodium in the temperature range of 1075-1525 K, Proc. Symp. High Temp. Lamp Chem., E.G. Zubler, Ed., p.237 (1985).

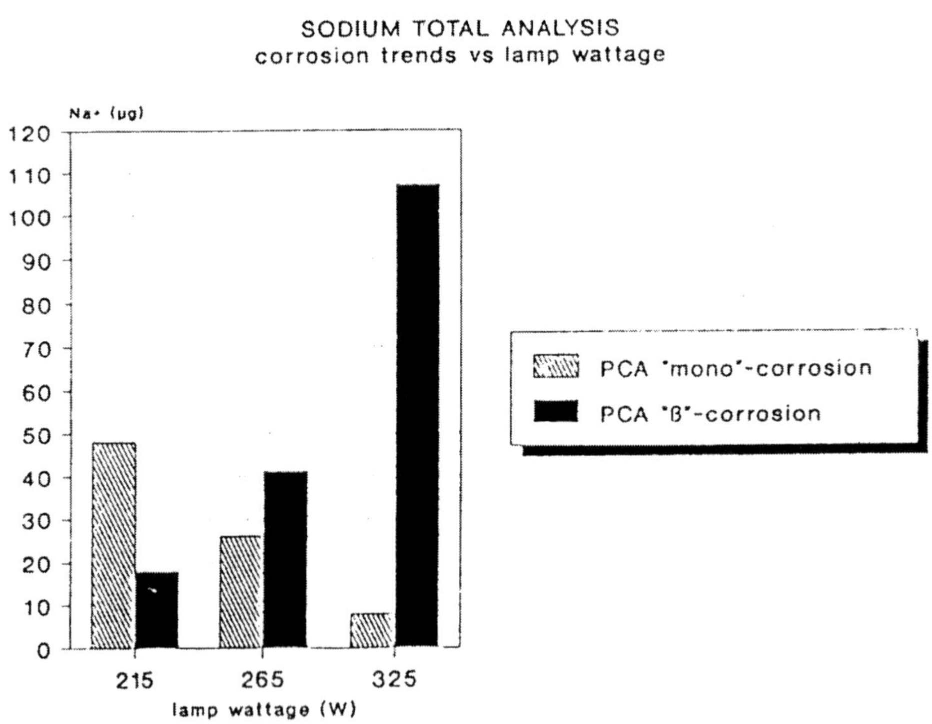

Fig. 1

Gap in pagination due to unavailable papers.

Pages 295-298

EFFECTS OF COLD SPOT TEMPERATURE AND OPERATION FREQUENCY ON THE SPECTRAL PROPERTIES OF METAL HALIDE LAMPS FOR LCD PROJECTORS

K. Tomokiyo, Y. Kitahara, T. Watanabe and T. Uemura
Light R&D Center, Matsushita Electronics Corp.
1-1, Saiwai-cho, Takatsuki, Osaka, 569, JAPAN

Short arc metal halide lamps are suitable for light sources of LCD projectors because of their high efficacies, high luminance and spectral properties. However, color separation due to radial and/or axial segregation causes the color difference on the screen.

In this paper, the effects of cold spot temperature and operation frequency on the color separation of the arc are discussed.

Lamp used for the measurements consists of discharge tube and reflector. Discharge tube, filled with Dy-Nd-Cs iodides, has the electrode distance of 7mm and is vertically operated at 250W. Discharge arc is positioned at the focal point of the parabolic glass reflector coated with dichroic layers. Screen characteristics were measured using reference optics, simulating the LCD projector.

Discharge arc can be divided into three zones at the cold spot temperature of 690℃: arc core (Zone A), area surrounding the arc core (Zone B) and outer shell of the arc (Zone C). Emission spectrum from Zone A is dominated by the mercury radiation, whereas, that from Zone B consists of quasi-continuous radiation of rare-earth atoms and/or ions. From Zone C, continuous radiation, probably due to dysprosium iodide molecules, is observed in longer wavelength region.

Color difference on the screen corresponds to the radial color separation of the arc: greenish white at the center, due to mercury radiation, in contrast to reddish white at the corner.

In order to reduce the color difference on the screen, mercury radiation at arc core should be supressed.

In Fig. 2, cold spot temperature dependence of the emission spectra from arc core (Zone A) is shown. At higher cold spot temperature, radiation of rare-earth elements increases and contribution of mercury radiation is reduced. This is due to vapor pressure increase of the iodides with lower excitation potentials. Above 850℃, Zone A is hardly observed and thus color difference on the screen can be minimized.

Operation frequency dependence on the emission spectra from Zone A were measured on square waveform between 50 to 500Hz. Relative intensity of mercury radiation is reduced with operation frequency rapidly upto 125Hz, then kept almost constant above 200Hz. The reason of this phenomenon has not been clarified yet, however, supposed reason is arc temperature decrease at higher frequency operation.

Uniform color characteristics of the screen can be realized by controlling the cold spot temperature and operation frequency.

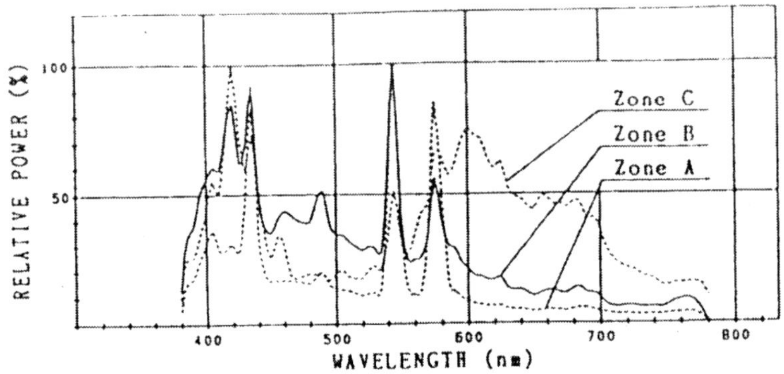

Fig. 1 Emission spectra from each zone of the arc

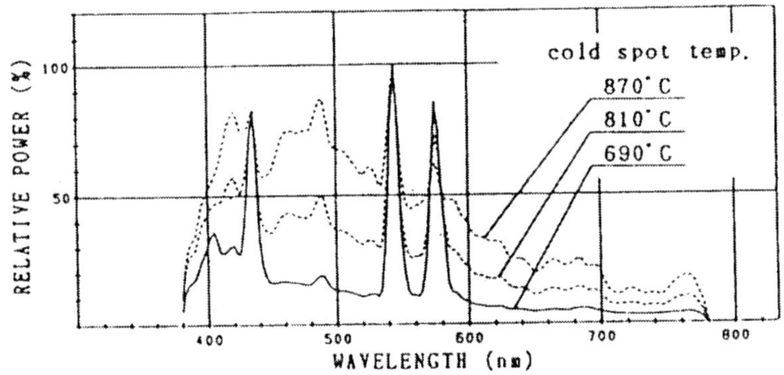

Fig. 2 Cold spot temperature dependence of the emission spectra from arc core

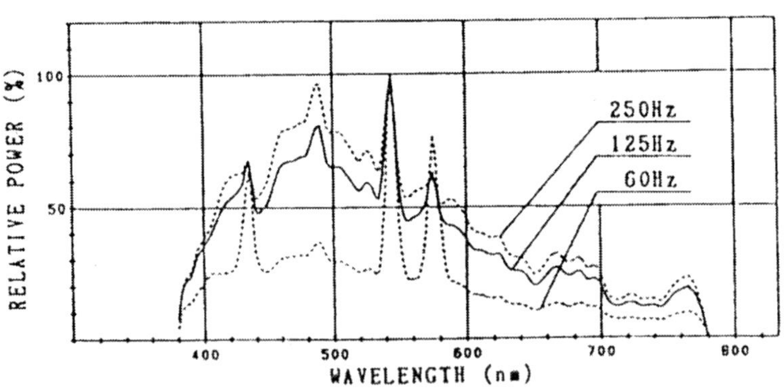

Fig. 3 Emission spectra from arc core under various operation frequencies

A NOVEL HIGH EFFICACY MICROWAVE POWERED LIGHT SOURCE

James T. Dolan, Michael G. Ury and Charles H. Wood
Fusion Systems Corporation, Rockville, MD 20855 USA

A unique plasma system has been discovered which uses high pressure sulfur or selenium vapor powered by microwaves. These lamps combine high efficacy and good color with high brightness and excellent spectral stability over life. The radiation mechanism is molecular transitions between the $B^3\Sigma$ and $X^3\Sigma$ states of S_2 or Se_2 dimers. Spectra from low pressure lamps show a strong UV component[1,2] which is reabsorbed at multiatmospheric pressures, resulting in a lamp with output predominantly in the visible. Radiative efficiencies (microwave power to optical radiation) of over 50% are reported.

Research using high pressure sulfur bulbs in one of Fusion's production microwave lamp systems is discussed. Figure 1 shows the spectrum of this type of bulb. The properties of the optimized lamp are summarized below:

Bulb Size	28 mm ID (sphere)
Microwave Power	3400 W
Luminous Flux	410 klumens
Efficacy (Microwave)	120 lpw
Color Coordinates	x = .31, y = .37
Correlated Color Temperature	6500 K
Color Rendering Index	R_a = 86

Variation of color with fill quantity has also been investigated, with the result that the output spectrum shifts to the red with increasing pressure while retaining its distinctive shape. Preliminary life test data are discussed, with the conclusion that bulb life may be very long. Sulfur is apparently inert with respect to the quartz envelope, as there was no evidence of sulfur loss or reaction over 9,000 hrs of operation. In consequence, the spectral output is very stable, with no measurable shift in color over life.

Smaller bulbs with higher brightness will also be described. Lamps attaining approximately 350 lum/mm^2 have been produced, and have been operated for over 800 hrs with no signs of degradation.

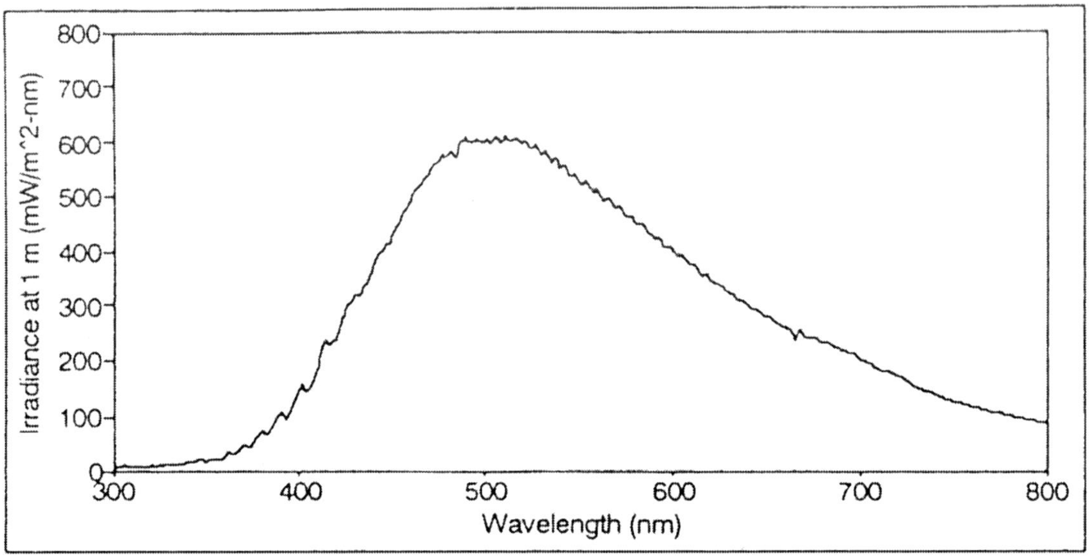

Figure 1. Spectrum of microwave powered high pressure sulfur lamp. 3400 W microwave power, 28 mm ID bulb

Acknowledgements
The authors gratefully acknowledge the guidance and encouragement of Dr. John Waymouth in preparing this paper

References

1. R. W. B. Pearse and A. G. Gaydon, *The Identification of Molecular Spectra*, 4th Edition (Chapman and Hall, London, 1976)

2. D. A. Peterson and L. A. Schlie, Journal of Chemical Physics 73, 1551 (1980)

NON-PASSIVE FAILURE OF HIGH PRESSURE GAS DISCHARGE LAMPS

75:L

K.S. Haagsma
Philips Lighting Division, P.O. Box 80020, 5600 JM Eindhoven, The Netherlands

INTRODUCTION

Most high intensity discharge lamps use an arc tube which operates at an over pressure. Especially for metal halide lamps a possibility exists that the arc tube could rupture at the end of life. Therefore most of those lamps are operated in enclosed fixtures, whose requirements are covered in UL 1572. Presently, there is an increasing demand for HID lamps which can be operated in open fixtures (better optical performance, easier to handle) [1]. Therefore, precautions must be taken to prevent the hot arc tube fragments penetrating the outer bulb. One such method is to employ a safety shield within the lamp which may consist of glass and metal parts. As a consequence, testing methods must be developed for evaluating the chosen protection. A well known laboratory method to cause an operating lamp to explode, is the so-called Capacitor Discharge test [2]. Within Philips Lighting other tests are available such as the Critical Operation Pressure test and the Bullet Impact Resistance test.

MEASURING THE CONTAINMENT SAFETY

Three different tests are currently being used within Philips Lighting:

a. *the Capacitor Discharge test*: this is a well known force fail test to make HID arc tube explode by discharging a certain amount of energy from a capacitor over the lamp electrodes while the lamp is in normal operation. The test results of the force failed arc tube depend on the (mechanical) strength of the seals (pinches). Therefore, it is very important to control the amount of discharged energy in order to assuring a realistic simulation (or approximation) of an end of life rupture.

b. *the Critical Operation Pressure test*: the arc tube of a specially prepared test lamp is pressurized via the exhaust tube until it ruptures. A range of bursting pressures at burst is realized by making a controlled scratch on the outside of the arc tube. The damage to the outer bulb or protective shield depends on the arc tube pressure at burst. This test is realistic in the sense that we know the lamp operation pressure and we have measured the (critical) pressure range where the outer bulb is damaged.

c. *the Bullet Impact Resistance test*: a quartz glass Fragment Simulating Projectile (FSP) is shot at the protective shield of the lamp as if the FSP was a shard from a ruptured arc tube. The probability of outer bulb damage depends on the velocity and weight of the FSP. Since the real fragment velocity and weight of a ruptured arc tube is generally unknown, this test can be used mainly to rank the impact resistance of different protective shields, independent of the type of arc tube.

QUANTIFYING THE CONTAINMENT SAFETY OF THE PROTECTED LAMP

The statistical tool which is used to analyze the test data is called the *probit analysis technique* [3]. As an example, measurement results of containment safety of the protected MHN-T 70W lamp are given in the following table.

Table: summary of containment safety test results of the MHN-T 70W lamp

Test	Results
Capacitor Discharge	No damages (n>100) at 145 J energy
Critical Operation Pressure	Critical pressure = 45 bar Lamp operation pressure ≈ 18 bar
Bullet Impact Resistance	Critical velocity of Bullet = 60 m/s Measured velocity of arc tube fragments = 25-35 m/s

Collectively, the three tests give a good indication of the containment safety of the HID lamp. It is clearly shown in the table that in the three tests the MHN-T 70W is containment safe in the case of a non-passive arc tube failure.
Although close examination of an arc tube rupture is possible, e.g. using high-speed photography during the Capacitor Discharge test, careful attention must be paid to compare the test results to a spontaneous end of life arc tube rupture. When comparing lamp designs, the other tests give further insights into the efficacy of the protection against non-passive arc tube failure.

CONCLUSIONS

The protection against non-passive failure of high pressure gas discharge lamps can be measured to a satisfactory degree. The test methods are well suited to characterize containment safety of HID lamps. When all results of the above discussed tests are taken together, it is shown that the MHN-T 70W lamp is well protected against non-passive arc tube failure.

REFERENCES

1. Tillman, N.J., v. Kemenade, J., v. Meel, F., v. Vliet, J., *High Quality Accent Lighting with Low Wattage Metal Halide Lamps*, Proc. IES Conference, Baltimore USA, August 1990

2. Parrott, R., Nortrup, E., Wyner, E., *Low Wattage Metal Halide Lamps with Enhanced Containment*, Proc. IES Conference Montreal, August 1991

3. Finney, D.J., *Probit Analysis, a Statistical Treatment of the Sigmoid Response Curve*, Cambridge Univ. Press, 1952, 2nd ed.

Session 7

Chairman: J. F. Waymouth

RF Discharge/Novel

76:I **J. Schlejen**
Inductively coupled fluorescent lamps: "The QL lighting system"

77:L **V. A. Godyak, R. B. Piejak, B. M. Alexandrovich**
Radio frequency driven miniature fluorescent lamps

78:L **C. N. Stewart, M. E. Duffy, J. T. Dakin, V. D. Roberts, S. El-Hamamsy, H. L. Witting, A. Inoye, K. Shimizu, K. Araki**
Inductively Coupled HID Lighting System

79:P **C. von Arx, U. Kogelschatz**
Intense silent discharge excimer UV source operating at room temperature

80:P **M. E. Duffy, J. T. Dakin, G. E. Duffy, M. M. Secen**
Diagnostics and model of an inductive HID Hg Discharge

81:P **V. A. Godyak, R. B. Piejak, B. M. Alexandrovich**
Matching network losses for capacitive RF discharge

82:P **A. T. Rowley, D. O. Wharmby**
Power dissipation and light generation in surface wave discharges

83:P **M. Shinomiya, M. Toho, M. Kawaguchi**
Impedance characteristics and its equivalent circuit of the electrodeless fluorescent lamp excited by the air-cored coil

84:I **R. Scholl, B. Weber, C. Weijtens**
New light generation mechanism: continuous radiation from clusters

INDUCTIVELY COUPLED FLUORESCENT LAMPS: 'THE QL LIGHTING SYSTEM'.

J. Schlejen
Philips Lighting, Central Development Laboratories.
PO Box 80020, 5600 JM Eindhoven, The Netherlands.

1. INTRODUCTION.

Electrodeless discharges offer many attractive features over existing light sources, both for manufacturers and end-users. Among others, the most important features due to the absence of electrodes, are:

I the exclusion of the main life-time determining factor of the lamp and, thus, a large extension of lamp life.

II a better lamp maintenance, due to lack of material sputtered-off from the electrode.

III freedom in lamp (and luminaire) design and the possibility to create new innovative light sources.

From a physical point of view, electrodeless discharges can be classified into four types: (1) capacitive E discharges, (2) inductive H discharges, (3) microwave discharges and (4) travelling wave discharges. A very usefull overview and short explanation on the four electrodeless discharge types has been given by Wharmby [1].

Although the idea of electrodeless lamps by Tesla [2] is 100 years old, upto now, no electrodeless general lighting was available[1]. The main problems which prevented the commercial introduction were size (temperature), electronics, electro-magnetic interference (EMI) and price aspects. However, exactly

---

[1]) However, electrodeless lamps for special purposes have been reported by Philips [3], Boumans [4] and Shinomiya et al. (the MEW 'EverLight' [5]).

100 years from Tesla, due to the large developments in electronics and modern phosphors, Philips Lighting has been able to present a commercially available light source, based on the inductive electrodeless discharge with internal excitation coil: the QL induction lighting system.

The 85 W QL lamp system is a high quality and a high system efficacy (>65 lm/W) general light source with an extremely long lamp life (60,000 hours) and good lamp maintenance; issues such as size, thermal problems, electronics and EMI are solved.

The QL system consists of a lamp, power coupler and HF generator (see fig. 1), which is combined with a luminaire. An overview will be given on the QL induction lamp and power coupler, QL HF generator, EM interference and finally possible future developments in induction lighting.

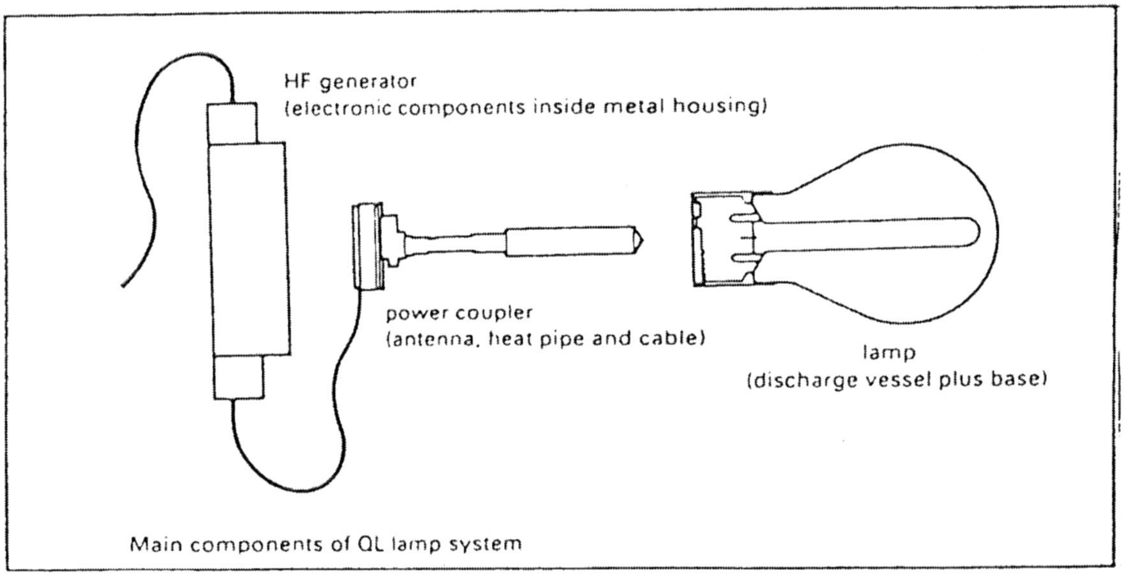

Figure 1: The QL lamp system.

2. THE ELECTRODELESS QL INDUCTION LAMP.

In figure 2a, the QL power coupler is shown. This power coupler contains an induction coil, winded around a ferrite core. The current through the coil induces a magnetic field H: the ferrite core intensifies the field[2]. The alternating magnetic field induces an alternating electrical field (for the QL at 2.65 MHz, higher frequencies are possible also), perpendicular to the magnetic field lines. Both the magnetic and electrical fields are shown in fig. 2a.

After ignition, in the discharge vessel, the electrons are accelerated in the electrical field to energies of ca. 1 eV and collide with the gas particles. This results in a circular electrical current in the plasma (see fig. 2b). The phenomena which then occur in the plasma are similar to those in a common low pressure Hg fluorescent lamp.

The gas plasma contains Hg atoms (as active medium) and a noble gas as buffer gas. By the electron collisions, the Hg atoms are excited to a higher energy level (some are ionized, also), from which they fall-back by emission of an UV photon. These UV photons are converted into visible light by the phosphor on the bulb wall and stem.

The electrical operation of the power coupler and lamp can be considered as a transformer (see fig. 2). The induction coil in the power coupler acts as the primary coil. The secondary coil is formed by the circular current in the discharge; this 'plasma' coil has one turn only. As in the QL lamp the primary coil has 15 turns and a ca. 1 A current, the current through the plasma is in the order of 10 A.

The induction coil and the plasma coil are coupled strongly: varying the coil geometry, gives rise to large variations in the plasma conditions of the lamp. The dependence of plasma current on the induction coil parameters is roughly as in a transformer.

---

[2]) Instead of an internal coil, also an external coil can be choosen. This reduces thermal problems for the coil and, for sufficient high frequencies, no ferrite is needed. However, such systems generate more EMI and have lower system efficacy.

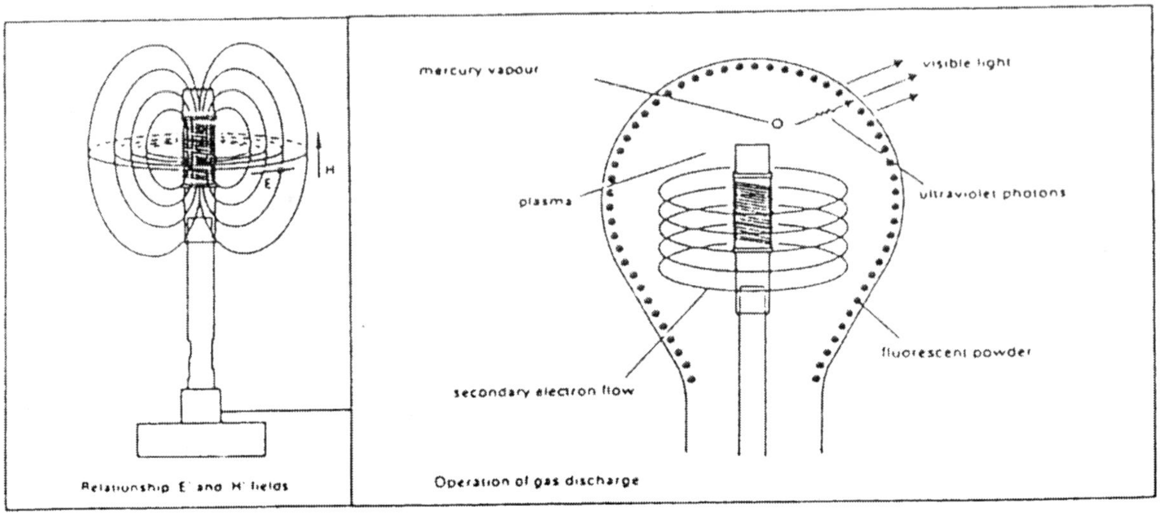

fig. 2a fig. 2b

Figure 2: Schematic view of the power coupler (a) and gas discharge (b) for QL lamps. The secondary electron flow (plasma current) and electric 'E' and magnetic 'H' fields are shown also.

Since the power coupler is located inside the centre of the QL lamp it heats up. Therefore, its function is not only to couple HF power into the plasma, but also to conduct the dissipated heat away from the centre. This is done by filling-up the air gaps in the power coupler with temperature resistant silicon rubber (to increase radial heat conduction towards the power coupler centre) and by having a heat-pipe in the power coupler (to increase axial heat conduction towards the bottom flange). Finally, the QL luminaire is essential to conduct the heat away from the bottom flange.

Moreover, due to the high thermal loading of the QL lamp, a BiIn amalgam is used in the lamp to obtain a suffuciently low (optimum) mercury pressure (also used in the 'Everlight' [5]). To speed-up the run-up of the QL lamp, an auxiliary amalgam is placed in the lamp.

To prevent lamp glass darkening and so obtain a high lamp maintenance, the QL lamp comprises a SiO2 protective coating

('Everlight': TiO_2 [5]) between glass and phosphor. In this way, the QL lamp still has a maintenance of more than 70% after 60,000 hours.

Starting of inductive lamps is done by supplying a high voltage over the induction coil. Ignition of the inductive discharge in the QL lamp is always preceeded by a capacitive discharge over the length of the induction coil (see fig. 3). The QL lamp starts within several tens of milliseconds.

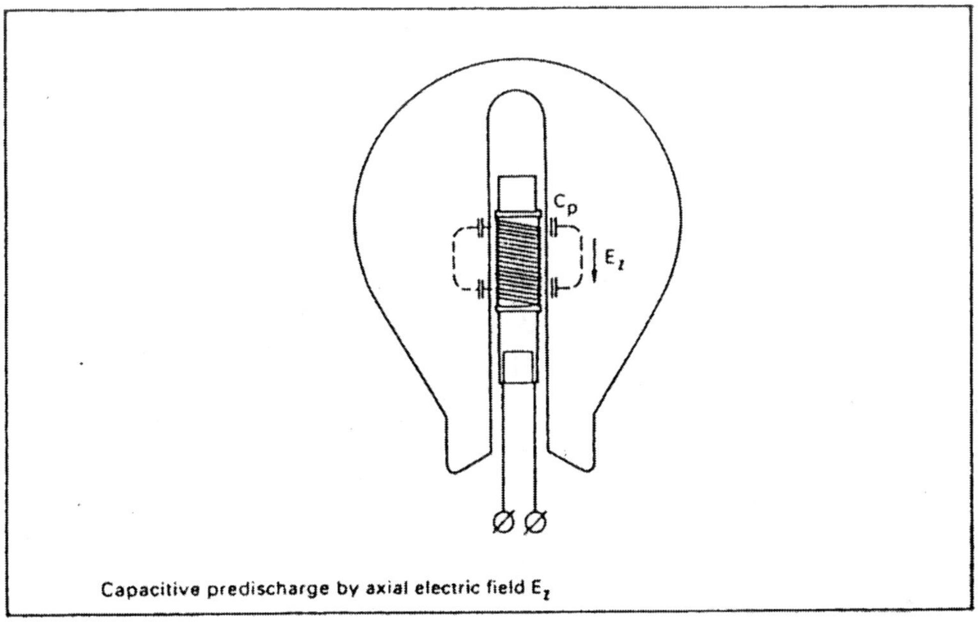

Figure 3: Schematic view of the capacitive discharge Ez along the induction coil (over parasitic capacitances Cp).

3. THE QL HF GENERATOR.

The QL HF generator (85 W) consists of the following main parts, which are indicated in the block diagram of fig. 4:

i) Radio Frequency Interference (RFI) filter.

ii) Preconditioner (including buffer capacitor).

iii) Power oscillator at 2.65 MHz.

The lamp and power coupler are part of the circuit design.

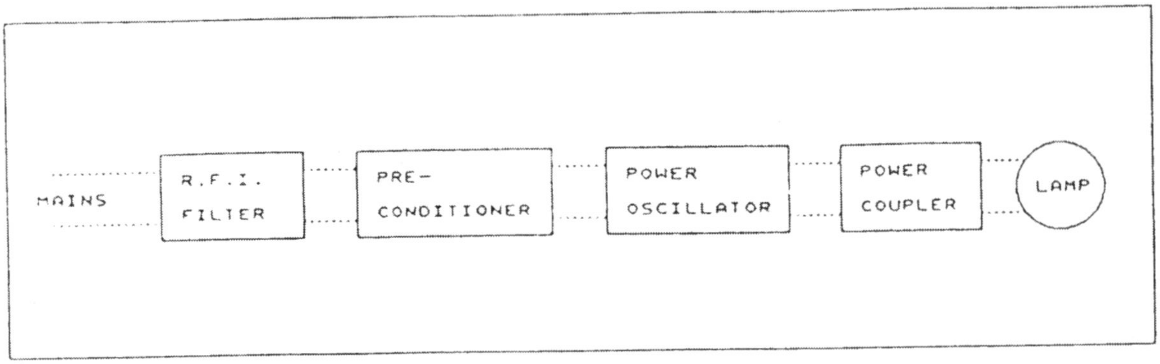

Figure 4: Block diagram of the QL HF generator.

The **RFI filter** is one of the measures which ensures that the QL system complies to the EMC demands (EN 55015).

The **Preconditioner** is designed similar to HF fluorescent ballast and provides a constant DC output voltage (independent of mains voltage variations) and a low mains current distortion (according to IEC 555-2). Besides these functions the preconditioner blocks mains transients and reduces the influence of line interuptions.

The **HF power oscillator** converts the DC voltage of the up-converter into a 2.65 MHz AC voltage. It generates and controls the HF power supplied to power coupler and lamp. Basically, this circuit is a self-oscillating half bridge circuit.

During starting (when the lamp is not yet ignited), the circuit operates near resonance and the induction coil merely acts as a coil over which a high voltage is generated. This voltage ignites the QL lamp (see chapter 2).

The efficiency of the 85 W HF generator is about 90%. The generator design has been such to ensure a low component stress (i.e. low DC voltage, ripple current and ambient temperature), in order to achieve a low failure rate and high lifetime of 60,000 hours.

4. EMC.

The suppression of unacceptable EM interference is one of the main technical barriers to overcome in the development of induction lighting. For the QL, the frequency chosen (2.65 MHz) is between the SW bands (>4MHz) and the MW bands (<1.6 MHz)[3].

The EM interference can be divided in two catagories:

i) Radiated electric and magnetic dipole radiation.

ii) Conducted electric interference (conducted through mains).

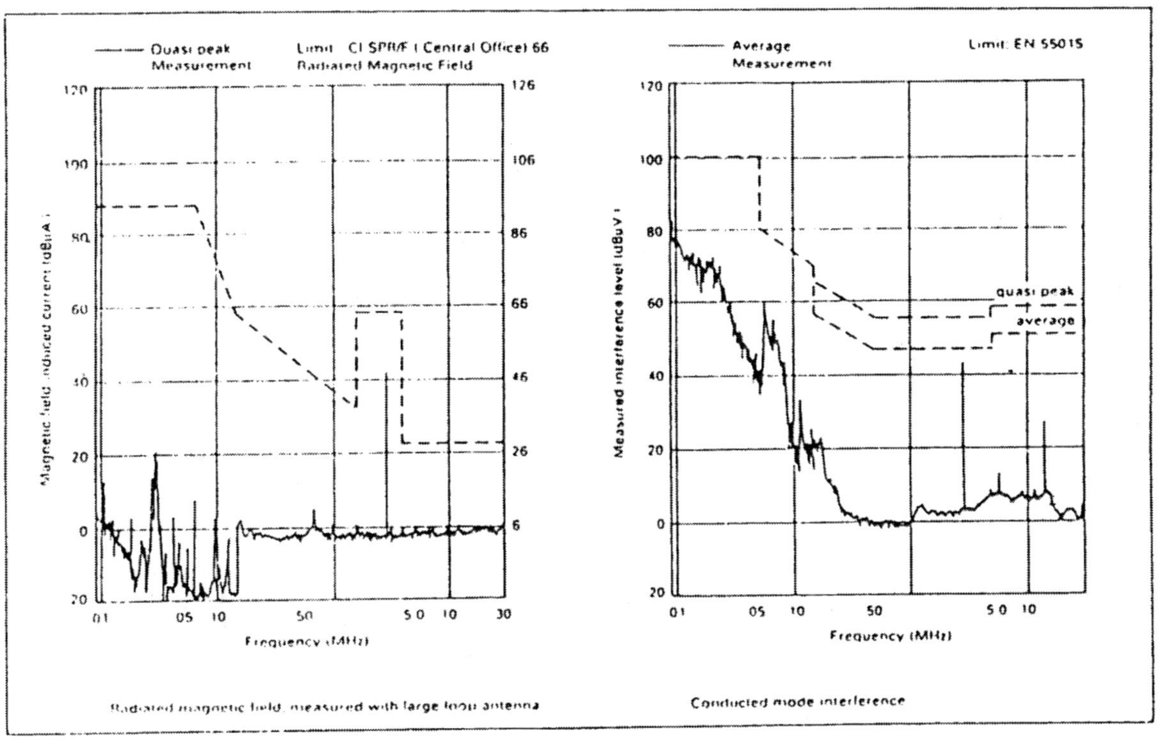

Figure 5: Typical EM interference spectra for a QL lighting system.

---

[3]) Another option is to choose for an ISM band (f.i. 13.56 MHz). Problem with these higher frequencies is the stronger generation of unacceptable higher harmonics. Moreover, the effiency of practical HF generators drops for higher frequencies. At lower frequencies (c. < 1 MHz) it becomes difficult to couple the energy efficiently into the discharge.

In the QL system, several measures have been taken to comply with the legal limits for both interference types. The main contributions to suppress the interference in the QL system are (most measures reduce both types of interference):

i) Shielding of the EM fields (generated by the internal induction coil) by the plasma on the outside.

ii) Compensation of the electrical parasitic currents by a bifilarly winded induction coil in the power coupler.

iii) Shielding of the EM fields by the grounded metal in the luminaire.

iv) Special design of the HF generator (shielding of the HF power oscillator, RFI filter).

Using these measures the QL lighting system complies with the legal limits (see fig. 5).

5. FUTURE DEVELOPMENTS.

As has already been shown by some Lighting Companies in their laboratories (Philips Lighting, GTE/Sylvania), the induction lamp principle can be used to develop Compact Fluorescent Lamps (CFL). Major problems to solve in this field will be (i) the thermal load of the integrated electronic circuit and, thus, circuit lifetime and (ii) the suppression of unwanted EM interference without the aid of a luminaire.

For the high lumen end, a luminous flux of roughly 10 klm seems to be a practical limit: above this limit, thermal problems in lamp and electronics are difficult to solve with the current state of the art in fluorescent induction lamp technology.

Finally, due to the absence of electrodes, all kinds of sophisticated voltage and current waveforms may be applied to power coupler and lamp in order to make special lighting effects possible (lamps with variable colour point ?) or to reduce the colour point shift during dimming.

REFERENCES:

[1] Wharmby D.O. (1989), 'Review of Electrodeless Discharges for Lighting', 5th Int. Symp. of Science ad Techn. of Light Sources, York.

[2] Tesla N. (1891), Elec. Engineer, 7, 549-550.

[3] Philips R. (1983), 'Sources and Applications of UV Radiation', New York: Academic Press.

[4] Boumans P.W.J.M. (1987), 'Inductively Coupled Plasma Emission Spectroscopy', New York: Wiley.

[5] Shinomiya M., Kobayashi K., Higashikawa M., Ukegawa S., Matsuura, J. and Tanigawa K. (1991), 'Development of the Electrodeless Fluorescent Lamp', J. Ill. Eng. Soc., Winter 1991, p44-49.

Radio Frequency Driven Miniature Fluorescent Lamps

V.A. Godyak, R.B. Piejak and B.M. Alexandrovich
GTE Laboratories Incorporated, Waltham, MA 02254, USA

77:L

Miniature fluorescent lamps (MFL) are widely used as a back light source for liquid crystal displays (LCD) in computers, automotive and aircraft instrumentation clusters. Typically, an MFL is composed of a glass tube with an inner diameter (ID) of a few mm's and a length of 10 to 20 cm, containing a cold cathode, longitudinally oriented discharge in a rare gas-mercury mixture. Due to the nature of a cold cathode (glow) discharge, a significant portion (about half) of the applied discharge voltage is across the cathode sheath. The power dissipated in the cathode sheath is wasted, as far as light production is concerned, since light output is mainly produced by the positive column of the discharge. Attempts to increase the light output of the MFL by increasing the discharge power lead to an additional drop in the MFL efficiency (LPW), since an increase in the discharge current (power) results in an increase in the cathode sheath voltage and a decrease in the voltage across the positive column. Moreover, the discharge operation at higher cathode voltage results in significant enhancement in the cathode sputtering rate and drastically reduces MFL life expectancy. Because of this, the cold cathode MFL operates at relatively small discharge currents with very limited light output.

Driving an MFL with RF instead of high frequency (HF, typically 40 KHz) introduces the possibility of increasing the MFL's efficiency while also increasing its absolute light output. This possibility stems from the predominant capacitive impedance of the RF electrode sheaths that separate the RF plasma body (positive column) from the RF electrodes, or from the glass surface (when RF is externally coupled, as in electrodeless RF discharges). For an RF driven discharge, the RF sheaths are a capacitive analog of the ballasting inductor of traditional, line frequency operated fluorescent lamps. In both cases the power loss in the ballasting element (an inductor and/or an RF sheath) can be made significantly less than the power loss in the plasma body of the discharge, while the differential impedance of the ballasting element may be considerably larger than that of the plasma body.

In the present report we discuss the results of a comparative study of the electrical characteristics of RF driven MFL's. The RF driving voltage V_d, the discharge current I_d, the total discharge power P and the relative light output L have been measured at three frequencies: 13.56, 27.12, and 40.68 MHz and at three argon pressure: 5, 20 and 40 Torr. All measurements were performed in an MFL having a 7 mm OD, 5 mm ID and a 20 cm length, in a discharge current interval between 15 and 100 mA. The lamp geometry and phosphor is comparable to a commercial cold cathode MFL (Sylvania F1.5/7mm/C/D850) which is conventionally driven by a 40 KHz supply with a discharge current of 5.5 mA and a power of 1.5 watts shared nearly equally between the cathode sheath and the plasma body. From measurements of the external RF discharge characteristics and the RF electric field along the plasma body, we have inferred the value of the internal discharge parameters as functions of the discharge current based on procedures developed in Ref [1]; these parameters are: the voltage drop across the plasma, the RF and DC components of the RF sheath voltage and the RF power dissipated in the RF sheath and in the plasma (for three RF frequencies and three argon pressures).

An example of the result from such analysis is shown in Fig. 1 where the ratio of electrode sheath power to total discharge power (P_{sh}/P) is given as a function of P for two gas pressures. Fig. 1 clearly demonstrates the advantage of driving the MFL with RF rather than HF. A commercial MFL driven at 40 KHz, with p = 20 Torr and P_d = 1.5W has a $P_{sh}/P \approx 0.5$ while the same type of lamp driven electrodelessly at 13.56 MHz demonstrates a considerably lower P_{sh}/P while operating at a higher discharge power.

Measurements show that P_{sh}/P is falling with rising frequency and gas pressure. Reduction in sheath power with RF operation of MFL also reduces the DC voltage across the sheath, thereby reducing ion energy and ion flux to the wall (or electrode) and should result in improved lamp maintenance. Thus, a switch from cold cathode HF operation to RF electrodeless operation allows one to increase lamp efficiency, wattage and lifetime of the MFL while reducing manufacturing cost because of the absence of electrodes.

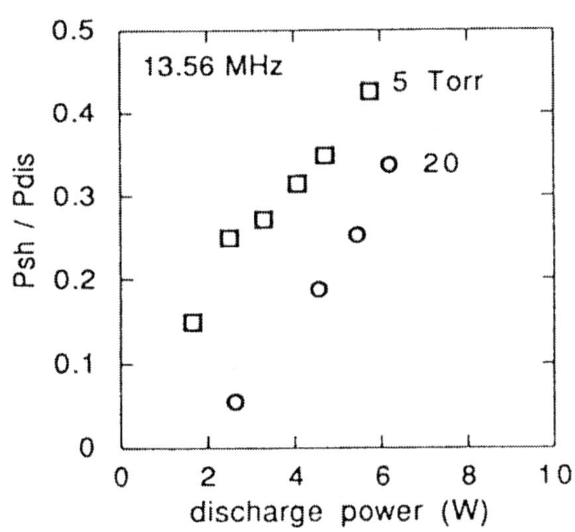

Fig. 1 Ratio of sheath power to total discharge power for RF driven MFL.

References

1.) V.A.Godyak, R.B.Piejak and B.M. Alexandrovich, IEEE Trans. Plas. Sci., 19, 660 (1991).

INDUCTIVELY COUPLED HID LIGHTING SYSTEM

Charles N. Stewart, Mark E. Duffy, James T. Dakin
GE Lighting, Nela Park, Cleveland, Ohio 44112, USA

Victor D. Roberts, Sayed-Amr El-Hamamsy, Harald L. Witting
GE Corporate Research & Development
Box 8, Schenectady, NY 12301, USA

Akihiro Inouye, Keiichi Shimizu, Kenji Araki
Toshiba Lighting & Technology Corporation
1-201-1 Funakoshi, Yokosuka, 237 Japan

The utilization of inductively coupled radio frequency energy to excite high intensity discharges has long been recognized as a possibility for significantly increasing the efficiency of lighting systems. Only recently, however, has this concept been reduced to a practical system [1,2]. We will describe a lamp and ballast system that shows considerable promise in the application of inductively coupled discharges to lighting. Arc efficacies of >200 lumens per watt have been achieved in lamp-ballast systems with efficacies of >135 lumens per watt. The electrical efficiency of the power supply, matching network and coil is >70%.

Within the ballast a resonant circuit is employed to ensure sufficient coil current to cause glow to arc transition. The RF output stage is driven by a 13.56 MHz crystal controlled oscillator. Both half bridge and full bridge class D power stage design have been evaluated. The full bridge approach requires superior balance of the gate drive, while the half bridge requires superior power devices. An auxiliary high voltage low current supply is used for starting. Heat generated by the circuits, especially the power devices, is conducted to the outer fixture surface to allow operation in a 55° C ambient. Calculations of the radiated EMI indicate regulatory limits are achieved; initial measurements support this.

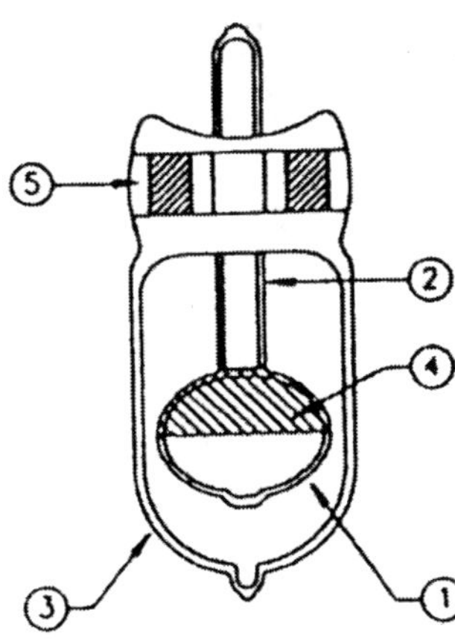

A multi-turn coil couples RF energy into the lamp. Analytic and finite element models have been developed to quantify the effects of arc tube position and coil shape and size on coil electrical and optical efficiency. Both the coil and the arc tube coating (4) can be used to optimize light from the luminaire.

Inductive coupling allows symmetry and simplicity of lamp construction, shown in the figure. The arc tube (1) is an oblate spheroid with an alumina or zirconia coating (4) and without metal seals. The starting probe (2), used to start the lamp is attached to the arc tube and held in position by a pinch seal (5) to the outer jacket (3). When the lamp is started a low pressure discharge, established in the starter probe by the high voltage supply of the ballast, couples energy to the arc tube and ensures quick glow to arc transition of the main discharge. Starting at or below the operating power of the lamp is easily achieved. The compact symmetric arc tube results in more rapid warm up and improved color uniformity, when compared to standard HID lamps; the figure shows a set of sixteen lamps run at constant power whose color spread is well less than three steps. Dimming to low levels (<50%) is possible because of the inductive coupling and the lack of electrodes.

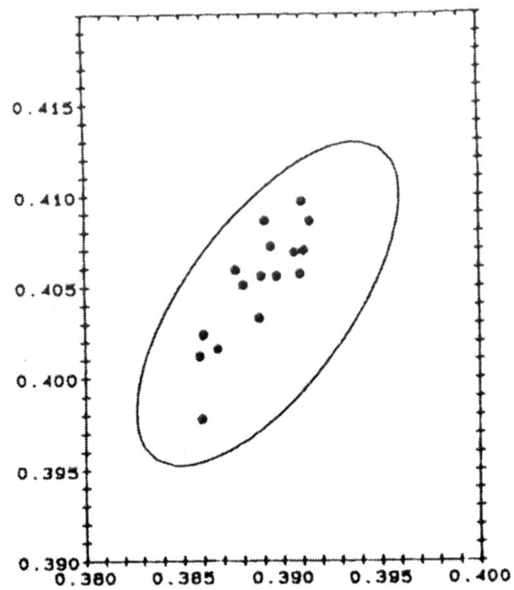

The inductively coupled discharge allows much greater freedom in the range of doses that can be utilized. A wide range of halides have been studied, alone and in binary and ternary combinations; all of these operate on the same ballast without any modification. The life limiting mechanism is believed to be reaction with the wall:

$$4REI_3 + 3SiO_2 \leftrightarrow 2RE_2O_3 + 3SiI_4$$

$$4REI_3 + 7SiO_2 \leftrightarrow 2RE_2Si_2O_7 + 3SiI_4$$

The decomposition reaction of silicon tetraiodide at the electrode surface and subsequent dissolution into the tungsten bulk is absent. Further, the absence of electrodes eliminates wall darkening and improves color shift with burning.

The advantages of this compact lighting source in optical systems and the environmentally favorable chemistry will also be covered.

1. J.M. Anderson, US patent 4,180,763, December 25, 1979

2. P.D. Johnson, J.T. Dakin, and J.M. Anderson, US patent 4,810,938, March 7, 1989.

INTENSE SILENT DISCHARGE EXCIMER UV SOURCE OPERATING AT ROOM TEMPERATURE

C. von Arx, U. Kogelschatz
Asea Brown Boveri Corporate Research, 5405 Baden / Switzerland

Incoherent excimer radiation can be efficiently generated in silent discharges (dielectric-barrier discharges) [1-4]. Powerful UV and VUV sources for surface modification and photochemical processes can be based on this principle [5]. Figures 1 and 2 show a typical set-up of an excimer lamp.

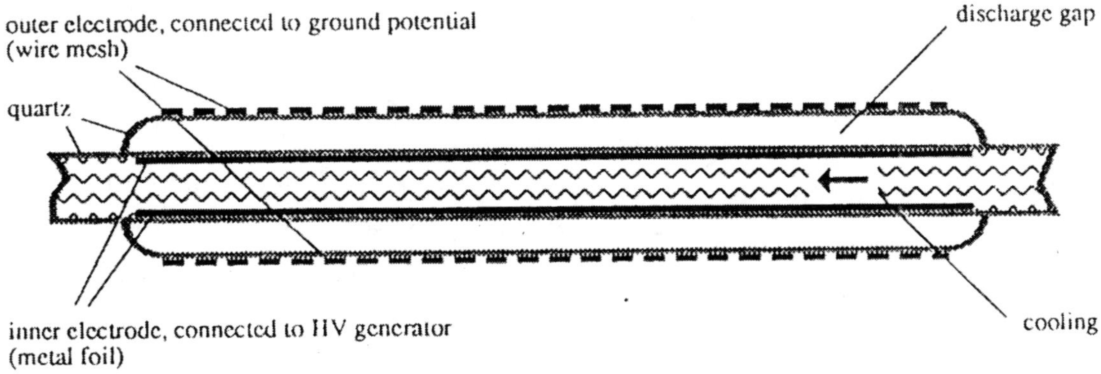

Fig. 1. Sealed cylindrical silent discharge configuration with annular discharge gap.

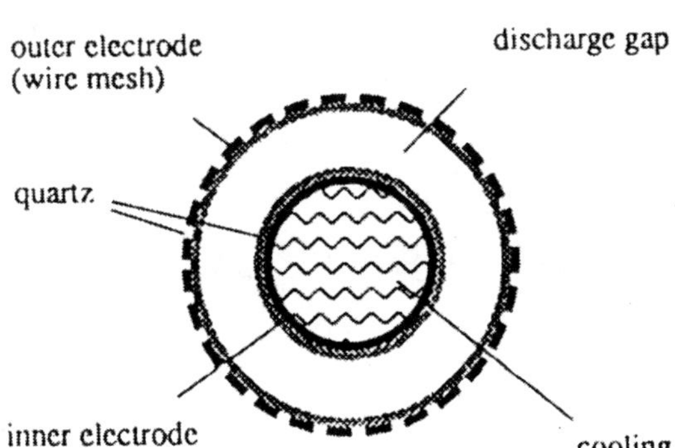

Fig. 2. Cross-section of a standard configuration.

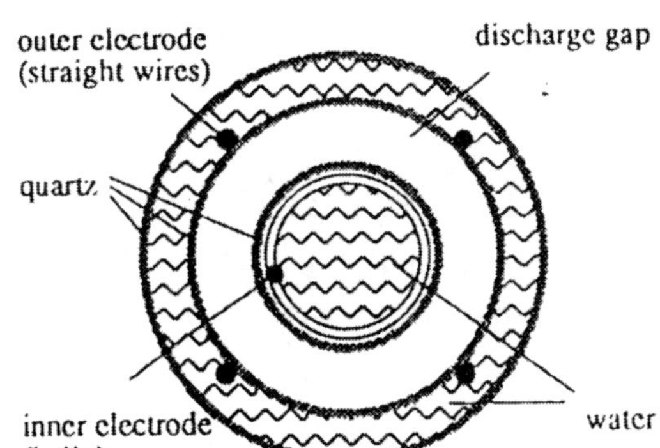

Fig. 3. Cross-section of a double-cooled set-up.

Unlike most other light sources excimer UV sources can be operated at fairly low temperatures, e.g. room temperature. Excimer formation requires excited and ionic precursor species that are formed by electron collisions. The gas temperature or heavy particle temperature has little influence on this process. However, high power densities are compatible with the excimer formation kinetics, as is known from excimer laser research [6]. In order to get low surface temperatures, excimer lamps operated at high powers must be efficiently cooled. In this article we describe a configuration with a cooling system using distilled water inside and outside the sealed coaxial discharge tube. A cross-section of this set-up is shown in Fig. 3. Four important properties of distilled water are made use of. First, water is excellent for cooling purposes. Second, water is transparent in the visible and ultraviolet spectral range down to 200 nanometers. Third, water absorbs infrared radiation from the plasma. Fourth, water has a high permittivity ($\varepsilon_{rel} \approx 81$). This unique property allows special and simple electrode designs.

The inner electrode is a solid wire, formed to a helix of a relative large pitch (roughly 2 cm) and a diameter that fits that of the water cooled inner tube. A high transmission of the inner section results which is useful in combination with a rear reflector that directs the radiation into one half-space. This is a common requirement for the irradiation of planar surfaces. The outer electrode consists of a few straight wires immersed in water (Fig. 3). For safety reasons these wires are usually connected to ground potential while the helix is connected to the high voltage supply operating at 55 kHz. The straight wires are positioned in such a way that only negligible blockage of the UV radiation occurs. This electrode configuration is based on the fact that the high permittivity of water tends to concentratethe electric field in the annular discharge space ($\varepsilon_{rel} = 1$). The experiment showed that these electrodes behave in a similar way to cylindrical solid metal electrodes ($\varepsilon_{rel} = \infty$), which would completely confine the electric field.

Silent discharges are characterized by the presence of a large number of constricted microdischarges. In this electrode configuration with lumped electrodes a homogeneous distribution of filaments in the discharge space is observed as in the standard configuration (Fig. 1 and 2). This is a consequence of the high ε of water surrounding the electrodes. The electrical properties of silent discharges result in a parallelogram-like shape of the voltage vs. charge Lissajous figure (U-Q-diagram) [7]. A recorded U-Q diagram (Fig. 4) of this special electrode configuration is comparable to U-Q-diagrams obtained with the standard electrode configuration (Fig. 1 and 2).

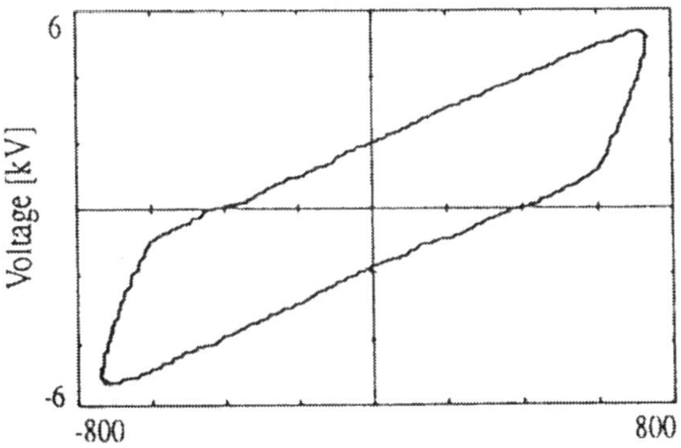

Fig. 4. Recorded U-Q-diagram of a XeCl* excimer lamp cooled inside and outside.

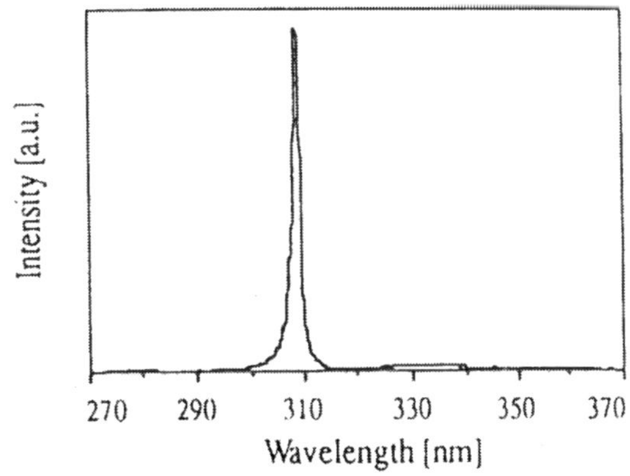

Fig. 5. Spectrum of a XeCl* excimer lamp.

The experiments were performed with a double-walled coaxial quartz tube filled with Xe, Cl_2 and Ar. Sealed tubes could be operated for several hundred hours without much loss of UV intensity. In this gas mixture strong fluorescence at 308 nm is observed, which corresponds to the B → X transition of the XeCl* excimer which is also made use of in excimer lasers. In an optimized gas mixture a narrow-band emission feature of 2 nm halfwidth is obtained with practically no additional radiation between 200 and 800 nm. A spectrum is shown in Fig. 5. The input power was 300 W for a discharge tube of 30 cm length and 3 cm outer diameter (gap spacing 5 mm). Even at this power level surface temperatures could be kept at room temperature. This intense narrow-band UV-B source operating at low temperatures opens up new possibilities for photopolymerization processes on heat sensitive substrates like thin polymer foils.

References:

[1] B. Eliasson, U. Kogelschatz, Appl. Phys. B 46 (1988) 299
[2] B. Gellert, U. Kogelschatz, Appl. Phys. B 52 (1991) 14
[3] H. Müller, M. Neiger, V. Schorpp, K. Stockwald, Proc. LS:5 York (1989) 171
[4] U. Kogelschatz, Pure & Appl. Chem. 62 (1990) 1667
[5] U. Kogelschatz, Appl. Surf. Sci. 54 (1992) 410
[6] Ch. K. Rhodes, Ed., Excimer Lasers (Springer, Berlin 1979 and 1984)
[7] U. Kogelschatz, Topical Invited Lecture, XX International Conference on Phenomena in Ionized Gases (ICPIG XX), Pisa (1991)

Diagnostics and Model of an Inductive HID Hg Discharge

M. E. Duffy, J.T. Dakin, G. E. Duffy, and M. M. Secen
GE Lighting, Nela Park, Cleveland, OH 44120

Experimental diagnostics and a numerical model are used to investigate an inductive HID Hg discharge.[1,2] Light sources based on similar discharges offer the potential of long life, high efficiency and good lumen maintenance due to the absence of electrodes and glass to metal seals.[3] The system considered here provides an excellent vehicle for developing diagnostics and models of inductive electrodeless discharges due to the well known behavior of high pressure Hg vapor.

The discharge system is shown schematically in Fig. 1. The arctube is a right circular cylinder having height, diameter and wall thickness of 20, 30 and 1 mm respectively. The arctube is filled with 9 mg of Hg and 3 Torr of Ar starting gas before sealing. The arctube is excited by a 33 mm diameter, 2 turn induction coil. The coil is driven by a 13.56 MHz power source through a 50 ohm transmission line and a capacitive matching network. When the drive power is in the range 200 to 400 Watts, the Hg is fully vaporized, and the pressure estimated to be 1 atm.

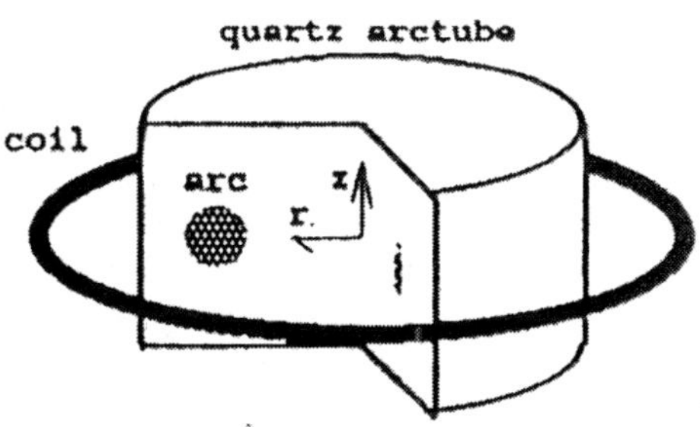

Figure 1: Diagram of discharge system

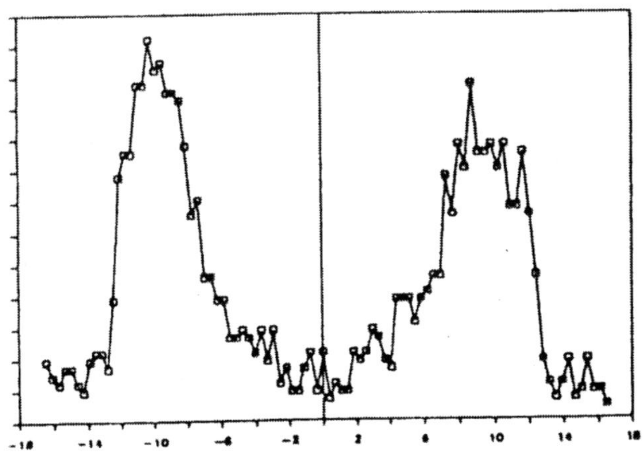

Figure 2: Radial scan of imaged Hg 577 nm radiation

A variety of diagnostic techniques are used to characterize the state of the discharge system. Measurements of the coil impedance under a variety of load conditions provide information on the inductive coupling coefficient and on the discharge impedance. Spectroscopic measurements provide information on the discharge temperature and shape. The radial scan of Hg 577 nm radiation shown in Fig. 2, for instance, shows the discharge to have a hot annular core surrounding a somewhat cooler center. Photometry provides information on the luminous efficacy and energy balance. Optical pyrometry provides the quartz wall temperature.

The 2-D axisymmetric model provides a solution to the coupled energy and momentum equations for the discharge fluid.[4] The source term for the energy equation includes ohmic heating in the self consistent field created by the coil and the discharge. It also includes radiative cooling, determined by integrating the radiation transfer equations for the collision broadened Hg lineshapes. The source terms for the momentum equations include gravity and the $\mathbf{J} \times \mathbf{B}$ force. In determining the fluid properties and the wavelength dependent radiation absorption, the discharge fluid is assumed to exhibit local thermodynamic equilibrium. The model indicates a strong convective flow driven by the $\mathbf{J} \times \mathbf{B}$ force, as shown in Fig. 3. The peak convective velocities are of order 40 cm/sec.

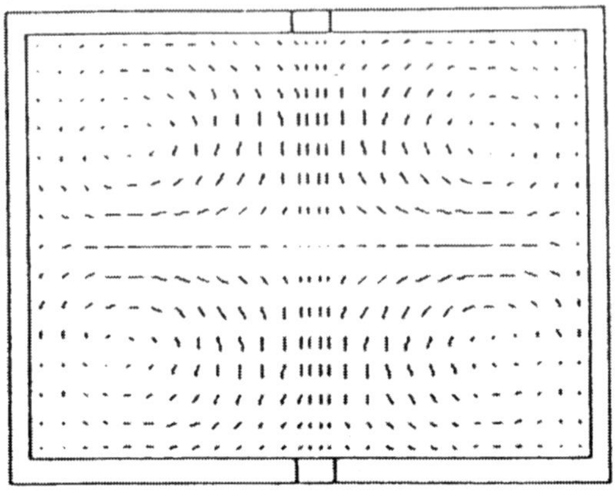

Figure 3: Flow field induced by $\mathbf{J} \times \mathbf{B}$ force

The model and diagnostics provide a rather complete description of the inductive discharge, which in many ways is similar to an equivalent discharge in an electroded arctube. Significant differences between the two cases include the role of convection, and the replacement of "end losses" in the electroded case with "coil losses" in the inductive case.

References

[1] J.M. Anderson, paper 7, LS4, Karlsruhe (1986).

[2] D. Hollister et al., paper 43:p, LS5, York (1989).

[3] C.N. Stewart et al., paper at LS6, Budapest (1992).

[4] J.T. Dakin et al., J. Appl. Phys., vol. 56, p. 4074 (1989).

Matching Network Losses for Capacitive RF Discharge

81:P

V.A. Godyak, R.B. Piejak and B.M. Alexandrovich
GTE Laboratories Incorporated, Waltham, MA 02254, USA

Proper design of the matching network is an important issue in any rf discharge application. Along with impedance transformation, the matching network should be energy efficient, provide discharge stability (ballasting), be tuneable and develop a high enough voltage to initiate the discharge (breakdown). In this work we consider matcher efficiency, by analyzing the relationship between power dissipated in the matcher and power delivered to a capacitive rf discharge.

A schematic diagram of the matching circuit and a capacitive rf discharge is shown in Fig. 1. Representing a capacitive rf discharge by an equivalent series RC network is convenient since over a wide range of discharge power both the discharge capacitance C and the ohmic component of the discharge voltage V_p (plasma voltage) remain essentially constant[1] ($V_p = IR = V\cos\phi \approx$ const) where V, I and ϕ are the discharge voltage, current and phase shift between them, respectively. In a typical fluorescent lamp with gas pressure p about a few Torr and a driving frequency of 13.56 MHz, provided that the external parameters are organized in such a way as to make ion acceleration losses in the sheaths negligibly small, the rms value of V_p can be estimated to be equal to the voltage drop across the positive column of a dc discharge with an equivalent gas mixture and geometry.

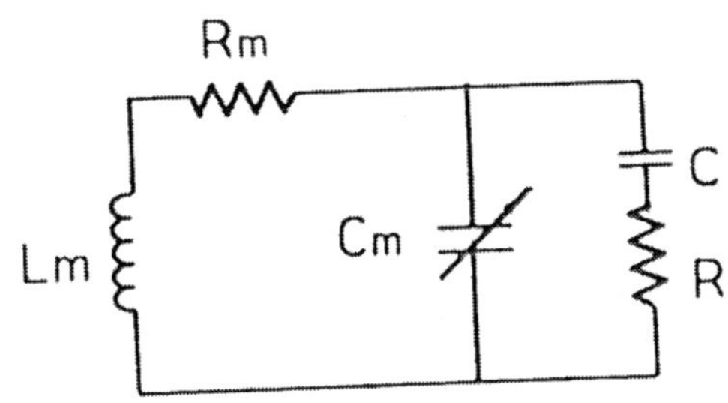

Fig. 1

The rf power delivered to a matcher from the output of an rf generator can be split into two parts: the discharge power P_d and the matcher loss P_m. Assuming a resonant condition, calculations based on the equivalent circuit in Fig. 1 yield:

$$QP_m/P_d = \tan\phi + C_m\,[C\cos\phi\,\sin\phi]^{-1}$$
$$= (V^2/V_p^2 - 1)^{1/2} + C_m V^2\,[CV_p^2\,(V^2/V_p^2 - 1)^{1/2}]^{-1}$$

where Q is the unloaded Q-factor of the matching network.

From this equation (shown in Fig. 2) it follows that the relative power lost in the matcher P_m/P_d is inversely proportional to its Q-factor (that is trivial) and it depends on the ratios of C_m/C and V/V_p. Note that the capacitive rf discharge can exist only when $V>V_p$ and thus $V/V_p > 1$ and $\cos\phi < 1$. On the other hand, the discharge capacitance C which is due to the rf sheaths is nearly proportional to the driving frequency and the square root of the gas pressure[1], thus $C \propto \omega p^{1/2}$. The matcher losses are minimal when C_m is zero and the matcher inductance is resonant with the discharge capacitance only. For this case:

$$Q\, P_m/P_d = \tan\phi = (V^2/V_p^2 - 1)^{1/2} = Q\, R_m/R$$

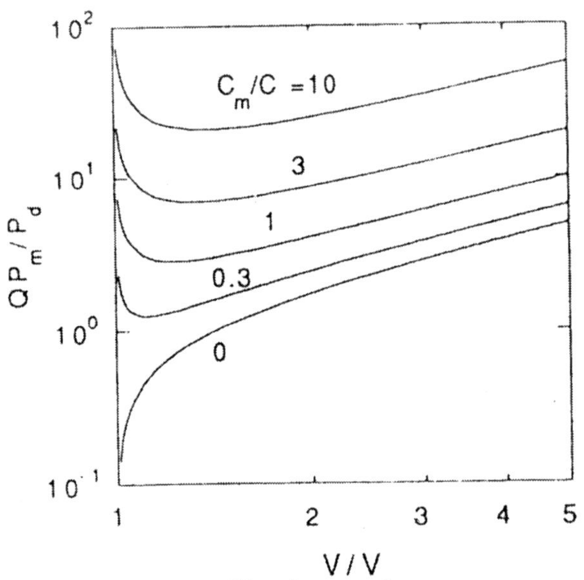

Fig. 2

with P_m/P_d tending to zero as V approaches V_p. This corresponds to discharge extinction which is accompanied by an unlimited rise in the plasma resistance. For the opposite case, when $V^2 > V_p^2$, $Q\, P_m/P_d = (1+C_m/C)\, V/V_p$ and $P_m \propto V^2$ while $P_d \propto V$. The different behavior of $P_m(V)$ and $P_d(V)$ is a consequence of discharge nonlinearity ($V_p \approx$ const and $R \propto I^{-1}$) while the matcher itself is a linear device for which $P_m \propto V^2$.

It is interesting to note that for $C_m \neq 0$ there is an optimal voltage V_1 where QP_m/P_d is minimal. This voltage can be found from extremum condition: $\tan^2\phi_1 = C_m[C+C_m]^{-1}$ where $\cos\phi_1 = V_p/V_1$. For a realistic case $C_m \geq C$ and ϕ_1 is close to $\pi/4$ which corresponds to $V_1^2 = 2V_p^2$. Fortunately, this case coincides with the condition for optimal self ballasting of the capacitive rf discharge (rf voltage across the sheaths and plasma are equal), and in this condition the rf power losses P_i due to ion acceleration in the sheath are relatively small[1,2]: $P_i/P_d \approx (m/M)^{1/2}\, \upsilon_{Te}/\upsilon_{\sim} << 1$. Thus, for minimal loss in the matcher one must arrange the capacitive rf discharge so that: $V_1 = \sqrt{2}V_p$ or $\cos\phi_1 \approx 0.7$. In this case, with sufficient accuracy for practical applications, the matcher efficiency is found to be: $P_m/P_d \approx Q^{-1}[1+2C_m/C]$.

References

1) V. Godyak, R. Piejak and B. Alexandrovich, IEEE Trans. Plas. Sci., 19, 660 (1991).
2) V. Godyak, R. Piejak and B. Alexandrovich, J. Appl. Phys. 69 3455 (1991).

POWER DISSIPATION AND LIGHT GENERATION IN SURFACE WAVE DISCHARGES

A T Rowley* and D O Wharmby
GE Thorn Lamps Ltd., Melton Road, Leicester, UK
*now at EA Technology, Capenhurst, Cheshire, UK

We are concerned here with surface wave discharges(1) (SWD) in mercury rare-gas systems in the range from 1 MHz to a few hundred MHz. We used discharge tubes of 11mm internal diameter and various lengths. These were filled with argon at 1 torr and mercury with a pressure determined by the cool spot temperature. Figure 1 shows how the discharge was operated. The annular gap in the launcher provides, at the surface of the discharge vessel, an electric field in the direction of propagation of the surface wave which has the correct symmetry. If the field is large enough, plasma is created. The plasma and the tube wall then form a waveguide down which the surface wave propagates, forming more plasma: thus the wave forms the plasma, and the plasma guides the wave.

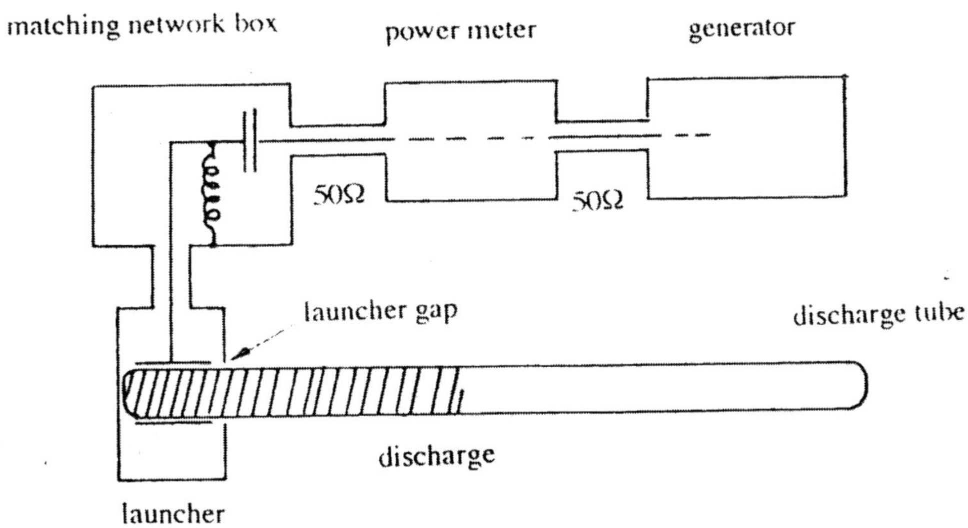

Figure 1 shows the experimental arrangement. The matching network adjusts the impedance of the cylindrically symmetric launcher with its lamp load and intrinsic capacitance to that of the 50 Ω generator. In this picture, the 'natural length' of the discharge (shown shaded) is less than the tube length.

A characteristic property of SWD is that the 'natural length' of the discharge depends on the power applied to the launcher. The higher the power, the longer the discharge; and the further from the launcher, the lower the intensity of light generation. The surface wave fields are attenuated as power is gained by the plasma from the wave: at some point the plasma cannot be maintained. Another characteristic is that the plasma power per unit length increases with applied frequency; low frequency, high power discharges are long. In experiments in which we measured both the power and light output per unit length of phosphor coated SWD[2], we found that the peak value of column efficacy was comparable to that in standard fluorescent lamps. The overall SWD efficacy was lower than that of fluorescent lamps because optimum conditions were not maintained throughout the whole SWD column. We have therefore investigated ways of achieving more uniform power dissipation.

If the power is increased in a short tube, the wave is 'reflected' from the end of the tube and the discharge becomes much more uniform. We have measured the power balance of such discharges using thermal imaging to compare the radiance per unit length of the discharge in the infra-red with that of a standard fluorescent lamp column of known power per unit length [2]. In addition we have measured the power dissipation in the matching network box by comparing the radiance of the surface of the box with that of an identical box containing resistances in which known powers were dissipated. We therefore measured the power dissipated in the part of the lamp outside the launcher, and the power in the matching network. Subtracting these from the power input from the generator yields the power dissipated in the launcher: radiated RF power is much less than 1% of total input power and is ignored in the power balance.

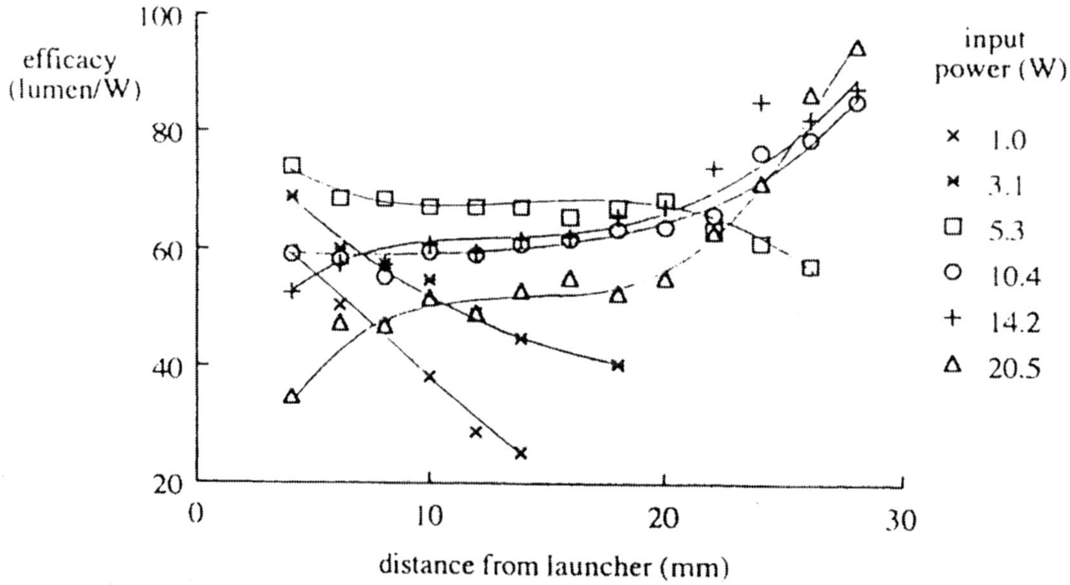

Figure 2 shows the efficacy along the length of a SWD as the input power is varied at optimum cool spot temperature. Discharge tube length is 300mm, launcher length 36mm and frequency 43MHz.

Light output and power measurements show that the natural length becomes equal to the 300mm tube length for a power of about 5W. Figure 2 shows that between about 5 and 10W, corresponding to an average power density of 15 to 30 W/m, the efficacy becomes relatively constant. Light output also becomes noticeably uniform. Highest efficacy occurs at the point in the discharge where the power density is around 20 W/m. As the power is increased further, the power in the plasma near the launcher increases, the efficacy in that region decreases, and the power wasted in the launcher increases sharply.

We have also investigated the use of a uniform thin wire helix [3] of a few turns wound around the tube, close to the surface and connected at the launcher end to RF ground. As well as reducing the RF interference, the helix can in some circumstances increase the efficacy of the lamp by up to 30%. Power balance measurements show that the helix increases the power dissipation in the discharge whilst reducing the power lost in the launcher. It is possible that the helix reduces the surface wave wavelength, since it produces a similar effect that of raising the frequency. Experiments on tubes having a 3 turn helix show that the power density can be adjusted somewhat. For example, with all turns of the helix bunched up near the end of the lamp furthest from the launcher, the power dissipation may be made more uniform than it is with a uniformly wound 3 turn helix, or with no helix at all.

CONCLUSIONS

SWD are not naturally uniform discharges, but there are ways in which uniformity may be increased, making them more suitable for lighting. One way is to increase the power so that the natural discharge length is greater than the length of tube so that the surface wave is 'reflected' from the end of the tube. If this is carried to excess, it leads to excessive power dissipation in the launcher. A second way is to wind a thin wire helix around the tube. This increases the power dissipation in the lamp and reduces the power lost in the launcher. Careful positioning of the turns of the helix can increase light output uniformity.

REFERENCES

(1) Moisan, M and Zakrzewski, Z in 'Radiative Processes in Discharge Plasmas', eds. J M Proud and L H Luessen, (New York: Plenum), 381, 1986

(2) Rowley A T, 'Measurement of power density distribution within an operating surface wave discharge lamp', 43rd Gaseous Electronics Conference, Champaign, Illinois, 1990

(3) Rowley A T and Linden-Smith N A, US Patent 5,063,333

IMPEDANCE CHARACTERISTICS AND ITS EQUIVALENT CIRCUIT OF THE ELECTRODELESS FLUORESCENT LAMP EXCITED BY THE AIR-CORED COIL

M.Shinomiya, M.Toho and M.Kawaguchi*
Central Research Laboratory, Matsushita Electric Works Ltd., Kadoma, Osaka 571, JAPAN
*Department of Industrial and Technical Education, Mie University, Tu, Mie 514, JAPAN

The electrodeless fluorescent lamp was developed as a new compact light source [1]. The discharge envelope is mainly spherical and a several turns coil is wound around the envelope. High frequency current is led into the coil, and discharge occurs in the envelope. Consideration of impedance characteristics for the system of the excitation coil and the plasma sphere is important for designing good system. The authors measured impedance of the electrodeless fluorescent lamps under various input power. An equivalent circuit is discussed and the circuit constants are estimated.

The lamp was spherically shaped and the excitation coil was wound tightly around the lamp. Power into the coil is varied. The current and the voltage of the excitation coil was measured by an oscilloscope. The complex impedance measured at the end of the coil is plotted in Fig.1 for various input power.

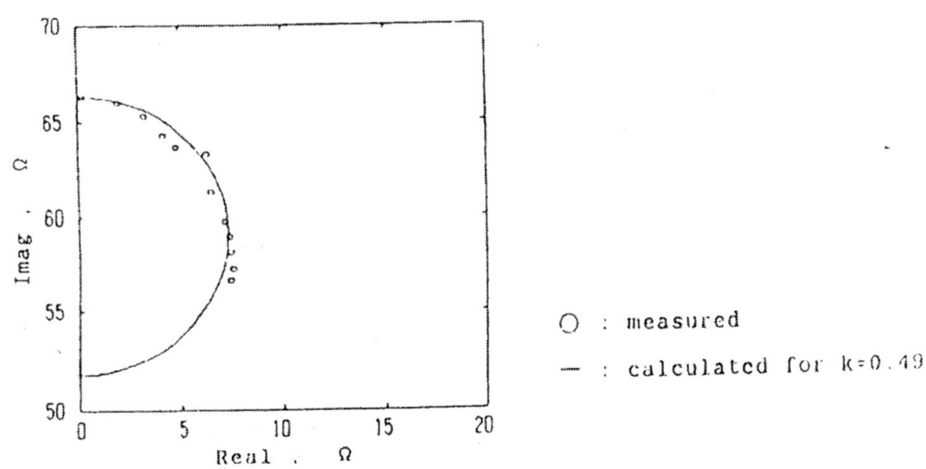

Fig.1: Impedance measured at the end of the coil
Lamp diameter: 45mm, Ar 133Pa, Hg pressure: not controlled.
Excitation coil: 3turns, wire diameter: 1.6mm pitch: 2mm,
coil impedance at 13.56MHz(@25C): $r + jX = 0.16 + j\ 65.4$ Ohms.
Input power: 5-30 W. Operating frequency: 13.56MHz.

Equivalent circuit is considered as an air-cored transformer which consists of the excitation coil as the primary winding and the plasma as the secondary [2]. Figure 2 shows the equivalent circuit, where a unity coupled transformer (n:1) is deduced and plasma impedance is converted to the primary of the unity transformer.

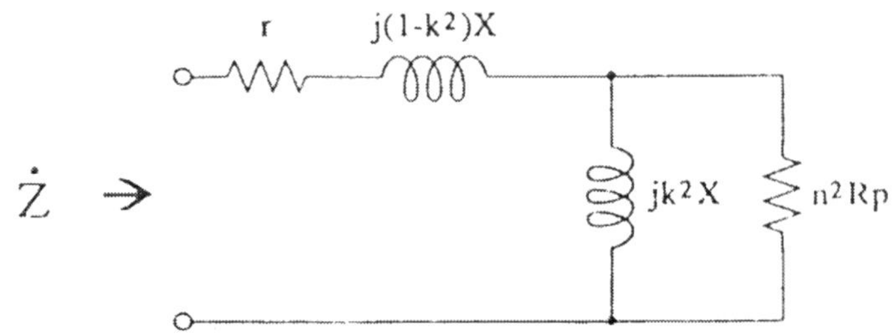

Fig.2: Equivalent circuit of the coil and plasma

By using the above circuit, the impedance of the system is written as,

$$\dot{Z} = \frac{n^2Rp \cdot k^4X^2}{n^4Rp^2 + k^4X^2} + r + j\left(\frac{n^4Rp^2 \cdot k^2X}{n^4Rp^2 + k^4X^2} + (1-k^2)X\right). \quad (1)$$

X, r : reactance and resistance of the coil, respectively.
Rp : equivalent resistance of plasma.
k : coupling coefficient.
n = kN , N : number of the turn of the coil.

With the plasma resistance(Rp) varied, the locus of eq.(1) makes a half circle on a complex impedance plane, when the coupling coefficient(k) is assumed to be constant. The solid half circle in Fig.1 is drawn for k=0.49, which coincides well with the experimental result. This indicates that the value of k is 0.49(const.) in our experimental condition. As shown in ref.[3], when the plasma density changes, both of the self inductance and the mutual inductance of a spherical uniform plasma remain constant, as long as skin depth(δ) is larger than sphere radius(a). The authors suppose that the coupling coefficient is constant in our experimental condition, because the skin depth remains in that range(δ/a>1).

The coil insertion loss can also be estimated from the measurement.

$$Pc/P = r/real(Z) \quad (2)$$

where, Pc: power loss of the coil, P: total power consumption . The coil loss is less than 3% of the total power consumption, when the lamp is operating at 20-30W input power.

References

1. Shinomiya,M. et al. 1991. Development of the electrodeless fluorescent lamp. J of the IES 20(no.1):44
2. Babat,G.I. 1947. Electrodeless discharges and some allied problems. Proceedings of IEE part3 94:27
3. Kawaguchi,M., Toho,M. and Shinomiya,M. 1989. Circuit constants of an electromagnetically excited plasma sphere. 19th Inter Conf on Phenomena in Ionized Gases

NEW LIGHT GENERATION MECHANISM: CONTINUOUS RADIATION FROM CLUSTERS

R.Scholl, B.Weber, C.Weijtens
Philips GmbH Forschungslaboratorium Aachen
P.O.Box 1980
D-5100 Aachen, Federal Republic of Germany

Introduction

Edison's invention of the carbon-filament lamp (1879) was the starting point of modern lighting research. A lot of developing steps led to the well known incandescent lamp with tungsten filament. The incandescent lamp is a Planckian radiator with a continuous spectrum and therefore favourable colour rendering properties. The luminous efficiency is limited by the maximum filament temperature of about 3400 K, some hundred degrees below the melting point of tungsten. Due to the Wien's displacement law an incandescent radiator of 3400 K mainly emits in the IR region and there is only little overlap with the eye sensitivity curve. Hence low efficiencies are characteristic for incandescent lamps.

It was the efficiency problem that led the lighting industry to alternatives for the incandescent lamp. The innovation of low and high pressure discharges offered lamps with much higher efficiencies but with a non-continuous emission and therefore poor colour rendering properties. It is important to recognize that the low efficiency of the continuous radiator was the reason to introduce selective radiators causing problems we still have today.

Cluster radiation which will be described in this article represents a new development in the field of light generation: it combines the high efficiencies of discharge lamps with the good light quality (continuous spectrum) of incandescent radiators. The idea is to create and to heat small particles (clusters, particles consisting of typically 1000 atoms) in a high pressure discharge and to use the thermal emission spectrum of these clusters for light generation. The thermal emission spectrum guarantees a continuous spectrum with good colour rendering properties. On the other hand it is possible to heat the clusters above their melting points and to shift the maximum of the emission into the visible range, the reason why the luminous efficiency increases. (There is also another effect due to the small size of the clusters which shifts the emission maximum into the visible range called Rayleigh effect).

The principle of light generation by glowing clusters is very old. The candle-light is based on such a mechanism [1]. By incomplete combustion of paraffin atomic carbon is formed, which condensates to form carbon clusters of 10 - 200 nm magnitude. These clusters are heated by the surrounding plasma and emit incandescent radiation. The carbon clusters only reach a temperature of about 2000 K and therefore the process of light generation in a candle is of low efficiency.

The cluster lamp

A cluster lamp has been realized at the Philips research laboratory Aachen (Germany) in form of a microwave excited high pressure discharge lamp [2],[3]. The cluster lamp consists of a cylindrical TM_{010} resonant microwave cavity in which microwave power is coupled in via a couple antenna (frequency ≈ 2.45 GHz). This microwave power (≈ 100 Watt) is used to support a high pressure discharge in an electrodeless quartz vessel with a typical volume of 0.25 cm^3 (see fig.1). The light of the cluster discharge leaves the cavity via a reflector and through a wire grid. The experimental set-up has been described in more detail in reference [4].

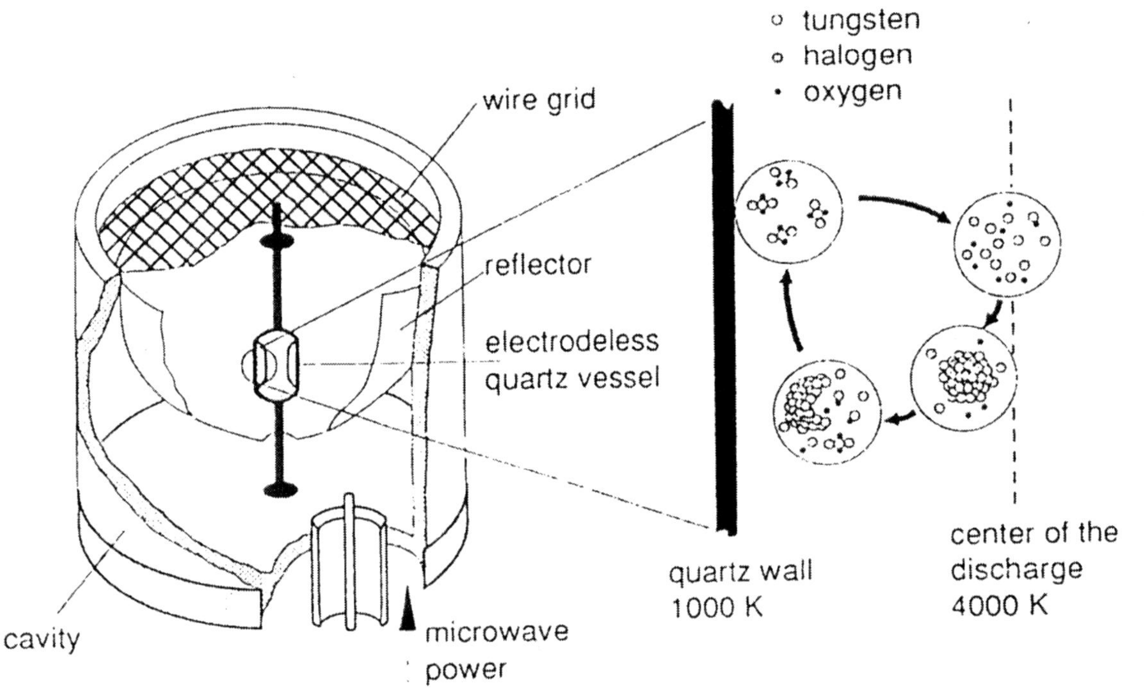

Fig.1: Cluster lamp together with a schematic sketch showing the formation of tungsten clusters

The cluster lamp is based on the following principle. A compound like WO_2Br_2 is filled in an electrodeless quartz vessel. These WO_2Br_2 molecules completely evaporate at typical wall temperatures of 1000 K. In the center of the discharge at temperatures of about 4000 K these molecules dissociate and atomic tungsten is created in such high amounts that the partial pressure of tungsten exceeds its vapour pressure (assuming sufficient WO_2Br_2 filling). So the tungsten vapour condenses in the hot region of the discharge and tungsten clusters are formed. The clusters are heated by the surrounding plasma and emit incandescent radiation until they are transported out of the condensation region. They are finally evaporated in hotter zones of the discharge or chemically dissolved in cooler parts (see fig.1). This cycle of cluster formation and decomposition proceeds stationary without any consumption of the substances filled into the discharge tube.

We use an electrodeless quartz vessel (and the excitation mechanism by microwaves) because WO_2Br_2, but especially the generated oxygen and bromine are very aggressive elements which would destroy an electrode in a few minutes. The transport mechanism which is essential to produce tungsten clusters (tungsten is transported via oxygen and bromine into the hot region of the discharge) is therefore not compatible with conventional electrodes.

Summarising the cluster lamp is based upon two ideas: first the light generation by hot clusters (similar to a candle) and secondly the continuous generation of these clusters by a chemical cycle.

The Regenerative Chemical Cycle

Let us have a more detailed look to the condensation of tungsten. At the Philips research laboratory in Aachen a computer tool was developed which is able to calculate temperature-dependent chemical equilibria on the basis of thermodynamics [5]. The computer program minimizes the Gibb's free enthalpy ΔG for a given temperature and a total gas pressure as a function of the partial pressures of the dissociation products of the filled compounds. Fig.2 shows an example of such a calculation. The calculation includes 21 solid, 4 liquid and 23 gaseous species, but only the most important species are plotted. The filled compound WO_2Br_2 is stable up to about 2500 K. Going to higher temperatures we see that a first bromine atom is separated and the partial pressure of WO_2Br increases. At about 3000 K the second bromine atom is separated and so on. At very high temperatures atomic tungsten dominates in the discharge. If we compare this partial pressure of tungsten (——) with its vapour pressure (·····) we obtain in a broad temperature region a tungsten supersaturation, that means that tungsten vapour condenses and tungsten clusters are created in the hot region of the discharge.

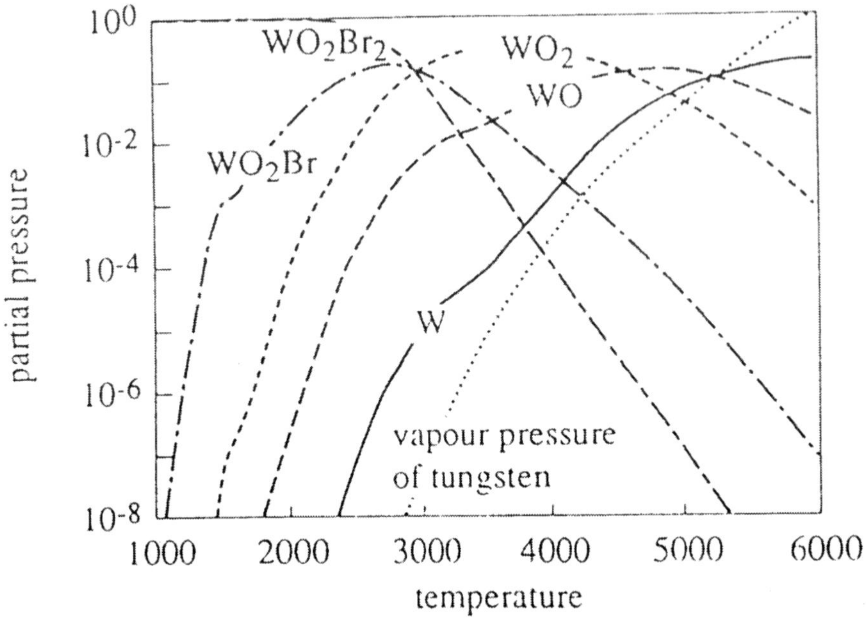

Fig.2 : Calculated chemical equilibrium composition in a WO_2Br_2 discharge

However, a more accurate description of the the cluster formation requires the incorporation of homogeneous condensation. For this purpose we have compared the tungsten supersaturation in our discharge (——) with the required critical supersaturation to form critical nuclei (·····) based upon a model of Becker and Döring [6].

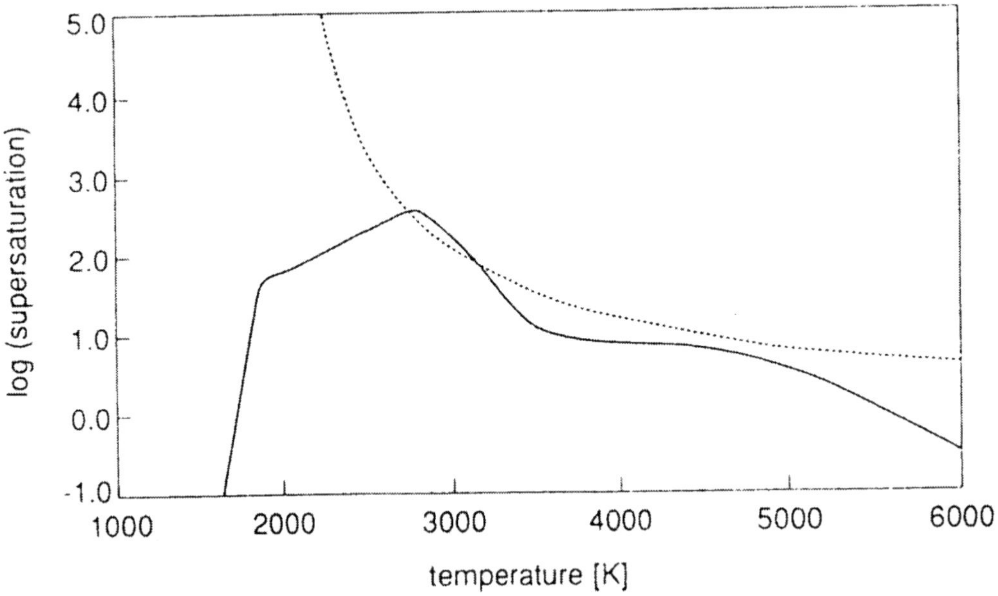

Fig.3 : Calculated tungsten supersaturation in a WO_2Br_2 discharge (——) and needed critical supersaturation to form critical nuclei (very small clusters) (·····)

Only in a small temperature range around 3000 K the critical supersaturation is exceeded and critical nuclei are formed. This leads us to the following picture of the cluster formation. The critical nuclei (one can calculate a size of 11 tungsten atoms for these critical nuclei) diffuse into hotter and colder regions of the discharge. Especially in hotter regions they will grow because there is a high concentration of atomic tungsten. If these clusters reach a size of about 250 atoms they diffuse due to the thermodiffusion into colder regions. On the way to the wall they first grow up to a size of some nanometers until they are chemically dissolved by oxygen and bromine. At the wall the thermodynamically stable compound is again WO_2Br_2.

Experimental Results

The so created tungsten clusters have the following properties. The temperature of the tungsten clusters is about 3500 K, which could be detected by measuring the temperature of the surrounding plasma. A temperature of about 3500 K is also consistent with the position of the maximum of the clusters' emission (In the case of very small particles compared to the wavelength of visible light the Wien's displacement law has the form $\lambda_{max} \cdot T = 2.404 \cdot 10^{-3}$ mK).

We have proven the existence of clusters in the discharge by laser scattering. An 1 W cw argon ion laser beam is focused into the discharge and the scattered laser light is detected under 90°. Light coming from spontaneous emission of the discharge or laser light scattered by the vessel walls has been reduced by lock-in-techniques. The scattered laser light is linearly polarized perpendicular to the scattering plane, a fact from which we deduce that the clusters are spheres [7]. From the scattering coefficient we estimate a mean radius of the clusters to about 2 nanometers.

Beside these nanometer size clusters some larger clusters appear in the discharge which are visible with the bare eye in a projection of the discharge.

The emission spectrum of the discharge consists of 3 contributions (see fig.4a): first identified atomic lines, secondly molecular band radiation in the range 350-600 nm originating from WO molecules and thirdly a broad dominating continuum with a maximum at about 700 nm which is the thermal emission spectrum of the hot tungsten clusters.

The different contributions of the emission spectrum originate from different regions of the discharge. Fig.4b shows the spectral radiance observed on lines of sight through the discharge axis and 1.0 mm beside it (the diameter of the discharge vessel is 5.0 mm). The emission from the axis is dominated by the molecular band spectrum of the WO molecules and the off-axis regions emit a broad structureless continuum, the incandescent radiation of tungsten clusters.

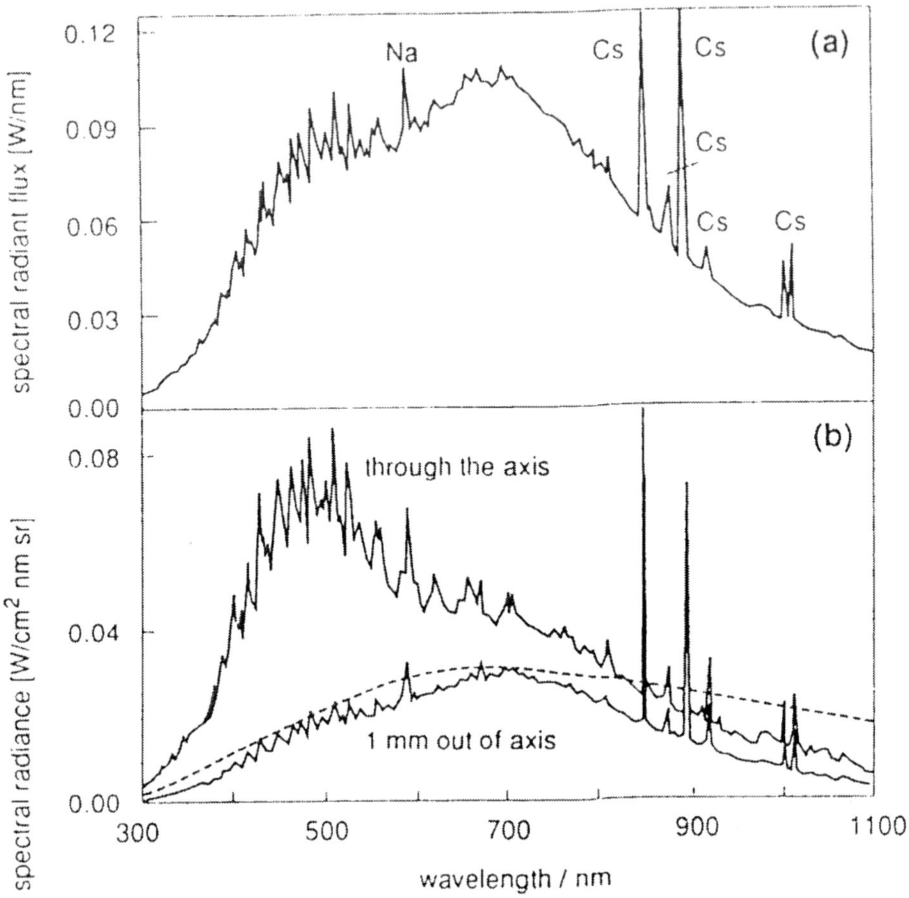

Fig.4: (a) Measured emission spectrum of a WO_2Br_2/CsBr/Ar discharge.
(b) Spectral radiance observed on lines of sight through the discharge axis and 1.0 mm beside it. The dashed line shows a calculated cluster spectrum at 3500 K which has been normalized to fit the observed continuum.

To describe the clusters' emission spectrum (see lower curve in fig.4b) theoretically we use the Rayleigh formula [7], an approximation of the Mie theory for objects much smaller than the wavelength of the emitted radiation:

$$L(\lambda) \sim \frac{1}{\lambda} \cdot B_{Planck}(\lambda,T)$$

The above formula describes the spectral radiance $L(\lambda)$ of an ensemble of non-interacting particles of temperature T with wavelength independent emissivity (grey clusters). $B_{Planck}(\lambda,T)$ is the spectral radiance of a black body of temperature T.

With a cluster temperature of 3500 K a rough description of the clusters' emission spectrum is obtained (see fig.4b: ----).However, the cluster emission in the IR decreases more rapidly than predicted by the Mie theory. This fact does not change if we use a wavelength dependent emissivity or a more realistic temperature dependent cluster distribution in our model. Obviously, there are still some open questions concerning the theoretical understanding of the incandescent radiation of hot clusters.

Properties of cluster discharges

Beside the tungsten clusters in tungstenoxihalide discharges we have also found rhenium clusters in rheniumoxide discharges (Re_2O_7 or ReO_3). In both cases the emission spectrum is dominated by the incandescent radiation of the hot clusters (see fig.5). It is therefore not astonishing that we obtain high to very high colour rendering indices of $R_a \approx 95$.

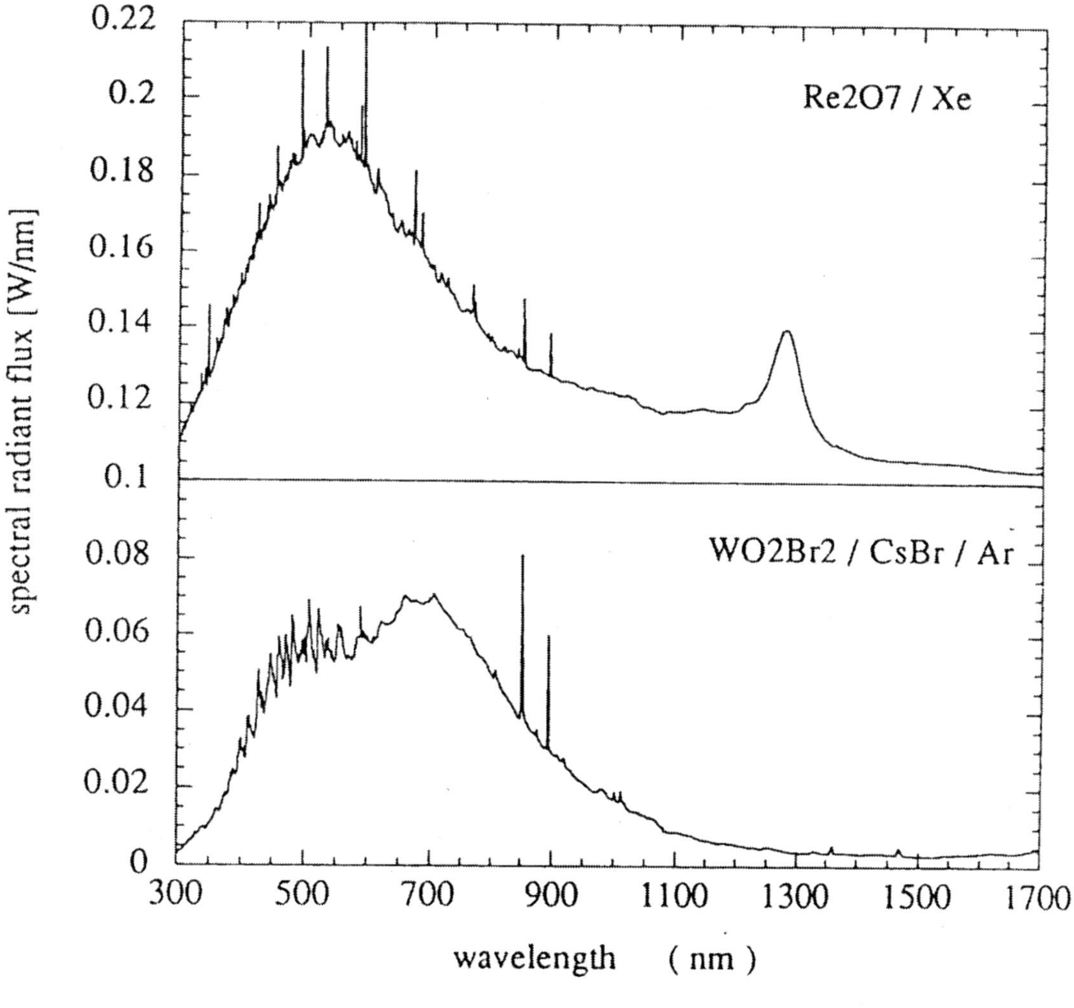

Fig.5 : Measured spectral radiant flux $\Phi(\lambda)$ of a Re_2O_7/Xe and a WO_2Br_2/CsBr/Ar discharge.

A power balance of the cluster discharge yields a conversion of 60 % microwave power into radiation (30 % visible, 30 % IR) resulting in luminous efficiencies of about 60 lm/W (rhenium: 62 lm/W; tungsten: 56 lm/W). The remaining 40 % are losses due to heat conduction, convection, chemical transport and radiation transport (we obtain an optical thickness $\tau=\int\kappa_{abs}ds$ of 0.1 - 0.4).

An important feature of cluster discharges is the possibility of dimming. From fig.6 it can be seen that the colour temperature and the chromaticity coordinates are approximately constant over a broad power range. To understand this fact we have to notice that the temperature of the clusters (in the case of tungsten: 3500 K) is mainly dominated by the point where the critical supersaturation is exceeded (see fig.3). This temperature value is determinated by the chemistry in the lamp, i.e. by the decomposition of tungstenoxihalide molecules, and is independent of the temperature profile of the discharge. If we increase the input power and therefore the temperature of the plasma the clusters will appear in more outer regions of the discharge and they will reach again a temperature of about 3500 K (in the case of tungsten clusters). Therefore the emission spectrum of the clusters remains constant. Only the ratio of cluster to molecular radiation changes causing to a small change in colour temperature.

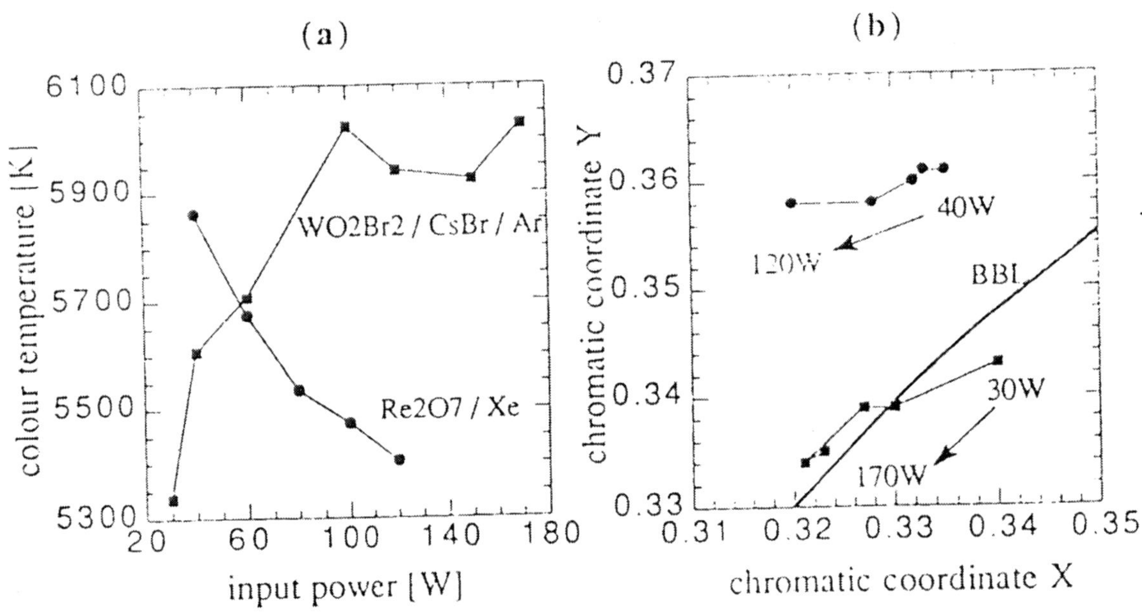

Fig.6 : Colour temperature (a) and chromaticity coordinates (b) of cluster discharges as a function of the input power.

Summary

We have found a new light generation mechanism. Via a chemical cycle it is possible to produce a stationary 'cloud' of continually regenerating hot clusters. Due to their high temperature the clusters emit a thermal emission spectrum which we use for light generation. Discharges with rhenium or tungsten clusters exhibit a number of features that are attractive for lighting applications. The continuous spectra ensure perfect colour rendering and the characteristic cluster formation process leads to stable and reproducible colourmetric properties of a future cluster lamp.

References

[1] A.G. Gaydon, H.G. Wolfhard, "Flames, their Structure, Radiation and Temperature", Chapman & Hall, London 1953

[2] B. Weber, "Die mikrowellenangeregte Cluster-Entladung: eine neuartige Lichtquelle", Ph.D.thesis, Universität Karlsruhe 1990

[3] R. Scholl u. B.Weber, "Incandescent Radiation from 3500 K Hot Clusters" in Proc. Int. Sym. on the Physics and Chemistry of Finite Systems: From Clusters to Crystals, Richmond, USA, 1991, edit. by P. Jena et. al, Kluwer, Boston/London 1992

[4] S. Offermanns, "Electrodeless High Pressure Microwave Discharges", J.Appl.Phys. **67** (1990) 115

[5] E.Schnedler, "The calculation of complex chemical equilibria", CALPHAD **8** (1984) 265-279

[6] F.F.Abraham, "Homogeneous Nucleation Theory", Academic Press, New York/London 1974

[7] C.F.Bohren, D.R.Huffman, "Absorption and Scattering of Light by Small Particles", Wiley, New York 1983

Author index

Author Index

Foundation for the Advancement of the Science & Technology of Light Sources
FAST-LS
Belmayne House
99 Clarkehouse Road
Sheffield, United Kingdom
S10 2LN

ISBN 978-1-5108-4110-9

CURRAN ASSOCIATES INC.
proceedings
.com